T0331625

INTRODUCTION TO THE NETWORK APPROXIMATION METHOD FOR MATERIALS MODELING

In recent years the traditional subject of continuum mechanics has grown rapidly and many new techniques have emerged. This text provides a rigorous, yet accessible introduction to the basic concepts of the network approximation method and provides a unified approach for solving a wide variety of applied problems.

As a unifying theme, the authors discuss in detail the transport problem in a system of bodies. They solve the problem of closely placed bodies using the new method of the network approximation for partial differential equations with discontinuous coefficients.

Intended for graduate students in applied mathematics and related fields such as physics, chemistry and engineering, the book is also a useful overview of the topic for researchers in these areas.

Encyclopedia of Mathematics and Its Applications

This series is devoted to significant topics or themes that have wide application in mathematics or mathematical science and for which a detailed development of the abstract theory is less important than a thorough and concrete exploration of the implications and applications.

Books in the **Encyclopedia of Mathematics and Its Applications** cover their subjects comprehensively. Less important results may be summarized as exercises at the ends of chapters. For technicalities, readers can be referred to the bibliography, which is expected to be comprehensive. As a result, volumes are encyclopedic references or manageable guides to major subjects.

All the titles listed below can be obtained from good booksellers or from Cambridge University Press. For a complete series listing visit www.cambridge.org/mathematics.

ENCYCLOPEDIA OF MATHEMATICS AND ITS APPLICATIONS

Introduction to the Network Approximation Method for Materials Modeling

LEONID BERLYAND
Pennsylvania State University

ALEXANDER G. KOLPAKOV
Università degli Studi di Cassino e del Lazio Meridionale

ALEXEI NOVIKOV
Pennsylvania State University

CAMBRIDGE
UNIVERSITY PRESS

CAMBRIDGE
UNIVERSITY PRESS

Shaftesbury Road, Cambridge CB2 8EA, United Kingdom

One Liberty Plaza, 20th Floor, New York, NY 10006, USA

477 Williamstown Road, Port Melbourne, VIC 3207, Australia

314–321, 3rd Floor, Plot 3, Splendor Forum, Jasola District Centre, New Delhi – 110025, India

103 Penang Road, #05–06/07, Visioncrest Commercial, Singapore 238467

Cambridge University Press is part of Cambridge University Press & Assessment, a department of the University of Cambridge.

We share the University's mission to contribute to society through the pursuit of education, learning and research at the highest international levels of excellence.

www.cambridge.org
Information on this title: www.cambridge.org/9781107028234

© Leonid Berlyand, Alexander G. Kolpakov and Alexei Novikov 2013

First published 2013

A catalogue record for this publication is available from the British Library

Library of Congress Cataloging-in-Publication data
Berlyand, Leonid, 1957–
Introduction to the network approximation method for materials modeling / Leonid Berlyand, Pennsylvania State University, Alexander G. Kolpakov, Universit`a degli Studi di Cassino e del Lazio Meridionale, A. Novikov, Pennsylvania State University.
pages   cm. – (Encyclopedia of mathematics and its applications)
Includes bibliographical references and index.
ISBN 978-1-107-02823-4 (hardback)
1. Composite materials – Mathematical models.   2. Graph theory.   3. Differential equations, Partial.   4. Duality theory (Mathematics)   I. Kolpakov, A. G.   II. Novikov, A. (Alexei)   III. Title.
TA418.9.C6B465   2013
620.1´18015115 – dc23   2012029156

ISBN   978-1-107-02823-4   Hardback

"To my mother and great supporter, Mayya Berlyand".
L. Berlyand

"With fond memories of my wonderful time at Penn State".
A. Kolpakov

"To my mother".
A. Novikov

Contents

Preface

In the natural sciences, there exist two main classes of models – continuum and discrete. The continuum models represent media occupying a volume in space and they are described by real-valued functions of spatial variables. The discrete models correspond to systems with a finite number of components and they are described by real-valued functions of discrete variables. Typically, continuum models are based on partial differential equations whereas discrete models are described by systems of algebraic equations (or systems of ordinary differential equations for evolutionary problems).

The main objective of this book is to indicate physical phenomena and specific properties of solutions of continuum boundary-value problems, which make it possible to describe them by discrete models based on the so-called *structural* approximation. We now briefly outline the idea of this approximation and its advantages in qualitative and quantitative analysis of particle-filled composites, which are widely used in modern technology.

Compare two types of discrete approximation: a numerical approximation and a structural approximation. Here the numerical approximation means the finite-difference, finite-element and similar approximations of the original continuum problem (partial differential equation), where the discretization scale (mesh size) is adjustable depending on the desired precision.

The structural approximation is that which is based on a "physical discretization", when edges and vertices of the discrete network correspond to material objects (e.g., particles in particle-filled composites or beams in a framework). In particular, in the structural approximation the scale of discretization is determined by a natural scale representing the size of inhomogeneities (e.g., particle size). For a wide class of problems, such a structural approximation leads to discrete (finite-dimensional) *network* (graph) models.

We now present a well-known example from structural engineering to illustrate the advantages of structural discretization. Consider a framework, which consists of a finite number of beams (like a framework of a building or a bridge). Each beam is described by an ordinary differential equation whose solution contains unknown constants. The boundary conditions at the junctions of the beams provide a finite system of linear algebraic equations for these constants. This algebraic system corresponds to the structural discretization of the framework.

In contrast, let us now apply a numerical discretization to the same problem. It is possible (and, in fact, would be more accurate) to consider this framework as a system of thin continuum elements. Then we can write the corresponding partial differential equations of elasticity theory for each element (beam) and carry out a numerical discretization for the system of elements. As a result, we obtain another discrete model. While this model looks more precise, its solution requires resolving a very large system of algebraic equations with an ill-conditioned matrix (the ill-conditioning is a result of the small thickness of the structural elements), which leads to significant computational difficulties. The structural discretization in this example results in an algebraic system of much smaller dimension and solves the problem of the framework of beams with high accuracy.

In the above example the structural discretization is clearly better than the numerical discretization. However, there are many problems where a numerical approximation works very well. For example, the numerical discretization obviously must be used in the analysis of the stress–strain state in the joints of a framework of a bridge.

Another well-known example of the structural approximation is the "spring" model used to describe a great variety of phenomena from molecular dynamics to composite materials. This model became classical after the book by M. Born and K. Huang (1954) and has been intensively used ever since.

The two models mentioned above, belonging to the same class of structural models, demonstrate the strong difference in their justification. The finite-dimensional model of a framework is a model at the mathematical level of rigor. At the same time, "spring" models are usually considered as "the physical level of rigor" models, which can be used to describe the corresponding phenomenon qualitatively but not quantitatively.

Structural models are naturally related to graphs (networks) describing relations between the interacting elements of the system under consideration. This is why they are often referred to as *network* models.

Various network models have been widely used in the physics and engineering literature for the construction of simplified analogs of continuum problems. Such analogs are usually obtained based on some kind of an intuitive consideration, and the relation between the original continuum model and the corresponding network is not derived in any systematic way. The problem of the approximation of a given continuum model by a corresponding discrete network has received far less attention. Here by approximation we mean a rigorous justification of the "closeness" in some sense (e.g., asymptotically) between two models with a controlled error estimate. Such justification includes a systematic theoretical derivation of a network from a given continuum model. The approximation issue plays a crucial role in determining the limits of validity of network models. If the issue of validity is not addressed, then a network model may not approximate the continuum problem and therefore may provide misleading results.

The specific feature of the problems presented in this book is that the mathematical methods are intimately related to physical phenomena and the development of our mathematical approach and the analysis of specific problems for various particle-filled composites is done hand in hand. This feature determines our style of presentation, which is a unity of mathematics and applications. Since typically the number of particles is large, the study of particle-filled composites involves large-scale networks. As a result, we face the specific issues of large networks, such as percolation and homogenization.

The problem considered in this book is described by simple constitutive laws (Ohm's, Stokes', Fourier's) which correspond to well-known linear equations. The main difficulty is due to the complex geometry of the domain where these equations are considered. This geometry includes the shape of particles, non-periodicity of the particle locations and close to maximal concentration.

A number of recent results available only in journal publications are treated within the framework of this approach, which makes it convenient for the reader.

This book is an introduction to a wide and dynamically developed field of applied mathematics. Its aim is to present, simultaneously with mathematical rigor and in the simplest possible way, the basic concepts of the network approximation method. For this reason, the authors have selected for detailed exposition one of the existing approaches to developing the network approximation (which can be characterized as the direct construction of trial functions providing asymptotically exact two-sided estimates) and one example of the application of the network approximation method to a real-world problem (computation of the transport property of a disordered high-contrast high-filled composite, including the solution of the well-known problem of the influence of polydispersity on the transport property of the composites).

The authors restrict the presentation to a two-dimensional problem and bodies, which have the shape of circular disks. The reason for this restriction is that this is the modern state of network approximation theory. For example, in this case the network approximation is developed not only for a finite number of bodies but also for an infinite number of bodies (the homogenization type network approximation). Some applied results were obtained by using the method of complex variables. The modern state of the general framework of approximation theory (the theory for arbitrary dimensions and bodies of arbitrary shape) can be characterized as a *work in progress*. Significant progress has been made in some directions in these fields. At the same time, some problems resolved for the two-dimensional problem and bodies, which have the shape of circular disks, remain open in the general case. So, the most complete theory of the network approximation now exists for the particular case described above. The authors decided to write an introduction to the network approximation using this particular problem rather than wait for the completion of investigations of the general case.

The structure of the book is the following:

- Chapter 1 contains a review of mathematical notions used in the analysis of transport problems in composite materials.
- Chapter 2 contains the formulation of the problem and its brief history.
- Chapter 3 presents the network approximation method for linear problems in the presence of a finite number of inclusions.
- Chapter 4 discusses polydispersed composites. This illustrates how the network approximation method is used for the solution of applied problems.
- Chapter 5 expands the network approximation method to an infinite system of disks.
- Chapter 6 expands the network approximation method to the nonlinear problem. We use this chapter also to demonstrate an approach to the construction of a network approximation based on the idea of a perforated medium.
- Chapter 7 demonstrates that the network model provides us with an approximation not only for overall characteristics (such as the total flux, energy) but nodal values approximate potentials of inclusions.
- Chapter 8 presents some results on the conductivity of polydispersed materials obtained by using the method of variables complex.

This book is intended for graduate students in applied mathematics, materials science, physics, mechanics, chemistry and engineering as an introduction to the mathematically rigorous theory of network approximations for partial differential problems or, alternatively, the theory of network models for continuum problems. The prospective applications of network approximation theory could be of interest to industrial engineers, who design composite materials, suspensions, powders, etc.

We thank Vladimir Mityushev, Brian Haines, Oleksandr Misiats, Alexander A. Kolpakov, Sergeiy I. Rakin, Vitaliy Girya and Mykhailo Potomkin for useful comments and suggestions which led to an improvement of the manuscript. A.G. Kolpakov's activity was supported through Marie Curie actions FP7: project PIIF2-GA-2008-219690.

L. Berlyand and A. Novikov gratefully acknowledge NSF support.

The authors offer this book in the hope that it may prove useful to researchers, professors and students working in the areas of pure and applied mathematics, physics, chemistry and material science, especially dealing with problems accounting for spatial inhomogeneity.

LEONID BERLYAND
State College, PA, USA

ALEXANDER G. KOLPAKOV
Cassino, Italy

ALEXEI NOVIKOV
State College, PA, USA

1

Review of mathematical notions used in the analysis of transport problems in densely-packed composite materials

In this chapter, for the convenience of the reader, we briefly recall several basic notions used in the network approximation method which will be used throughout the book.

The network approximation method intensively addresses three branches of mathematics: graph theory, the theory of partial differential equations and duality theory (convex analysis). These branches are very wide and we present here the minimal necessary information. Detailed explanations can be found in the literature referred to in the corresponding sections of our review.

1.1 Graphs

In this section, we present basic notions related to networks. The study of networks, in the form of mathematical graph theory (see, e.g., West (2000)) is the fundamental cornerstone of discrete mathematics.

1.1.1 General information about graphs (networks)

A network is a set of items, which we call vertices or nodes, with connections between them, called edges, see Figure 1.1. In the mathematical literature networks are often called graphs.

Two vertices $\mathbf{x}_i$ and $\mathbf{x}_j$ connected by an edge e_{ij} are called adjacent. The connections in a network can be represented by the collection of the edges $\{e_{ij}\}$ or by the connectivity matrix G_{ij} determined as

$$
G_{ij} = \begin{cases} 1 & \text{if the } i\text{-th and } j\text{-th vertices are connected,} \\ 0 & \text{otherwise; } i, j = 1, \ldots, N. \end{cases}
\tag{1.1.1}
$$

A graph can be described by the set $\mathbf{X} = \{\mathbf{x}_i; i = 1, \ldots, N\}$ of its vertices and the set $\mathbf{E} = \{e_{ij}; i, j = 1, \ldots, N\}$ of its edges or by the set $\mathbf{X}$ of vertices and the

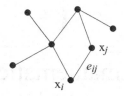

Figure 1.1 A graph.

connectivity matrix $\mathbf{G} = \{G_{ij};\ i, j = 1, \ldots, N\}$. Thus, a graph can be represented in the form $\mathcal{G} = (\mathbf{X}, \mathbf{E})$ or $\mathcal{G} = (\mathbf{X}, \mathbf{G})$.

Both vertices and edges may have a variety of properties associated with them. In many cases some numbers (called the weights of the edges) are assigned in correspondence to the edges. A graph with weights is called a weighted graph. For a weighted graph, the definition (1.1.1) is modified as follows

$$G_{ij} = \begin{cases} g_{ij} \neq 0 & \text{if the } i\text{-th and } j\text{-th vertices are connected,} \\ 0 & \text{otherwise; } i, j = 1, \ldots, N. \end{cases} \tag{1.1.2}$$

The weights g_{ij} (1.1.2) include complete information about the connectivity matrix G_{ij} (1.1.1).

The edges of a graph can also be directed. Graphs composed of directed edges are called directed graphs. Directed weighted graphs are described by (1.1.2) with the weights taking positive or negative values. In specific problems the weights g_{ij} will represent fluxes in networks such as fluxes of fluid, electric current, heat, etc. A weighted graph can be described by the set $(\mathbf{X}, \mathbf{G}) = \{x_i,\ g_{ij}; i, j = 1, \ldots, N\}$.

The size of the network is defined as the total number N of vertices in the network. In this book, we consider networks of finite (possibly large) size.

We say that a network (graph) is connected if every two vertices of the network are connected by a path consisting of edges in this network. A loop in the network is a path, which begins and ends at the same vertex.

1.1.2 Delaunay–Voronoi graphs in modeling of disordered structures

The key condition for structural modeling of disordered particle-filled composites is the high intensity of physical fields in the gaps between closely spaced neighboring particles. That is why the notion of neighboring particles plays an important role. While for periodic arrays of particles the notion of neighbors is obvious, for disordered (non-periodic) arrays the formal definition of neighbors requires an effort.

One way of introducing such a notion is by employing the Delaunay–Voronoi method (Aurenhammer and Klein, 2000; Sahimi, 2003). We now briefly outline this method. Recall that the Voronoi tessellation (also known as Dirichlet tessellations and sometimes referred to as Wigner–Seiz tessellations (Sahimi, 2003) of a

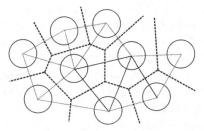

Figure 1.2 Voronoi polygons (tessellations) and Delaunay graph in dimension two.

collection of geometric objects, called the generating points (e.g., a collection of centers of particles in our case), is a partition of the space into Voronoi cells, each of which consists of the points of the space, which are closer to one particular object than to any others.

It is well known that for a given set of generating points in a plane (space) the Voronoi cells are polygons (polyhedra). The generating points which share a common edge (face) in such a partition are called *neighbors*.

While the faces (edges) of Voronoi cells in two dimensions form a graph, in dimension three, this tessellation does not define a graph. However, in all dimensions a graph, called a *Delaunay graph*, can be introduced based on a given Voronoi tessellation. The edges of the Delaunay graph are obtained by connecting neighboring generating points in the Voronoi tessellation. Sometimes the Delaunay graph is referred to as a Voronoi graph (Sahimi, 2003).

Finally, we remark that the Delaunay graph is connected; that is, every two vertices of this graph are connected by a path, which consists of edges of this graph, see Figure 1.2.

1.2 Functional spaces and weak solutions of partial differential equations

In this section we briefly recall several basic notions from the theory of partial differential equations. Detailed explanations can be found in (Kato, 1976; Kolmogorov and Fomin, 1970; Lions and Magenes, 1972; Mizohata, 1973; Schwartz, 1966; Yosida, 1971).

1.2.1 Distributions and distributional derivatives

A complete normed linear space is called a Banach space. A Banach space H with the norm defined as $||x||_H = \sqrt{(x, x)_H}$, where $(x, y)_H$ means the scalar product in H, is called a Hilbert space.

Denote by $\mathcal{D}(Q)$ the set of infinitely differentiable functions with compact support (a function has compact support if it is equal to zero outside a compact set $K \subset Q$), where Q is a bounded domain in $\mathbb{R}^N$.

Topology (convergence) on $\mathcal{D}(Q)$ is introduced as follows: let $\Theta^i \in \mathcal{D}(Q)$, $i = 1, 2, \ldots$; $\Theta^0 \in \mathcal{D}(Q)$. Then the convergence $\Theta^i \to \Theta^0$ in $\mathcal{D}(Q)$ as $i \to \infty$ means that the supports of all functions belong to the same compact set from Q, and Θ^i and all its derivatives converge uniformly to Θ^0 and its corresponding derivatives.

Now, let T be a linear continuous functional determined on $\mathcal{D}(Q)$, i.e., a map (a rule) that assigns to every $\Theta \in \mathcal{D}(Q)$ a number $\langle T, \Theta \rangle$, which is the value of the functional T on this function $\Theta \in \mathcal{D}(Q)$ (in other words $\langle , \rangle$ means the dual coupling). This map is linear with respect to Θ and $\langle T, \Theta^i \rangle \to \langle T, \Theta^0 \rangle$ if $\Theta^i \to \Theta^0$ in $\mathcal{D}(Q)$ as $i \to \infty$. Such functionals are called distributions on Q. The set of distributions is denoted by $\mathcal{D}'(Q)$.

Every locally integrable function $f(\mathbf{x})$ on Q generates a distribution $\tilde{f} \in \mathcal{D}'(Q)$ defined as follows

$$\langle \tilde{f}, \Theta \rangle = \int_Q f(\mathbf{x}) \Theta(\mathbf{x}) d\mathbf{x}.$$

If $T \in \mathcal{D}'(Q)$, its distributional derivative $\dfrac{\partial T}{\partial x_i} \in \mathcal{D}'(Q)$ is determined by the equality

$$\left\langle \frac{\partial T}{\partial x_i}, \Theta \right\rangle = -\left\langle T, \frac{\partial \Theta}{\partial x_i} \right\rangle \tag{1.2.1}$$

for every $\Theta \in \mathcal{D}(Q)$.

Formula (1.2.1) of the distributional derivative reminds us of the integration by parts identity. This is not a coincidence. Historically, the notion of distributional derivative was inspired by this formula, see (Schwartz, 1966; Sobolev, 1937, 1950).

1.2.2 Sobolev functional spaces

We consider a function $f(\mathbf{x})$ defined on the domain Q such that $|f(\mathbf{x})|^p$ (the p-th power of the function) is Lebesgue integrable (Burkill, 2004; Rudin, 1964). This set of such functions is denoted by $L_p(Q)$. When supplied with the norm

$$\|f\|_{L_p} = \left(\int_Q |f(\mathbf{x})|^p d\mathbf{x} \right)^{1/p}$$

$L_p(Q)$ is Banach functional space.

The functional space $L_\infty(Q)$ is introduced as the set of integrable functions bounded almost everywhere (Rudin, 1964). The norm in $L_\infty(Q)$ is introduced as

$$\|f\|_{L_\infty} = \operatorname*{ess\,sup}_{\mathbf{x} \in Q} |f(\mathbf{x})|$$

or

$$\|f\|_{L_\infty} = \lim_{n \to \infty} \|f\|_{L_n}.$$

These two definitions are equivalent (Rudin, 1964).

For $p = 2$, $L_2(Q)$ becomes a Hilbert space with the scalar product

$$(f, g)_{L_2} = \int_Q f(\mathbf{x})g(\mathbf{x})d\mathbf{x}.$$

The Sobolev functional space $W^{m,p}(Q)$ is the set of distributions which (with all its distributional derivatives of order less than or equal to m) are generated by functions from $L_p(Q)$. This is a Banach space with the norm

$$||f||_{W^{m,p}} = \left(\int_Q \sum_{0 \leq m_1 + \ldots + m_N \leq m} \left| \frac{\partial^{m_1 + \ldots + m_N} f}{\partial x_1^{m_1} \ldots \partial x_N^{m_N}} \right|^p d\mathbf{x} \right)^{1/p}. \tag{1.2.2}$$

For $p = 2$, $W^{m,p}(Q)$ is a Hilbert space, which is denoted by $H^m(Q)$, with scalar product

$$(f, g)_{H^m} = \int_Q \sum_{0 \leq m_1 + \ldots + m_N \leq m} \frac{\partial^{m_1 + \ldots + m_N} f}{\partial x_1^{m_1} \ldots \partial x_N^{m_N}} \cdot \frac{\partial^{m_1 + \ldots + m_N} g}{\partial x_1^{m_1} \ldots \partial x_N^{m_N}} d\mathbf{x}.$$

For $m = 1$, the scalar product in $H^1(Q)$ is

$$(f, g)_{H^1} = \int_Q \left(f(\mathbf{x})g(\mathbf{x}) + \nabla f(\mathbf{x}) \nabla g(\mathbf{x}) \right) d\mathbf{x}, \tag{1.2.3}$$

where

$$\nabla = \left(\frac{\partial}{\partial x_1}, \ldots, \frac{\partial}{\partial x_n} \right).$$

The norm in $H^1(Q)$ is

$$||f||_{H^1} = \sqrt{\int_Q \left(f^2(\mathbf{x}) + |\nabla f(\mathbf{x})|^2 \right) d\mathbf{x}}. \tag{1.2.4}$$

The functional space $H^m(Q)$ may also be introduced as the closure of the functional space $C^\infty(Q)$ in the norm (1.2.2) ($p = 2$) (Adams, 1975). The closure of $\mathcal{D}(Q)$ (the space of $C^\infty(Q)$ functions with compact support) in the norm (1.2.2) or (1.2.4) is denoted by $W_0^{m,p}(Q)$ or $H_0^1(Q)$, respectively (Adams, 1975). In particular, this means that the sets $C^\infty(Q)$ and $\mathcal{D}(Q)$ are dense subsets of the respective functional spaces $H^m(Q)$ and $H_0^m(Q)$.

It is possible to introduce the space $H^s(Q)$ for real (not necessary integer or even positive) values of s. In the special case where $Q = \mathbb{R}^N$, $p = 2$, the space $H^s(\mathbb{R}^N)$ can be introduced by using the Fourier transform. $H^s(\mathbb{R}^N)$ is the set of all functions $f \in L_2(\mathbb{R}^N)$ such that their Fourier transforms

$$\hat{f}(\zeta) = (2\pi)^{-\frac{N}{2}} \int_{\mathbb{R}^N} f(\mathbf{x})e^{-i(\mathbf{x}, \zeta)} d\mathbf{x}, \quad \zeta = (\zeta_1, \ldots, \zeta_N) \in \mathbb{R}^N$$

satisfy the condition $|(1 + |\zeta|^2)^{s/2} \hat{f}| \in L_2(\mathbb{R}^N)$. The norm of $f(\mathbf{x})$ in $H^s(\mathbb{R}^N)$ is introduced as follows

$$\|f\|_{H^s(\mathbb{R}^N)} = \|(1 + |\zeta|^2)^{s/2} \hat{f}(\zeta)\|_{L_2(\mathbb{R}^N)}.$$

This way of introducing of non-integer order derivatives is based on the well-known property of the Fourier transform of the derivative (see, e.g., Rudin (1992))

$$i\zeta_n \hat{f}(\zeta) = (2\pi)^{-\frac{N}{2}} \int_{\mathbb{R}^N} \frac{\partial f}{\partial x_n}(\mathbf{x}) e^{i(\mathbf{x}, \zeta)} d\mathbf{x}, \quad n = 1, \ldots, N \tag{1.2.5}$$

and Plancherel's theorem

$$\int_{\mathbb{R}^N} f^2(\mathbf{x}) d\mathbf{x} = \int_{\mathbb{R}^N} \hat{f}^2(\zeta) d\zeta. \tag{1.2.6}$$

From (1.2.5) and (1.2.6) we have

$$\int_{\mathbb{R}^N} |\nabla f|^2(\mathbf{x}) d\mathbf{x} = \int_{\mathbb{R}^n} |\zeta|^2 \hat{f}^2(\zeta) d\zeta. \tag{1.2.7}$$

For the second, third and higher order derivatives

$$(i)^2 \zeta_n \zeta_m \hat{f}(\zeta) = -\zeta_n \zeta_m \hat{f}(\zeta) = (2\pi)^{-\frac{N}{2}} \int_{R^N} \frac{\partial^2 f}{\partial x_n \partial x_m}(\mathbf{x}) e^{-i(\mathbf{x}, \zeta)} d\mathbf{x},$$

and so on, and we conclude that the power m of the factor ζ in the expression $|\zeta|^m \hat{f}(\zeta)$ is the order of the derivatives of the original function $f(\mathbf{x})$.

While the classical derivatives are defined for integer orders only, the expression

$$\int_{\mathbb{R}^N} |\zeta|^s \hat{f}^2(\zeta) d\zeta$$

is determined for various s (integer and real, positive and negative).

In particular, from (1.2.6) and (1.2.7), it follows that the norm in $H^1(\mathbb{R}^N)$ can be defined in two equivalent ways:

$$\|f\|_{H^1} = \sqrt{\int_{\mathbb{R}^N} (f^2(\mathbf{x}) + |\nabla f(\mathbf{x})|^2) d\mathbf{x}} = \sqrt{\int_{\mathbb{R}^n} (1 + |\zeta|^2) \hat{f}^2(\zeta) d\zeta}.$$

The functional space $H^{-s}(Q)$, $s > 0$, can be associated with the dual space of $H_0^s(Q)$ (Lions and Magenes, 1972).

1.2.3 Traces of functions from $H^m(Q)$

The material properties of particle-filled composite materials are often described by piecewise constant functions, which is why the classical formulation of the corresponding boundary-value problems does not apply. Then we will use weak formulations (see next section) in the space $H^1(Q)$. In particular, we will need to assign boundary values for the functions from $H^1(Q)$. These functions are not in general continuous and moreover they are defined almost everywhere (but not everywhere) in the domain Q. Thus assigning boundary values on ∂Q for such functions is not straightforward and requires special techniques.

Recall that the set $C^\infty(\bar Q)$ of infinitely differentiable functions in $\bar Q = Q \bigcup \partial Q$ (we assume ∂Q is a C^∞-smooth surface) is dense in $H^1(Q)$ (Lions and Magenes, 1972).

Consider an $(n-1)$-dimensional surface $\Gamma \subset \partial Q$. For a function $f \in H^1(Q)$, the trace ("boundary values" of f on Γ) $f|_\Gamma : \Gamma \to \mathbb{R}$ can be defined as follows. For a given $f \in H^1(Q)$ choose $f_i \in C^\infty(Q), i = 1, 2, \ldots,$ such that $f_i \to f$ in $H^1(Q)$ as $i \to \infty$. For $f_i \in C^\infty(Q)$ the notion of trace is defined in the classical sense as the value of f_i on Γ. The sequence $f_i|_\Gamma$ has a limit in $H^{1/2}(\Gamma)$ (Lions and Magenes, 1972). This limit is called the trace of the function $f \in H^1(Q)$ on the surface Γ. Due to the density property the trace operator can be extended by continuity from $C^\infty(\bar Q)$ to $H^1(Q)$ and thus it becomes a linear bounded (continuous) operator from $H^1(Q)$ to $H^{1/2}(\Gamma)$.

Similarly, the trace operator can be defined as a bounded operator from $H^m(Q)$ to $H^{m-1/2}(\Gamma)$. This definition of the trace operator allows for integration by parts for functions from Sobolev functional spaces which follows immediately from the density of $C^\infty(\bar Q)$ in $H^m(Q)$. The space of $H^m(Q)$ functions with zero trace is $H_0^m(Q)$.

This trace theorem will often be used in the following generalized form (Lions and Magenes, 1972; Ekeland and Temam, 1976). If a domain Q has Lipschitz ($C^{0,1}$ smooth) boundary, and $\mathbf{v}$ is a three- or two-dimensional vector field, such that

$$\mathbf{v} \in L_2(Q) \text{ and div } \mathbf{v} \in L_2(Q), \tag{1.2.8}$$

then the trace $\mathbf{v}(\mathbf{x}) \cdot \mathbf{n}$ is determined on $\Gamma \subseteq \partial Q$ as a function from $H^{1/2}(\Gamma)$.

A particular (but very important) case of (1.2.8) is the case of divergence-free functions from $L_2(Q)$:

$$\mathbf{v} \in L_2(Q) \text{ and div } \mathbf{v} = 0. \tag{1.2.9}$$

A detailed proof of the trace theorem for divergence-free functions (1.2.9) can be found in Temam (1979).

1.2.4 Weak solutions of partial differential equations with discontinuous coefficients

An introduction to the theory of weak solutions of partial differential equations can be found in Lions and Magenes (1972). Here we present a weak formulation of an elliptic boundary-value problem.

Consider the following boundary-value problem (hereafter the summation of repeated indices is assumed if not indicated otherwise)

$$\frac{\partial}{\partial x_i}\left(a(\mathbf{x})\frac{\partial u}{\partial x_i}\right) = f(\mathbf{x}) \text{ in } Q, \tag{1.2.10}$$

$$a(\mathbf{x})\frac{\partial u}{\partial \mathbf{n}}(\mathbf{x}) = u_1(\mathbf{x}) \text{ on } \Gamma \subseteq \partial Q, \tag{1.2.11}$$

$$u(\mathbf{x}) = u_2(\mathbf{x}) \text{ on } \partial Q \setminus \Gamma, \tag{1.2.12}$$

with coefficients

$$a(\mathbf{x}) \in L_\infty(Q) \tag{1.2.13}$$

that satisfy the coercivity condition

$$a(\mathbf{x}) \geq \gamma > 0 \text{ for all } \mathbf{x} \in Q. \tag{1.2.14}$$

If the functions $a(\mathbf{x})$, $f(\mathbf{x})$, $u_1(\mathbf{x})$, $u_2(\mathbf{x})$ and the surface ∂Q are sufficiently smooth, the problem (1.2.10) has a classical solution belonging to $C^2(Q) \bigcap C^1(\bar{C})$ (Ladyzhenskaya and Ural'tseva, 1968). Multiplying the differential equation (1.2.10) by an arbitrary function $\Theta \in C^\infty(Q)$ such that $\Theta(\mathbf{x}) = 0$ on $\partial Q \setminus \Gamma$ and integrating by parts, we obtain

$$\int_Q a(\mathbf{x})\frac{\partial u}{\partial x_i}(\mathbf{x})\frac{\partial \Theta}{\partial x_i}(\mathbf{x})dx = -\int_Q f(\mathbf{x})\Theta(\mathbf{x})dx + \int_\Gamma u_1(\mathbf{x})\Theta(\mathbf{x})dx \tag{1.2.15}$$

for any $\Theta \in \mathcal{D}(Q)$.

The equalities (1.2.15) and (1.2.12) are defined for functions $a(\mathbf{x}) \in L_\infty(Q)$, $u_1(\mathbf{x}) \in H^{-1/2}(\partial Q)$, $u_2(\mathbf{x}) \in H^{1/2}(\partial Q)$, $u(\mathbf{x}) \in H^1(Q)$, and $\Theta(\mathbf{x}) \in H^1(Q)$.

Thus, it is possible to define the solution of the boundary-value problem (1.2.10) for non-differentiable (in particular, for piecewise continuous) coefficients $a(\mathbf{x})$. The solution of (1.2.15) is referred to as a generalized solution of the boundary-value problem (1.2.10)–(1.2.12) and it is understood in the sense of distributions.

1.2.5 Variational form of boundary value problems

Let H be a Hilbert space. A scalar function $a(u, v)$ determined on $H \times H$ is called a bilinear form on H if it is linear with respect to the first and the second variables. The bilinear form is called continuous if there exists a constant $C < \infty$ such that

$$|a(u, v)| \leq C||u||_H \cdot ||v||_H$$

for any $u, v \in H$.

A bilinear form is called Hermitian if $a(u, v) = a(v, u)$ for any $u, v \in H$. A bilinear form is called coercive if there exists a constant $c > 0$ such that

$$a(u, u) \geq c||u||_H^2 \tag{1.2.16}$$

for any $u, v \in H$.

For any bilinear continuous, coercive and Hermitian form $a(u, v)$ on H there exists unique operator $L : H \to H^*$ such that

$$a(u, v) = \langle Lu, v \rangle \tag{1.2.17}$$

for any $v \in H$.

If $u_2 = 0$, the problem (1.2.10)–(1.2.12) is associated with the bilinear form

$$a(u, v) = \int_Q a(\mathbf{x}) \nabla u(\mathbf{x}) \nabla v(\mathbf{x}) dx \tag{1.2.18}$$

defined on the functional space

$$H = \{v \in H^1(Q) \mid v(\mathbf{x}) = 0 \text{ on } \partial Q \setminus \Gamma\}.$$

The form (1.2.18) satisfies the conditions above if $a(\mathbf{x})$ satisfies the conditions (1.2.13) and (1.2.14).

Then the operator

$$L = \frac{\partial}{\partial x_i} \left(a(\mathbf{x}) \frac{\partial}{\partial x_i} \right)$$

can be treated as an operator acting from H to H^*. The equation (1.2.15) is now valid for any test function in H. Namely, it takes the form

$$\int_Q a(\mathbf{x}) \frac{\partial u}{\partial x_i}(\mathbf{x}) \frac{\partial v}{\partial x_i}(\mathbf{x}) dx = -\int_Q f(\mathbf{x}) v(\mathbf{x}) dx + \int_\Gamma u_1(\mathbf{x}) v(\mathbf{x}) dx \tag{1.2.19}$$

for any $v \in H$.

The problem (1.2.19) is called the variational form of the problem (1.2.10)–(1.2.12). The equation (1.2.19) is a necessary condition of the minimum of the quadratic functional

$$I(u) = \int_Q a(\mathbf{x}) |\nabla u(\mathbf{x})|^2 dx + \int_Q f(\mathbf{x}) v(\mathbf{x}) dx - \int_\Gamma u_1(\mathbf{x}) v(\mathbf{x}) dx \tag{1.2.20}$$

on H.

Under conditions (1.2.13) and (1.2.16) there exists a unique $u \in H$ such that (1.2.19) is satisfied (see, e.g., Ekeland and Temam (1976)).

If the solution of the problem (1.2.15) is sufficiently smooth, then it satisfies (1.2.10)–(1.2.12) with $u_2(\mathbf{x}) = 0$, see, e.g., Ladyzhenskaya and Ural'tseva (1968).

1.3 Duality of functional spaces and functionals

We present brief information about the duality of functional spaces and functionals, the Legendre transform and the minimax problem following Ekeland and Temam (1976). Information on convex analysis in finite-dimensional and functional spaces can be found in Rockafellar (1970, 1969) and Ekeland and Temam (1976) (see also the references in the cited books).

1.3.1 Legendre transform

Let H be a Hilbert space. The set of linear functionals defined on H is also a Hilbert space. It is called the dual (conjugate) space with respect to H and denoted H^*.

As usual (Ekeland and Temam, 1976), we denote by $\langle u, u^* \rangle$ the dual coupling of elements $u \in H$ and $u^* \in H^*$ (the value of the linear functional u^* on the element u).

Let H and H^* be two dual Hilbert spaces and F be a functional of H into R. the functional

$$F^*(u^*) = \sup_{u \in H}\{\langle u, u^* \rangle - F(u)\} \tag{1.3.1}$$

that defines a function from H^* is called the conjugate functional of F.

It is known that F^* is a convex functional and if F is a convex functional, $F^{**} = F$ (Rockafellar, 1970).

For $H = \mathbb{R}^N$ (in this case $H^* = \mathbb{R}^N$) the right-hand side of (1.3.1) is known as a Legendre transform. In this case two convex functions $\phi(\mathbf{x})$ and $\phi^*(\mathbf{y})$, related by the equality

$$\phi^*(\mathbf{y}) = \max_{\mathbf{x} \in \mathbb{R}^N}\{\mathbf{xy} - \phi(\mathbf{x})\}, \tag{1.3.2}$$

are called conjugate functions. In (1.3.2) $\mathbf{xy}$ means the scalar product of the vectors $\mathbf{x}$ and $\mathbf{y}$ in $\mathbb{R}^N$.

If two numbers α and α^* satisfy $\dfrac{1}{\alpha} + \dfrac{1}{\alpha^*} = 1$, the functions $\phi(\mathbf{x}) = \dfrac{1}{\alpha}|\mathbf{x}|^\alpha$ and $\phi^*(\mathbf{y}) = \dfrac{1}{\alpha^*}|\mathbf{y}|^{\alpha^*}$ are conjugate functions (Ekeland and Temam, 1976).

For $\alpha = 2$, (1.3.2) takes the form

$$\frac{1}{2}|\mathbf{x}|^2 = \max_{\mathbf{y} \in \mathbb{R}^N}\left\{\mathbf{xy} - \frac{1}{2}|\mathbf{y}|^2\right\}$$

and the functions $\phi(\mathbf{x}) = \dfrac{1}{2}|\mathbf{x}|^2$ and $\phi^*(\mathbf{y}) = \dfrac{1}{2}|\mathbf{y}|^2$ are conjugate functions.

The functions $\phi(\mathbf{x}) = \dfrac{a}{2}|\mathbf{x}|^2$ and $\phi^*(\mathbf{y}) = \dfrac{1}{2a}|\mathbf{y}|^2$ are conjugate functions for $a \neq 0$. The condition for a maximum of the function $\mathbf{xy} - \dfrac{a}{2}|\mathbf{x}|^2$ is $\mathbf{y} - a\mathbf{x} = 0$. Substituting $\mathbf{x} = \dfrac{\mathbf{y}}{a}$ in the analyzed function, we find that its maximum value is $\dfrac{\mathbf{y}^2}{2a}$.

1.3.2 The minimax problem

Let H and Z be Hilbert spaces. We consider a minimization problem

$$I(\phi) \to \min, \phi \in H. \tag{1.3.3}$$

We assume that all minima and maxima mentioned exist. Else they must be replaced by *inf* and *sup* (Ekeland and Temam, 1976).

Let $I(\phi)$ be written as a maximum of a functional $L(\phi, p)$ (called a Lagrangian function)

$$I(\phi) = \max_{p \in Z} L(\phi, p). \tag{1.3.4}$$

Then the minimum in the problem (1.3.3) is

$$\min_{\phi \in H} \max_{p \in Z} L(\phi, p). \tag{1.3.5}$$

It is convenient to introduce

$$\max_{p \in Z} \min_{\phi \in H} L(\phi, p) \tag{1.3.6}$$

and treat (1.3.5) and (1.3.6) as dual *minimax* and *maximin* problems.

In general

$$\sup_{p \in Z} \inf_{\phi \in H} L(\phi, p) \le \inf_{\phi \in H} \sup_{p \in Z} L(\phi, p). \tag{1.3.7}$$

In many cases of practical interest the non-strict inequality in (1.3.6) transforms into equality (see, e.g. Ekeland and Temam (1976)).

Definition 1.1 A pair $(\bar{\phi}, \bar{p}) \in H \times Z$ is a saddle point of $L(\phi, p)$ on $H \times Z$ if

$$L(\bar{\phi}, p) \le L(\bar{\phi}, \bar{p}) \le L(\phi, \bar{p}), \; \forall \phi \in H, p \in Z. \tag{1.3.8}$$

Proposition 1.2 *The function $L(\phi, p)$ defined on $H \times Z$ possesses a saddle point $(\bar{\phi}, \bar{p})$ on $H \times Z$ if and only if*

$$L(\bar{\phi}, \bar{p}) = \max_{p \in Z} \inf_{\phi \in H} L(\phi, p) \le \min_{\phi \in H} \sup_{p \in Z} L(\phi, p). \tag{1.3.9}$$

We present here the hypotheses concerning L, which guarantee the existence of saddle points and equality in (1.3.7) (Ekeland and Temam, 1976).

A function $F : H \to R$ is called lower semicontinuous if for any $\phi \in H$ $\underline{\lim} F(\phi) \ge F(\varphi)$ as $\phi \to \varphi$ in H. Here "$\underline{\lim}$" means the "lower limit" (Rudin, 1964, 1992).

A function $F(\phi)$ is called upper semicontinuous if $-F(\phi)$ is lower semicontinuous.

Let H and Z be reflexive (Ekeland and Temam, 1976) Banach spaces and let the function $L : H \times Z \to R$ satisfy the following conditions: for any $\phi \in H$, the function

$$p \to L(\phi, p) \text{ is concave and upper semicontinuous} \tag{1.3.10}$$

for any $p \in Z$ and the function

$$\phi \to L(\phi, p) \text{ is convex and lower semicontinuous.} \tag{1.3.11}$$

In addition, there exists $p_0 \in Z$, such that

$$\lim_{||\phi|| \to \infty} L(\phi, p_0) = +\infty \tag{1.3.12}$$

and there exists $\phi_0 \in H$, such that

$$\lim_{||p|| \to \infty} L(\phi_0, p) = -\infty. \tag{1.3.13}$$

Proposition 1.3 *Under the conditions (1.3.10)–(1.3.13), $L(\phi, p)$ possesses at least one saddle point $(\bar{\phi}, \bar{p})$ on $H \times Z$ and*

$$L(\bar{\phi}, \bar{p}) = \max_{p \in Z} \inf_{\phi \in H} L(\phi, p) = \min_{\phi \in H} \sup_{p \in Z} L(\phi, p). \tag{1.3.14}$$

The proofs of Propositions 1.2 and 1.3 can be found in Ekeland and Temam (1976).

Under conditions (1.2.13) and (1.2.16) the Lagrangian function for the quadratique functional

$$I(u) = \frac{1}{2} \int_Q a(\mathbf{x}) |\nabla u(\mathbf{x})|^2 d\mathbf{x}$$

exists and satisfies the conditions (1.3.10)–(1.3.13). The Lagrangian function has the form

$$L(u, \mathbf{p}) = \frac{1}{2} \int_Q \left(\nabla u(\mathbf{x}) \mathbf{p}(\mathbf{x}) - \frac{1}{2a(\mathbf{x})} \mathbf{p}^2(\mathbf{x}) \right) d\mathbf{x}.$$

1.4 Differentiation in functional spaces

The notion of derivatives can be introduced for functionals defined on functional spaces. It is a useful tool widely used both in pure mathematics and applications (Dieudonne, 1969; Ekeland and Temam, 1976; Haug *et al.*, 1986).

Let F be a functional of H into R. The limit (if it exists)

$$\lim_{\lambda \to 0} \frac{F(u + \lambda v) - F(u)}{\lambda} \tag{1.4.1}$$

is called the directional derivative of F at u in the direction $v \in H$ and we denote it by $F'(u, v)$.

If there exists $u^* \in H^*$ such that:

$$F'(u, v) = \langle u^*, v \rangle \tag{1.4.2}$$

for any $v \in H$, we say that F is Gâteaux differentiable at u and call u^* the Gâteaux differential at u and denote it by $F'(u)$.

The Gâteaux differential is characterized by the equation

$$\frac{F(u + \lambda v) - F(u)}{\lambda} = \langle F'(u), v \rangle \tag{1.4.3}$$

for any $v \in H$.

We will deal with the integral functional of the form

$$F(u) = \int_Q f(u(\mathbf{x})) d\mathbf{x}. \qquad (1.4.4)$$

If the function $f : \mathbb{R} \to \mathbb{R}$ is differentiable with respect to the variable u, then

$$\langle F'(u), v \rangle = \int_Q f'(u(\mathbf{x})) v(\mathbf{x}) d\mathbf{x}, \qquad (1.4.5)$$

where $f'(u)$ means the derivative of the function $f(u)$ with respect to the variable u. Formula (1.4.5) is known from the calculus of variations as the first variation of the functional $F(u)$ (1.4.4).

1.5 Introduction to elliptic function theory

In the present section, complex functions of a complex variable $z = x + iy$ are considered, where i is the imaginary unit. Elliptic functions (Akhiezer, 1990) are doubly-periodic meromorphic functions with periods ω_1, ω_2 whose ratio ω_2/ω_1 is not a real number (it should be mentioned that Jacobi's theorem says that there is no single-valued analytic function with more than two periods).

Let us recall some general properties of doubly-periodic functions. Let $f(z)$ be a single-valued analytic function with two periods ω_1, ω_2, $\text{Im}\,\omega_2/\omega_1 > 0$. Then

$$f(z + w) = f(z), \ \forall w = m_1\omega_1 + m_2\omega_2, \ m_1, m_2 \in \mathbb{Z},$$

where $\mathbb{Z}$ stands for the set of integers. Any parallelogram with points z_0, $z_0 + \omega_1$, $z_0 + \omega_1 + \omega_2$, $z_0 + \omega_2$ as its vertices is called a *parallelogram of periods* (and also a *fundamental cell*, whenever $z_0 = 0$).

A doubly-periodic function without singularities is a constant. The number of poles (calculating with their multiplicities) in the fundamental cell is called the *order of the elliptic function*.

The sum of the residues at the poles of an elliptic function in its parallelogram of periods is equal to zero. The derivative of an elliptic function is again an elliptic function. The number of zeros (as well as a-points) of an elliptic function (calculating with their multiplicities) in the fundamental cell is equal to the order of the elliptic function.

1.5.1 Weierstrass $\wp$-function

The Weierstrass $\wp$-function can be represented in the form of a series:

$$\wp(z) = \frac{1}{z^2} + \sum_{m_1, m_2}' \left[\frac{1}{(z - m_1\omega_1 - m_2\omega_2)^2} - \frac{1}{(m_1\omega_1 + m_2\omega_2)^2} \right], \qquad (1.5.1)$$

where $\sum\limits_{m_1, m_2}'$ means that summation is performed over all integers $m_1, m_2 \in \mathbb{Z}$ except $m_1 = m_2 = 0$. The following properties (a)–(e) follow directly from (1.5.1):

(a) $\wp(z)$ is a doubly-periodic function with the only pole $m_1\omega_1 + m_2\omega_2$.

(b) $\wp(z)$ is an even function of order 2:

$$\wp(z) = \wp(-z). \qquad (1.5.2)$$

(c) Its derivative $\wp'$ is an odd function of order 3.

(d) In the neighborhood of the origin its principal part is equal to $\dfrac{1}{z^2}$.

(e) $\wp(z) - \dfrac{1}{z^2}$ tends to zero as $z \to 0$.

(f) It is an inverse function to the elliptic integral:

$$z = \int_{\infty}^{\zeta} \frac{dx}{\sqrt{4x^3 - g_2 x - g_3}} \quad (\zeta = \wp(z)), \qquad (1.5.3)$$

where

$$g_2 = 60 \sum_{m_1,m_2}' \frac{1}{(m_1\omega_1 + m_2\omega_2)^4}, \quad g_3 = 140 \sum_{m_1,m_2}' \frac{1}{(m_1\omega_1 + m_2\omega_2)^6}. \qquad (1.5.4)$$

The functions $\wp(z)$ and $\wp'(z)$ are related by the following algebraic relation

$$\wp'^2(z) = 4\wp^3(z) - g_2\wp(z) - g_3. \qquad (1.5.5)$$

1.5.2 Weierstrass ζ-function

The Weierstrass ζ-function (which should not be confused with the well-known Riemann ζ-function playing an important role in the description of prime numbers) is defined by integration of the Weierstrass $\wp$-function, namely,

$$\zeta(z) = \frac{1}{z} - \int_0^z \left\{ \wp(z) - \frac{1}{z^2} \right\} dz, \qquad (1.5.6)$$

which leads to the following series representation

$$\zeta(z) = \frac{1}{z} + \sum_{m_1,m_2}' \left[\frac{1}{z - m_1\omega_1 - m_2\omega_2} + \frac{1}{m_1\omega_1 + m_2\omega_2} + \frac{z}{(m_1\omega_1 + m_2\omega_2)^2} \right]. \qquad (1.5.7)$$

Differentiating (1.5.6), we have $\zeta'(z) = -\wp(z)$. Therefore, the Weierstrass ζ-function is an odd function ($\zeta(-z) = -\zeta(z)$) having only the one pole at $z = 0$ in the parallelogram of periods with residue 1. Hence, the ζ-function cannot be elliptic; sometimes it is called *quasi-periodic* since the following relations are valid:

$$\zeta(z + \omega_1) = \zeta(z) + \eta_1, \quad \zeta(z + \omega_2) = \zeta(z) + \eta_2, \qquad (1.5.8)$$

where the constants η_1, η_2 are $\eta_1 = 2\zeta(\omega_1/2)$, $\eta_2 = 2\zeta(\omega_2/2)$. By calculating the line integral along the boundary of the fundamental cell, one can obtain the following formula relating these constants and periods:

$$\eta_1\omega_2 - \eta_2\omega_1 = \pi i. \tag{1.5.9}$$

1.5.3 Weierstrass σ-function

This function of the Weierstrass collection is obtained by line integration of $\zeta(z) - \dfrac{1}{z}$ along an arbitrary curve starting from the origin and not passing through any pole of the integrand. To avoid multi-valuedness, $\sigma(z)$ is defined as follows:

$$\log\frac{\sigma(z)}{z} = \int_0^z \left\{\zeta(z) - \frac{1}{z}\right\} dz. \tag{1.5.10}$$

Direct calculation immediately gives an infinite product representation for $\sigma(z)$:

$$\sigma(z) = z\prod_{m_1,m_2}' \left\{\left(1 - \frac{z}{m_1\omega_1 + m_2\omega_2}\right)\right.$$
$$\left.\times \exp\left(\frac{z}{m_1\omega_1 + m_2\omega_2} + \frac{z^2}{2(m_1\omega_1 + m_2\omega_2)^2}\right)\right\},$$

where the designation $\prod_{m_1,m_2}'$ is similar to $\sum_{m_1,m_2}'$. The differentiation formula has in this case the form $\dfrac{\sigma'(z)}{\sigma(z)} = \zeta(z)$. The σ-function is an odd function having no singularity in any bounded domain and only zeros at $z = m_1\omega_1 + m_2\omega_2$. Thus, it is not an elliptic function either. The following holds with the same constants as for the ζ-function $\sigma(z+\omega_1) = -e^{\eta_1(z+\omega_1)}\sigma(z)$, $\sigma(z+\omega_2) = -e^{\eta_2(z+\omega_2)}\sigma(z)$.

1.5.4 θ-function

In practice, it is often supposed that one of the periods of an elliptic function is real. This can be realized by performing the following changes of variables $v = \dfrac{z}{\omega_1}$, $\tau = \dfrac{\omega_2}{\omega_1}$. Then, in the variable v, the elliptic function will have periods 1 and τ, still supposing that $\mathrm{Im}\,\tau > 0$. In this variable, the θ-function is defined in the form of a series:

$$\theta(v) = i\sum_{-\infty}^{\infty}(-1)^n q^{\left(n-\frac{1}{2}\right)^2} e^{(2n-1)\pi vi}, \tag{1.5.11}$$

where $q = e^{\pi i\tau}$. There is a connection between the θ-function and the σ-function given by the formula:

$$\sigma(z) = \frac{\omega_1}{\theta'(0)} e^{\frac{\pi z^2}{\omega_1}} \theta\left(\frac{z}{\omega_1}\right). \tag{1.5.12}$$

Hence, the θ-function has no poles at any bounded domain (and is not an elliptic function). From the definition (1.5.11) of the θ-function, it follows that

$\theta(v+1) = -\theta(v)$, $\theta(v+\tau) = -\dfrac{1}{q}e^{-2\pi vi}\theta(v)$. The points $v = m_1 + m_2\tau$ are the only zeros of the θ-function.

1.5.5 Eisenstein–Rayleigh sums

It is convenient to use the elliptic functions in the form of the *Eisenstein series* introduced by Eisenstein in 1847 and described by Weil (1976). The classical lattice sums (the Eisenstein sums) were applied to the calculation of the effective conductivity tensor by Rayleigh (1892) (see also McPhedran (1986); McPhedran *et al.* (1988)) when a representative cell contains one inclusion. In the present subsection, we introduce the fundamental parameters of elliptic function theory following Weil (1976) and Akhiezer (1990).

The Eisenstein summation method is defined as follows:

$$\sum_{m_1,m_2} = \lim_{N\to\infty} \sum_{m_2=-N}^{N} \left(\lim_{M\to\infty} \sum_{m_1=-M}^{M} \right). \tag{1.5.13}$$

Using this summation, we introduce:

$$S_n(\omega_1, \omega_2) = \sum_{m_1,m_2}{}' (m_1\omega_1 + m_2\omega_2)^{-n}, \tag{1.5.14}$$

where m_1 and m_2 run over all integers except the pair $m_1 = m_2 = 0$, $n = 2, 3, \ldots$. The sum (1.5.14) with $n = 2$ is conditionally, hence, slowly convergent. The formula deduced in Mityushev (1997), $S_2(\omega_1, \omega_2) = \dfrac{2}{\omega_1}\zeta\left(\dfrac{\omega_1}{2}\right)$, is efficient in computations. Rylko (2000) deduced another efficient formula:

$$S_2(\omega_1, \omega_2) = \left(\dfrac{\pi}{\omega_1}\right)^2 \left(\dfrac{1}{3} - 8\sum_{m=1}^{\infty} \dfrac{mq^{2m}}{1 - q^{2m}} \right), \tag{1.5.15}$$

where $q = \exp(\pi i \tau)$.

The sums (1.5.14) with $n > 2$ are absolutely convergent. It is known that $S_n(\omega_1, \omega_2) = 0$ for odd n. For even n, the Eisenstein–Rayleigh sums (1.5.14) can be easily calculated through the rapidly convergent infinite sums (see (1.5.4))

$$g_2 = g_2(\omega_1, \omega_2) = \left(\dfrac{\pi}{\omega_1}\right)^4 \left(\dfrac{4}{3} + 320\sum_{m=1}^{\infty} \dfrac{m^3 q^{2m}}{1 - q^{2m}} \right), \tag{1.5.16}$$

$$g_3 = g_3(\omega_1, \omega_2) = \left(\dfrac{\pi}{\omega_1}\right)^6 \left(\dfrac{8}{27} - \dfrac{448}{3}\sum_{m=1}^{\infty} \dfrac{m^5 q^{2m}}{1 - q^{2m}} \right).$$

Then, $S_4(\omega_1, \omega_2) = \dfrac{1}{60}g_2(\omega_1, \omega_2)$ and $S_6(\omega_1, \omega_2) = \dfrac{1}{1400}g_3(\omega_1, \omega_2)$.

The sums $S_{2n}(\omega_1, \omega_2)$ $(n \geq 4)$ are calculated by the recurrence formula:

$$S_{2n} = \frac{3}{(2n+1)(2n-1)(n-3)} \sum_{m=2}^{n-2} (2m-1)(2n-2m-1) S_{2m} S_{2(n-m)}.$$

(1.5.17)

Remark. A formula for non-periodic (random) arrays to properly define S_2 is discussed in Mityushev (1999).

Weierstrass functions can be expressed as Taylor expansions:

$$\ln \sigma(z) = \ln z - \sum_{n=2}^{\infty} \frac{S_{2n}}{2n} z^{2n}, \quad \zeta(z) = \frac{1}{z} - \sum_{n=2}^{\infty} S_{2n} z^{2n-1}, \quad (1.5.18)$$

$$\wp(z) = \frac{1}{z^2} + \sum_{n=2}^{\infty} (2n-1) S_{2n} z^{2n-2}.$$

The formulas (1.5.18) are not used for calculating the Weierstrass functions. For instance, $\sigma(z)$ is better computed by (1.5.12), because the θ-function can be computed by the very fast formula (1.5.11). Formula (1.5.3) is also effective in computations when $\wp(z)$ is calculated as the function inverse to the elliptic integral.

1.5.6 Eisenstein series

In the following part, we summarize the main facts of the Eisenstein series theory following Weil (1976). The Eisenstein series are defined as follows:

$$E_n(z; \omega_1, \omega_2) = \sum_{m_1, m_2} (z - m_1 \omega_1 - m_2 \omega_2)^{-n}, \quad n = 2, 3, \ldots . \quad (1.5.19)$$

The Eisenstein summation method (1.5.13) is applied to $E_2(z; \omega_1, \omega_2)$. The series $E_n(z; \omega_1, \omega_2)$ for $n = 3, 4, \ldots$ as a function in z converge absolutely and almost uniformly in the domain $\mathbb{C} \backslash \bigcup_{m_1, m_2} (m_1 \omega_1 + m_2 \omega_2)$. Each of the functions (1.5.19) is doubly-periodic and has a pole of order n at $z = 0$. However, further it will be convenient to define the value of $E_n(z; \omega_1, \omega_2)$ at the point zero as follows:

$$E_n(0; \omega_1, \omega_2) = S_n(\omega_1, \omega_2). \quad (1.5.20)$$

The Eisenstein functions of even order $E_{2n}(z)$ can be represented in the form of the series:

$$E_{2n}(z) = \frac{1}{z^{2n}} + \sum_{k=0}^{\infty} \sigma_k^{(n)} z^{2(k-1)}, \quad (1.5.21)$$

where

$$\sigma_k^{(n)} = \frac{(2n+2k-3)!}{(2n-1)!(2k-2)!} S_{2(n+k-1)}. \quad (1.5.22)$$

The Eisenstein series and the Weierstrass function $\wp(z; \omega_1, \omega_2)$ are related by the identities

$$E_2(z; \omega_1, \omega_2) = \wp(z; \omega_1, \omega_2) + S_2(\omega_1, \omega_2), \tag{1.5.23}$$

$$E_n(z; \omega_1, \omega_2) = \frac{(-1)^n}{(n-1)!} \frac{d^{n-2}}{dz^{n-2}} \wp(z; \omega_1, \omega_2).$$

Laurent's series for $E_l(z)$ in $0 < |z| < \varepsilon$ has the form Weil (1976)

$$E_l(z) = \frac{1}{z^l} + (-1)^l \sum_{s=0}^{\infty} \frac{(l+s-1)!}{s!(l-1)!} z^s. \tag{1.5.24}$$

Example. Consider the square lattice generated by the fundamental translation vectors expressed by the complex numbers $\omega_1 = 1$, $\omega_2 = i$. Then $S_2 = \pi$, $S_4 = 3.15121$, $S_8 = 4.25577$, $S_{12} = 3.93885$ ($S_n = 0$ for integers $n > 2$ not divisible by 4); the function $\wp(z)$ for the square lattice can be represented in the form of the series

$$\wp(z) = \frac{1}{z^2} + \sum_{k=1}^{\infty} (4k-1) S_{4k} z^{4k-2}. \tag{1.5.25}$$

In particular, (1.5.25) implies that

$$\wp(iz) = -\wp(z). \tag{1.5.26}$$

The derivative $\wp'(z)$ is equal to zero only at the half-periods of the lattice; hence,

$$\wp'\left(\frac{1}{2}\right) = \wp'\left(\frac{i}{2}\right) = \wp'\left(\frac{i+1}{2}\right) = 0. \tag{1.5.27}$$

Moreover,

$$\wp\left(\frac{1}{2}\right) + \wp\left(\frac{i}{2}\right) + \wp\left(\frac{i+1}{2}\right) = 0. \tag{1.5.28}$$

Using (1.5.23) one can easily rewrite the relations (1.5.25)–(1.5.28) in terms of the Eisenstein functions.

1.6 Kirszbraun's theorem

Kirszbraun's theorem describes Lipschitz extensions, and we recall that a function u is said to be Lipschitz on Ω, if it is continuous and there exists constant $\mathcal{L}$ such that

$$|u(\mathbf{x}_1) - u(\mathbf{x}_2)| \leq \mathcal{L}|\mathbf{x}_1 - \mathbf{x}_2|$$

for any $\mathbf{x}_1, \mathbf{x}_2 \in \Omega$. The space of such functions is denoted as $C^{0,1}(\Omega)$. It is a Banach space with the norm

$$\|u\|_{C^{0,1}(\Omega)} = \|u\|_{C(\Omega)} + \sup_{\substack{\mathbf{x}_1, \mathbf{x}_2 \in \Omega \\ \mathbf{x}_1 \neq \mathbf{x}_2}} \frac{|u(\mathbf{x}_1) - u(\mathbf{x}_2)|}{|\mathbf{x}_1 - \mathbf{x}_2|}.$$

By Rademacher's theorem (see, e.g., Theorem 2 in Section 3.2.1 in Evans and Gariepy (1992)) if $\|u\|_{C^{0,1}(\Omega)} \leq K$, then $\|u\|_{W^{1,\infty}(\Omega)} \leq K$. Conversely, if Ω is convex, then $W^{1,\infty}(\Omega)$ consists of all bounded Lipschitz functions on Ω.

Theorem 1.4 (M.D. Kirszbraun) *Let $\Omega \in \mathbb{R}^n$, and let $u : \partial\Omega \to \mathbb{R}$ be a Lipschitz map with $\|u\|_{C^{0,1}(\partial\Omega)} \leq \mathcal{L}$. Then there exists a Lipschitz extension $\bar{u}$ in Ω such that $\|\bar{u}\|_{C^{0,1}(\Omega)} \leq \mathcal{L}$.*

Kirszbraun himself proved this theorem for vectorial problems. It was subsequently generalized for Lipschitz maps in Hilbert spaces. The above formulation will, however, be enough for our purposes.

2

Background and motivation for the introduction of network models

In this chapter we review several ways of applying network models to inhomogeneous continuum media and systems of inclusions.

Discrete networks have been used as analogs of continuum problems in various areas of physics and engineering for a long time (see, e.g., Acrivos and Chang (1986); Ambegaokar *et al.* (1971); Bergman *et al.* (1990); Curtin and Scher (1990b); Koplik (1982); Newman (2003); Schwartz *et al.* (1984)). However, as demonstrated in Kolpakov (2006a), such analogs may or may not provide a correct approximation. In recent decades, the problem of the development of network models as rigorous approximations of continuum models was posed and solved for certain physical problems.

The objectives of our book are two-fold. First, we will develop an approach that allows us to derive network models by structural discretization (structural approximation). The key feature of this approach is that it is based on a rigorous asymptotic analysis with controlled error estimates, and thus we obtain the limits of validity for the network approximation. Secondly, we show that our network models are efficient tools in the study and prediction of properties of disordered particle-filled composites of various kinds.

2.1 Examples of real-world problems leading to discrete network models

Our interest is motivated by real-word problems and we next present three examples of highly packed composites which can be modeled by networks.

2.1.1 Optimal design of electrical capacitors

It is well known that the capacitance of an electrical capacitor is proportional to the dielectric constant (Smythe, 1950). Most materials have a dielectric constant which is not large, usually no greater than 10 (in relative units in which the dielectric constant of the vacuum is 1). On the other hand, several kinds of ceramic materials have an extremely high dielectric constant (10^3–10^4 or even higher (Kuchling, 1980)). However, in many situations it is not practical to use a pure ceramic as a principal element of a capacitor because a ceramic has unsatisfactory mechanical properties (too brittle). Therefore, a composite of a highly elastic matrix with a low dielectric constant (which provides the desired mechanical properties) and ceramic particles (which increase the effective dielectric constant) is typically used in industry. Practical issues in the design of such composites include the selection of the filler identity, the size distribution and optimal array of the filling particles.

Application of network models to the analysis of electrical properties of composite materials can be found in Mertensson and Gafvert (2003, 2004) and Nettelblad *et al.* (2003).

2.1.2 Thermal management in the electronics industry

Modern integrated circuit design needs highly efficient composite packages for heat removal. This need is driven by the fact that modern circuits have more "elements" per volume (to miniaturize the devices and make them more powerful) and therefore they generate more heat per volume. A typical package is a compound of an epoxy-based polymer filled with ceramic (or other highly conductive material) powders where the ratio of the thermal conductivities is of the order of 10^2 (Hill, 1996; Hill and Supancic, 2002). The rationale behind using such composites is the same as for electrical capacitors; epoxy-based polymers provide the desired mechanical properties while the filler increases the heat conductivity.

2.1.3 Suspensions

The transport of slurries (highly concentrated suspensions) is an important problem arising in numerous industrial applications, ranging from construction engineering to combustion processes (Shook and Rocko, 1991), as well as in geophysics (mud-flow and debris flow rheology) and pharmacology (drug design) among others.

The rheological behavior of a suspension in a wide range of concentrations is described by the effective viscosity and the effective viscous dissipation rate. A wide range of experimental and numerical results on the effective properties of suspensions are available, see for example, the references in Berlyand and Panchenko (2007). Although the hydrodynamics of slurries is a well-studied subject with about a 100-year history (Einstein, 1906; Lamb, 1991), most of the research in this field is devoted to dilute suspensions. Suspensions of moderate concentrations are also well studied using a homogenization theory approach, e.g., Lévy (1986). However, it is often necessary to operate with highly concentrated suspensions

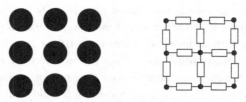

Figure 2.1 Models of a conducting composite: (left) continuum model and (right) discrete model (the resistor network).

(Nunan and Keller, 1984b). The network approximation allows us to describe the rheological properties of concentrated suspensions very efficiently (Berlyand and Panchenko, 2007).

2.2 Examples of network models

Discrete models usually describe systems consisting of a finite number of elements. The number of possible interactions between the finite number of elements is also finite. Thus, we can characterize such discrete models by a finite set of variables (which can take both real and discrete values). To each discrete model there corresponds a network constructed in a natural way: vertices correspond to the elements of the system and edges correspond to the interactions between the elements.

We distinguish two classes of discrete models. In the first class, a material is represented by a discrete set of atoms or molecules that interact with each other through interatomic potentials. Here the sizes of individual particles (atoms) are neglected. The models of the second class are called structural models (also called lattice models, see Kalamkarov and Kolpakov (1997); Sahimi (2003)). The parameters of structural models involve the sizes of elements forming the composite material and their shapes. Such elements could be grains, particles, cells, elastic and plastic regions, etc. In this book we deal with the second class of discrete models.

We now present examples of discrete models.

2.2.1 Resistor network models

The first example is a model of a conducting medium filled with a periodic array of absolutely conducting particles (Figure 2.1 (left)) widely used in applications (Sahimi, 2003; Yang and Hui, 1991). This medium is modeled by a periodic set of points (which correspond to particles) such that the neighboring points are connected by resistors of resistivity R (see Figure 2.1 (right)). This means that we prescribe the resistivity R to gaps between neighboring particles. One important question is how to choose the value of R which would model the conductivity of

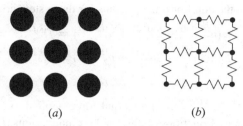

(a) (b)

Figure 2.2 Models of an elastic composite: (a) continuum model and (b) discrete model (the spring network).

the gaps. This will be addressed in Section 2.5.3. For the sake of simplicity, we consider a planar square array of points (Figure 2.1). Each point is indexed by two integers k and l (Cartesian coordinates in the plane); $k, l = 1, \ldots, N$ (N is the size of the array). The electric potential of a point with coordinates (k, l) is denoted by $\phi^{(k,l)}$. Each point interacts only with its four neighbors (Figure 2.1) (nearest-neighbor model). The Kirchhoff equations for such a system state that the algebraic sum of the currents entering each point is zero:

$$\frac{\phi^{(k+1,l)} - 2\phi^{(k,l)} + \phi^{(k-1,l)}}{R} + \frac{\phi^{(k,l+1)} - 2\phi^{(k,l)} + \phi^{(k,l-1)}}{R} = 0; \quad (2.2.1)$$

$$k, l = 2, \ldots, N - 1.$$

Equations (2.2.1) hold for inner points ($k, l = 2, \ldots, N - 1$). For the remaining boundary points the boundary conditions are prescribed. If the potentials $\phi_0^{(k,l)}$ are given for the boundary points, then

$$\phi^{(k,l)} = \phi_0^{(k,l)}; \quad (k, l) = (1, i), (N, i), (i, 1), (i, N); \quad i = 1, \ldots, N. \quad (2.2.2)$$

Equations (2.2.1), (2.2.2) provide a simple example of a discrete model. From the mathematical point of view, (2.2.1), (2.2.2) is a liner algebraic system, which can be solved by standard numerical procedures (see, e.g., Fox (1964)).

2.2.2 Spring network models

The second example is a model of an elastic medium filled with periodically arranged rigid particles (an analog of the well-known Cauchy–Born model (Born and Huang, 1954) used to describe a great variety of phenomena from molecular dynamics to composite materials). The elastic body is modeled by a periodic set of points such that neighboring points are connected by elastic springs of stiffness E. For the sake of simplicity, we again consider a planar square array of points (Figure 2.2). Each point is indexed by two integers k and l; $k, l = 1, \ldots N$. The displacement of a point with coordinates (k, l) is denoted by a vector $\mathbf{u}^{(k,l)} = \left(u_1^{(k,l)}, u_2^{(k,l)}\right) \in \mathbb{R}^2$. Each point interacts with only four neighbors (2.2) via the springs. If the elastic force of a spring is proportional to the relative elongation

(Hooke's law), then the equilibrium equations for the system are given by

$$E\left(u_1^{(k+1,l)} - 2u_1^{(k,l)} + u_1^{(k-1,l)}\right) = f_1^{(k,l)}; \tag{2.2.3}$$

$$E\left(u_2^{(k,l+1)} - 2u_2^{(k,l)} + u_2^{(k,l-1)}\right) = f_2^{(k,l)};$$

$$k, l = 2, \ldots, N-1,$$

where $\mathbf{f}^{(k,l)} = \left(f_1^{(k,l)}, f_2^{(k,l)}\right)$ is an external force applied to the (k, l)-th point (e.g., the gravity force). Equations (2.2.3) hold for inner points ($k, l = 2, \ldots, N - 1$). At the remaining boundary points certain boundary conditions are prescribed. For example, if the system is clamped at the boundary, then

$$u_k^{(1,i)} = u_k^{(N,i)} = u_k^{(i,1)} = u_k^{(i,N)} = 0; \quad i = 1, \ldots, N. \tag{2.2.4}$$

Equations (2.2.3), (2.2.4) are again a linear algebraic system.

More sophisticated examples of network models can be found in Dobrodumov and El'yashevich (1973); Sahimi (2003); Limat (1988); Feng (1985); Herrmann *et al.* (1989); Chung *et al.* (1996); Curtin and Scher (1990a); Kellomaki *et al.* (1996); Yan *et al.* (1989).

2.2.3 Beam network models

The next in the hierarchy of network models, after resistors and spring network models, are beam network models, where the edges are beams whose deformation (elongation and bending) is described by fourth-order ordinary differential equations. Beam network models were studied by numerous investigators (see, e.g., Kalamkarov and Kolpakov (1997); Roux and Guyon (1985); Pshenichnov (1993)) in problems related to, e.g., fracture analysis. The general homogenization approach to lattice solids, beams and plates based on the network approximation is presented in Kolpakov (2004). Beam framework models are another step in the network modeling and it is illustrated by the recent history of developing materials with negative Poisson ratio (Kolpakov, 1985; Almgren, 1985; Friis *et al.*, 1988; Lakes, 1991; Milton, 1992; Berlyand *et al.*, 1991, 1995) and negative thermal expansion.

2.2.4 Network models in physics

In the examples presented below the main problem is determining what to assign to the edges and vertices of the discrete model (e.g., which resistivity R must be prescribed to resistors in the first example and which stiffness E must be prescribed to springs in the second example). This is the main problem in most of the discrete models.

We emphasize that the majority of network models are derived as simplified analogs of corresponding continuum models, rather than their rigorous approximations. For instance, model (2.2.3), (2.2.4) is an analog of a continuum elasticity model, obtained by replacing partial derivatives with finite differences. The accuracy of this finite-difference approximation is not addressed here. Note that (2.2.3), (2.2.4) does not capture the shear deformation and therefore is not a proper approximation of the continuum problem. At the same time the model given by (2.2.1), (2.2.2) describes the continuum conductivity problem quite well. To explain this observation, we note that the problem (2.2.1), (2.2.2) is scalar (the potential ϕ is a scalar), while the problem (2.2.3), (2.2.4) is vectorial (the displacement $\mathbf{u}$ is a vector). In the process of the analysis developed below, it will be clear that there is a significant difference between scalar and vectorial problems. Our approach allows for the development of network approximations, which properly approximate both vectorial and scalar continuum problems.

We also mention here the problem of the continuum limit (Hrennikoff, 1941; Lenczner, 1997; Noor, 1988), when a discrete model is a starting point and a continuum model is derived in the limit as the lattice spacing goes to zero. This problem in a sense is inverse to our problem since now the continuum model approximates the discrete model derived from first principles.

Such discrete models (often called molecular dynamics models) became popular in physics after the seminal work (Born and Huang, 1954) on the Cauchy–Born rule for derivation of the continuum elasticity equations from spring lattice models. Here the main difficulty is to obtain adequate formulas describing the interaction of particles (the interparticle interaction potential). This continuum limit problem is beyond the scope of this book.

2.2.5 Network models in materials science, electrostatics and hydrodynamics

We now present a brief outline of network models in materials science. Network modeling is an interdisciplinary problem. It involves methods of natural sciences (physics, chemistry, biology), mathematical modeling of random media, and various techniques of partial differential equations and calculus of variations.

There exists a large number of studies of applied network models, e.g., Sahimi (2003); Willis (2002). We mention here two large groups of such models (that are not mutually exclusive): models of percolation theory (the study of phase transitions described by random graphs), e.g., Balberg (1987); Grimet (1992); Halperin *et al.* (1985); Kesten (1992); Meester and Roy (1992); Stauffer and Aharony (1992) and discrete models for composites, e.g., Broutman and Krock (1974); Chou and Ko (1989); Kelly and Rabotnov (1988); Sahimi (2003) (both lists of references are incomplete).

Percolation models are typically based on periodic networks (lattices) with material properties randomly assigned to the edges/vertices of the network (although there is a growing number of studies which use non-periodic networks, see, e.g., Balberg (1987); Meester and Roy (1992); Novikov and Friedrich (2005), most of the classical results in percolation theory are obtained for periodic networks). Many discrete models for composite materials are similar to the model (2.2.1) and are usually obtained as heuristic analogs of continuum models rather than their rigorous approximations (see the above example of an elastic composite).

Network models are widely used for modeling of failure and fragmentation of materials at all levels: macroscopic, mesoscopic and microscopic (including molecular and atomic levels). Earlier network models (Dobrodumov and El'yashevich, 1973) of fracture used spring lattices. Later more sophisticated models were developed, e.g., Curtin and Scher (1990a); Kun and Herrmann (1996); Sahimi (2003).

We now briefly review two classes of problems where modeling based on a rigorous network approximation has been significantly used in recent years.

We begin with a scalar problem of electrostatics, where the network models are applied to the problem of the electrical conductivity or capacitance of a particle-filled composite (or a collection of bodies) (Ambegaokar *et al.*, 1971; Bergman and Dunn, 1992; Bergman, 1983; Clerc *et al.*, 1990; Greengard and Lee, 2006; Greengard and Moura, 1994; Hinsen and Felderhof, 1992; Molyneux, 1970; Robinson and Friedman, 2001; Runge, 1925). An industrial problem of the optimal design of an electrical capacitor was discussed above. A review of problems in electrostatics and electrodynamics where network models have been extensively used can be found in Sahimi (2003). Recall the problem of the *electrostatic capacitance* (Wermer, 1974) of a collection of bodies. The stationary electric field of a system of perfect conductors is described by Laplace equations. The electrostatic capacitance of this collection is introduced as the electrostatic energy of this collection (i.e., the Dirichlet integral). For a pair of parallel plates the capacitance is easy to compute (the formula for the capacitance of a parallel plate capacitor is derived in high-school physics textbooks). However, calculation of the capacitance of a pair of disks or spheres is a much more difficult task whose solution can be found in college textbooks (Tamm, 1979; Brown, 1956). The solution of Laplace's equation for a body with a complex shape and, especially, for a collection of many bodies (which is equivalent to the calculation of the capacitance of this body or the capacitance of the collection of bodies) cannot be found analytically and can be obtained only numerically. For a large number of bodies, finding a direct numerical solution of this problem becomes a computational challenge. Then homogenization techniques and network approximations can be used for a significant reduction of its computational complexity.

Another traditional area for application of network models and similar approaches is the *hydrodynamics* of suspensions and slurries (Abbot *et al.*, 1991; Batchelor and Wen, 1972; Bürger and Wendland, 2001; Coussot, 2002; Ding *et al.*, 2002; Goto and Kuno, 1984; Graham, 1981; Happel, 1959; Koplik, 1982; Jabin and Otto, 2004; Ladd, 1997; Leal, 1992; Maury, 1999; Meredith and Tobias, 1960; Nott and Brady, 1994; Panasenko and Virnovsky, 2003; Shook and Rocko, 1991; Sierou and Brady, 2002; Subia *et al.*, 1998). Note that many papers were devoted to the description of the velocity field and calculation of the overall viscosity of concentrated suspensions (Brady, 1993; Brady and Bossis, 1985; Carreau and Cotton, 2002; Chang and Powell, 1994; Chen and Acrivos, 1978; Feng and Acrivos, 1985; Frenkel and Acrivos, 1967; Jeffrey and Acrivos, 1976; Leighton and Acrivos, 1987; Phillips *et al.*, 1992; Prager, 1963; Subia *et al.*, 1998). Recall that hydrodynamics problems are vectorial. There have been many generalizations of various results from scalar elliptic partial differential equations to vectorial systems (e.g., elasticity, Stokes). While in some instances such generalizations are straightforward, quite often new mathematical issues arise, which have no scalar analogs. Korn's inequality (see, e.g., Oleinik *et al.* (1962)), which plays a crucial role in the solvability of elasticity problems, is a prominent example. Another example is an attempt to develop a vectorial analog of the Keller–Dykhne geometric mean law (Dykhne, 1971; Jikov *et al.*, 1994; Keller, 1964) which led to asymptotic formulas for the zero and negative Poisson's ratio of two phase composites (Berlyand and Kozlov, 1992; Berlyand and Promislow, 1995). A straightforward generalization for vectorial (suspensions) problems of the scalar techniques developed in Berlyand and Kolpakov (2001) and Berlyand and Novikov (2002) did not work. The vectorial suspension problem required the development of a new fictitious fluid approach and led to the prediction of a new physical phenomenon of degenerate effective viscosity (Berlyand and Panchenko, 2007) which has no scalar analog.

2.3 Rigorous mathematical approaches

Although the idea of network modeling at first looks clear (maybe even trivial), a more careful investigation of the problem leads to the conclusion that it is not simple, especially from the mathematical point of view. In fact, the construction of a network model assumes an approximation of a *continuous* boundary-value problem with a *finite-dimensional* problem, and the approximation does not use the methods of finite-difference or finite-element approximations (in which approximation is achieved through mesh refinement).

In the pioneering works by Borcea (1998) and Borcea, Berryman and Papanicolaou (1999) the problem of an approximation or a relationship between a continuum model and a network model was formulated and addressed in the rigorous mathematical framework of approximations of Dirichlet-to-Neumann maps. It was further applied there for conductivity problems of composite media with continuously distributed properties arising, e.g., in imaging. The results of Borcea and Papanicolaou (1998) have been applied to the inverse problem of the recovery of the conductivity from boundary measurements, when there are regions of high contrast in the medium such that standard approximation methods (Born approximation) do not work. It has been shown in Borcea *et al.* (1999) that imaging of the conductivity of such a medium is asymptotically equivalent to the identification of a resistor network from voltage and current measurements at the boundary vertices.

The techniques of Borcea and Papanicolaou (1998) have also been generalized for the problem of quasi-static transport in high-contrast conductive media (Borcea, 1998). In Borcea (1998), Borcea *et al.* (1999) and Borcea and Papanicolaou (1998) the direct and dual variational principles were used to rigorously justify the discrete network approximation by constructing, up to the leading term, matching bounds for asymtotics of the effective resistivity. These bounds are given by discrete variational principles that can be interpreted in terms of networks. The key step in this duality approach is the construction of test functions. There is no general recipe for the construction of such functions. Therefore, implementation of this approach is a highly non-trivial analysis problem since a test function which will work for one problem may not work for another. Indeed, these functions essentially depend on the physical and geometrical features of the problem. In particular, in Borcea and Papanicolaou (1998) an original construction of a test function for the Kozlov model (Kozlov, 1989) (see (2.4.3) below) was developed.

The idea of the construction of test functions for particle-filled composites is similar to the construction of test functions for media with continuously distributed properties mentioned above. There are, however, some differences in the analysis of the direct and the dual variational bounds. In particular, in the latter the building block of the network is the Kozlov model (Kozlov, 1989), whereas in our approach the building block for the network is the Keller construction (Keller, 1963) for two closely spaced disks/spheres. We chose to describe in detail the latter case, because we were involved in developing it, and it is simpler to explain.

As was noted, composites of periodic structures with perfectly conducting particles of spherical and cylindrical shape (as well as with rigid particles) were investigated by Keller (1963) and Flaherty and Keller (1973). We also mention here the work of Frenkel and Acrivos (1967) and Graham (1981), among others. The analysis in these works was carried using "physical" arguments and a formal asymptotic for media with periodic structures. A rigorous mathematical approach to the problem of transport in high-contrast composites, including both periodic and disordered particle-filled composites, was later developed in a series of works: Berlyand and Kolpakov (2001); Berlyand and Novikov (2002); Novikov (2009); Berlyand and Mityushev (2005); Berlyand, Gorb and Novikov (2009); Berlyand, Borcea, and Panchenko (2005).

2.4 When does network modeling work?

We now discuss the key features of continuum problems, which allow for the use of structural disctetization described in the Introduction. Note that the network approximations described in this book are obtained by the structural network discretization procedure. For an arbitrary continuum problem that describes a heterogeneous medium one may not necessarily expect that this approximation would work because it requires specific features (e.g., high contrast in material properties, localization of physical fields) of a continuum problem. For instance, in the above examples of particle-filled composites (conducting particles in a dielectric matrix and suspensions), the interaction between particles is dominated by large physical fields inside the thin gaps between neighboring particles. Moreover, as will be shown below, in problems of this type the property of asymptotic localization of physical fields, such as the velocity field, the electrostatic potential, elastic stresses, etc., may take place. This property plays a central role in our structural discretization, and we next discuss the physical and geometrical conditions that result in such concentration.

We begin with two features of the problem that allow for network modeling and are easy to identify. First, the material properties of the particles and the medium should be vastly different – this property is usually referred to as the *high-contrast property* and it is needed for identifying the vertices of the network. Indeed, if the material properties of the particles and the matrix are of the same order of magnitude, then such a composite medium is described by slowly varying functions and typically no concentration of physical fields occurs. Thus, a physical condition which allows for discrete modeling is the high-contrast of the material properties of the particles and the matrix (background medium). The second condition is of a geometrical nature. It is called *dense packing*, when the distances between neighboring particles are much smaller than their radii, that is the concentration of particles is close to maximal. This is used to identify the edges of the graph.

These two conditions ensure that the dominant contribution to the overall material characteristics (e.g., effective conductivity, effective viscosity) comes from the physical fields between neighboring particles. Thus the particles may be represented as points (vertices of a network) and the gaps can be represented as edges connecting neighboring points. Integrals over the gaps of physical fields (e.g., the electrical or thermal energy of a gap) are assigned to the edges and we arrive at a discrete network model of a continuum problem obtained via the structural discretization procedure.

In this book we present a network modeling method for continuum problems that has two distinct aspects: modeling of material properties (e.g., constitutive equations for particles and the matrix in a composite, contrast in material properties, etc.) and modeling of geometry (e.g., particle shapes and locations).

2.4.1 Constitutive equation of scalar transport models

In this book we consider scalar constitutive equations. While they describe different physical phenomena (Ohm's law, Fick's law, Newton's law of cooling), they have the same mathematical form. We will refer to these as "transport" phenomena. Table 2.1 lists several transport problems. The equivalence of these problems makes it possible to treat these problems within a common mathematical framework.

Table 2.1 List of phenomena.

Phenomenon	Potential ϕ	Flux $\mathbf{J}$
Heat conduction	Temperature	Heat flux
Electrical conduction	Electric potentials	Current density
Diffusion	Density	Diffusion current density
Electrostatics	Electric potentials	Electric current

The driving force (the gradient of the potential ϕ) is related to the flux $\mathbf{J}$ by the corresponding constitutive equation. For all the phenomena listed in Table 2.1, the relationships between the flux and the driving force have the form:

$$\mathbf{J} = a(\mathbf{x})\nabla\phi, \tag{2.4.1}$$

where $a(\mathbf{x})$ is a material constant. In many cases equation (2.4.1) has the form $\mathbf{J} = -a(\mathbf{x})\nabla\phi$.

For all the phenomena listed, the balance equations have the form

$$\text{div }\mathbf{J} = 0 \tag{2.4.2}$$

or

$$\text{div}(a(\mathbf{x})\nabla\phi) = 0.$$

The network approximation approach, presented in this book, applies also to vectorial problems, e.g., highly packed suspensions as demonstrated in recent works Berlyand and Panchenko (2007) and Berlyand *et al.* (2009). However, because of the introductory nature of this book, we restrict our presentation to scalar problems since they are much simpler technically.

2.4.2 Periodic and disordered structures

Periodic and disordered structures are two main types of structures in nature and engineering. A periodic structure is obtained by periodic repetition of a typical element (called a periodicity cell or a basic cell (Bensoussan *et al.*, 1978; Sanchez-Palencia, 1980)). The disordered structures (also called topologically disordered structures (Gupta and Cooper, 1990)) are not periodic and are often random. The degree of disorder may vary from small disorder (e.g., small random perturbation of a deterministic, for example periodic, structure) to complete disorder (e.g., random structures) (Beran, 1968; Bonnecaze and Brady, 1991; Christensen, 1979; Kozlov, 1978; Novozilov, 1970; Torquato, 2002). Many natural and man-made materials are partially or completely disordered. Nevertheless, there are some solid three-dimensional composite materials with periodic structures that are presently manufactured, see, for example, Chou and Ko (1989) and McAllister and Lachman (1983). At the same time many two- and one-dimensional engineering

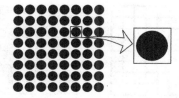

Figure 2.3 Periodic structure (left) and its periodicity cell (right).

structures and devices are ordered, usually periodic or quasi-periodic, e.g., Kalamkarov and Kolpakov (1997); Kolpakov (2004); Pshenichnov (1993).

Periodic systems are easier to model mathematically, since in such systems one can naturally select a basic periodicity element (a periodicity cell), see Figure 2.3, whose properties are representative of those of the entire system. This means that the local properties of the periodic system (a solution of the problem for the basic element, usually called a local or cell problem) completely determine the (overall) properties of the entire periodic system. In a disordered system, however, one cannot select a basic element that generates the entire system. For disordered systems, a so-called representative element is introduced. The representative element is determined as the minimal piece of the system or material, which contains the properties of the entire material. The problem of the choice of the representative element (e.g., its size) is still the subject of numerous discussions and is far from being completely solved, see, e.g., Ostoja-Starzewski (2006). There are experimental, numerical and theoretical results on the subject, predicting that the representative volume element must have a dimension along each direction not smaller than 10 characteristic dimensions of the basic elements (20–30 characteristic dimensions would be considered sufficient for sure). Elements along one direction are accepted as assuredly sufficient for the representative element. This means that the representative volume element must include more than 100 constitutive elements in the two-dimensional case and more than 1000 constitutive elements in the three-dimensional case. There is not a complete rigorously justified theory of the representative volume element yet. A disordered system can be naturally related to a graph (a network). In percolation theory often graphs of regular geometry are used for modeling of random materials or structures, see Figure 2.4(left). More realistic structures correspond to non-regular graphs Figure 2.4(right).

One can try to calculate the effective properties of random high-contrast composites using numerical methods by direct solution of the partial differential equations with rapidly oscillating coefficients (e.g., using finite elements or integral equations, see, e.g., Babǔshka *et al.* (1999); Cheng and Greengard (1998), or Fourier transforms, see, e.g., Michel *et al.* (2000, 2002); Vinogradov and Milton (2005); Peterseim (2010, 2012)). Then two questions arise. First, how large is the (computational) dimension of the problem (since the problem is highly inhomogeneous, the corresponding mesh has to have a large number of mesh

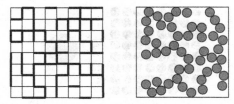

Figure 2.4 Disordered structures: random square lattice model (widely used in percolation theory) (left) and a system of randomly distributed disks (right).

points)? Second, how do we take into account the high contrast in the properties of the phases? In particular, it is necessary to address the issue of concentrated fluxes between closely spaced particles, which requires the use of a very fine mesh in the subdomains of concentration for an analytical study of the concentration phenomenon. It leads to a further increase in the computational complexity of the problem. Finally, it is necessary to repeat this computation many times in order to obtain an average over a large number of random configurations (i.e., to collect statistical information).

2.4.3 Media with continuously distributed characteristics. Kozlov model of high-contrast media

Kozlov (1989) proposed to describe the continuous distribution of high-contrast transport properties (conductivity, dielectric permittivity, complex impedance) in a simple geometric manner using a scalar oscillating function of the form

$$a(\mathbf{x}) = c(\mathbf{x})e^{S(\mathbf{x})/\epsilon^2} \qquad (2.4.3)$$

with $c(\mathbf{x}) > 0$, $S(\mathbf{x}) \geq 0$. Here $S(\mathbf{x})$ is a *smooth* function with isolated non-degenerate critical points. The small parameter ϵ models the high contrast.

Figure 2.5 displays a typical graph of the function (2.4.3). Figure 2.6 illustrates the behavior of the function (2.4.3) as ϵ tends to zero and $S(x) > 0$ in the one-dimensional case. For specific applications one must carefully select the function $a(\mathbf{x})$ in the Berryman–Borcea–Kozlov–Papanicolaou model.

In this book we consider particle-filled composites – a different class of problems in which the material properties cannot be described by a smooth function like equation (2.4.3). Furthermore, we consider materials with infinite contrast (corresponding to $\epsilon = 0$ in (2.4.3)) and our asymptotic analysis is carried out in the limit of dense packing, that is when distances between neighboring particles tend to 0.

2.4.4 Media with piecewise constant characteristics and systems of bodies. The Maxwell–Keller model

In this book, we consider media with piecewise constant characteristics, which correspond to particle-filled composites. Material properties of such composites

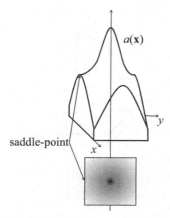

Figure 2.5 A typical graph and distribution over a periodicity cell of the functions $a(\mathbf{x})$ for two-dimensional composite described by Kozlov function.

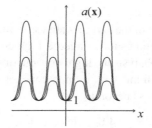

Figure 2.6 The evolution of the Kozlov function $a(\mathbf{x}) = c(\mathbf{x})e^{S(\mathbf{x})/\epsilon^2}$ as ϵ tends to zero in the one-dimensional case.

are described by discontinuous functions. If the material structure is periodic, we refer to such a model as a Maxwell–Keller model to distinguish it from a continuous Kozlov model. For materials with non-periodic structure we use the term "generalized Maxwell–Keller model".

A typical function describing material properties of a particle-filled composite material has the form

$$a(\mathbf{x}) = \begin{cases} a_i & \text{in particles,} \\ a_m & \text{in the matrix.} \end{cases} \qquad (2.4.4)$$

A typical graph of the function of the form (2.4.4) is presented in Figure 2.7. Functions of this form also describe systems of bodies. If we consider an electrostatic problem for a system of bodies, the distribution of the dielectric characteristics is described just by the function (2.4.4), where a_i means the dielectric constant of particles and a_m means the dielectric constant of the vacuum which is the background medium or the matrix in this case.

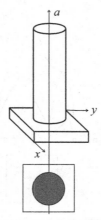

Figure 2.7 A typical graph and the distribution over the periodicity cell of the functions $a(\mathbf{x})$ for a two-dimensional composite with a periodic structure described by the function (2.4.4).

If $a_i \gg a_m$ (the properties of the particles are vastly different from the properties of the matrix), then the composite is called a high-contrast composite. For the purposes of mathematical analysis the highly conducting particles are replaced by a perfectly conducting medium (with infinite conductivity or, equivalently, zero resistivity). In this case (2.4.4) becomes

$$a(\mathbf{x}) = \begin{cases} 0 & \text{in particles,} \\ a_m & \text{in the matrix,} \end{cases} \qquad (2.4.5)$$

and $a(\mathbf{x})$ is proportional to the characteristic function of the matrix $\chi_m(\mathbf{x})$ defined as

$$\chi_m(\mathbf{x}) = \begin{cases} 0 & \text{in particles,} \\ 1 & \text{in the matrix.} \end{cases} \qquad (2.4.6)$$

This reflects the fact that the important parameter in this problem is the *dimensionless ratio* of resistivities of matrix and particles. The characteristic function of particles $\chi_i(\mathbf{x})$ is

$$\chi_i(\mathbf{x}) = \begin{cases} 1 & \text{in particles,} \\ 0 & \text{in the matrix.} \end{cases} \qquad (2.4.7)$$

The characteristic functions (2.4.6) and (2.4.7) satisfy the equation $\chi_i(\mathbf{x}) + \chi_m(\mathbf{x}) = 1$.

In the formula (2.4.5) there is no high-contrast parameter analogous to ϵ in (2.4.3). There is, however, another asymptotic parameter of a geometrical nature, which is implicitly present in (2.4.5). This parameter plays a key role in our considerations. It is called the interparticle distance parameter and is denoted by δ. It measures the typical distances between closely spaced neighboring particles.

It may be possible to derive model (2.4.5) as a degenerate limit of (2.4.3) for a suitably chosen family S_ϵ, that have saddle-points. Such an approximation requires extra effort in the analytical and numerical implementation. This question has not been addressed yet.

The interparticle distance parameter δ is a well-defined parameter of a physical nature, which is observed and measured experimentally. For highly packed composites, δ is small, which corresponds to the asymptotic limit $\delta \to 0$. Thus, the presence of this parameter allows one to apply asymptotic analysis.

Inspired by the successful rigorous justification (Borcea, 1998; Borcea and Papanicolaou, 1998; Borcea *et al.*, 1999) of the network approximation for media (2.4.3), we directly apply (bypassing the asymptotics in ϵ) their ideas of variational duality to highly packed particulate composites described by (2.4.5). The key difference between the two constructions is as follows. The variational test functions for (2.4.3) are based on saddle-points of $S(x)$ due to Kozlov (1989). The test functions for (2.4.5) are based on Keller's asymptotic solution for the flux between two closely spaced disks (Keller, 1963).

2.5 History of the mathematical investigation of overall properties of high-contrast materials and arrays of bodies

Network models have been widely used in applied sciences and engineering for a long time. However, the development of the mathematical foundation for these network models (in particular, the development of a justification, providing error estimates and limits of validity, for discrete network approximations of continuum problems) has lagged behind significantly. It would be incorrect to say that the problem of the mathematical justification of the network models did not attract the attention of researchers. On the contrary, many outstanding mathematicians and physicists have paid attention to the problems, which are, as we now understand, directly related to the network (structural) modeling. There exists a great variety of problems related to network (structural) modeling in one way or another. In this book, we briefly discuss the problems referred to as generalized Maxwell–Keller problems. Their formulation leads to asymptotic analysis of the overall transport properties of an arbitrary system of bodies under the condition that the distances between the bodies are much smaller than their sizes.

It is possible to distinguish three main historical stages in the analysis of the problem.

1873–1963. The history of the particle-filled composite conductor problems goes back to Maxwell (1873) who considered an electric field in a periodic array of bodies. Maxwell opened a stage in the mathematical study of the problem, which can be characterized as the analysis of transport properties of a periodic array of spheres and cylinders (disks). Another outstanding contribution in the mathematical analysis of the problem was made by Rayleigh (1892), who first studied the problem in a rigorous mathematical context. The solution of the problem of transport properties of composites with a periodic structure was completed when the framework of homogenization theory was developed in the last three decades of the twentieth century.

1963–1998. In 1963 Keller (1963) found that the Maxwell formula is not valid for densely packed bodies and derived formulas for transport properties of periodic arrays of circular disks and spheres different from the Maxwell formula. After Keller, various problems related to periodic arrays of closely placed spheres and cylinders (disks) were analyzed.

1998–1999. The next stage in the analysis of this problem is related to the idea of the approximation of continuum problems for high-contrast media with discrete (so-called network) models. These models were first introduced (in a rigorous mathematical context) in the work of Borcea (1998); Borcea *et al.* (1999); Borcea and Papanicolaou (1998), where a problem with periodic smooth high–contrast rapidly oscillating coefficients was investigated. The Berryman–Borcea–Papanicolaou approach used the Kozlov model of a high-contrast medium (see Section 2.4.3 for a description of the Kozlov model). The problem considered in Borcea (1998); Borcea *et al.* (1999); Borcea and Papanicolaou (1998) was first formulated in a periodic setting; however, the analysis of the cell problem by Berryman, Borcea and Papanicolaou clearly demonstrated that their approach works just as well for non-periodic problems, and it allowed them to approximate a wide variety of aforementioned Maxwell–Keller problems by the corresponding network models. Since 2001 we published several papers where the basic ideas and methods presented in this book have been introduced.

2.5.1 The problem of computation of overall properties of a periodic system of bodies

After Maxwell and Rayleigh, the problem of transport properties of periodic arrays of bodies attracted the attention of many investigators, see, e.g., Andrianov *et al.* (1996, 1999); Cheng and Greengard (1998); Doyle (1978); Hasimoto (1959); Keller (1987); Keller and Sachs (1964); Lions (1978); McPhedran (1986); McPhedran *et al.* (1988); Mityushev (1997); Nunan and Keller (1984b); Sangani and Acrivos (1983) (the list is not complete because the number of papers in the field is enormous and it keeps growing).

Later, the problem of transport properties of arrays of closely-placed bodies was analyzed by Doyle (1978); Drummon and Tahir (1984); Happel (1959);

Mityushev (1997); McKenzie *et al.* (1978); McPhedran and McKenzie (1978); Zuzovsky and Brenner (1977) (again the list is incomplete). A significant contribution to the problem was made by McPhedran and his co-authors in McPhedran (1986); McPhedran *et al.* (1988); Melrose and McPhedran (1991); Nicorovici and McPhedran (1996); Perrins *et al.* (1979); Yardley *et al.* (2001). Mityushev and co-authors (Berlyand and Mityushev, 2005, 2001; Makaruk *et al.*, 2006; Mityushev, 1997a; Mityushev and Adler, 2002b; Mityushev, 2001; Mityushev and Adler, 2002a; Mityushev and Rogosin, 2000) made a significant contribution to the investigation of transport properties of two-dimensional periodic arrays of bodies using the method of complex variables. We emphasize two specific features of the problems considered in the above papers:

(i) the bodies form a periodic array;
(ii) the bodies have the shape of spheres or cylinders (disks).

The transport properties of an array of elliptic cylinders and cylinders having square cross-section were considered recently by Andrianov *et al.* (2002, 1999); Mityushev (2009); Nicorovici and McPhedran (1996); Yardley *et al.* (2001).

Keller (1963) noted that while the classical dilute limit approximation by Maxwell and Rayleigh works up to relatively high concentration (about 30%), it fails to describe the effective properties when the particles are close to touching. In this work, asymptotic formulas for the effective conductivity of a periodic dense array of perfectly conducting spheres (cylinders) were derived. Analogous problems for the effective properties of solid/solid and fluid/solid composites with a periodic array of particles were considered by Drummon and Tahir (1984); Flaherty and Keller (1973); Happel (1959); Hasimoto (1959); Nunan and Keller (1984a,b); Sangani and Acrivos (1983). In the paper by McPhedran *et al.* (1988) a problem for closely spaced, highly conducting cylinders was considered using techniques from the theory of functions of complex variables. In Batchelor and O'Brien (1977) the conduction through a granular material was investigated using ensemble averaging and approximate solutions for closely packed spheres.

The problem of the effective viscosity for fluid/solid composites (suspensions) was also studied by Frenkel and Acrivos (1967) by formal asymptotic analysis for periodic arrays and more recently by Brady (1993) and Sierou and Brady (2002) by advanced numerical techniques of Stokesian dynamics for arbitrary arrays. In the work of Flaherty and Keller (1973) and Keller (1963, 1987) the contrast in material properties of the particles and the matrix was infinite. Taking into account the finite but high contrast is a very difficult task and it was done by McPhedran *et al.* (1988) (see also Batchelor and O'Brien (1977)).

In the monograph by Movchan *et al.* (2002), dynamic problems for densely-packed periodic composites were studied. Since periodic structures can be reduced to a periodicity cell for one particle, there was no need to construct a discrete network.

The need for such a construction arose naturally in the analysis of disordered materials. Continuum models for such materials were developed by many authors. We refer the reader to the recent monograph by Torquato (2002), where a comprehensive survey of the literature can be found. The history of the problem is presented by Landauer (1978), Markov (2000) and Milton (2002).

2.5.2 Homogenization theory for periodic and random structures

A general approach to finding transport properties of materials with periodic structure is provided by homogenization theory. Note that typically homogenization theory is restricted to problems with finite contrast in the material properties of components and it does not take into account the singularities due to neighboring particles close to touching.

There exists an extensive literature on the analysis of periodic structures in the framework of homogenization and multi-scale approaches (see, e.g., Bensoussan *et al.* (1978); Jikov *et al.* (1994); Panasenko (2005) and references therein). Homogenization theory establishes a relationship between the macroscopic and microscopic properties of heterogeneous materials. The multi-scale approach specifies the details of this relationship. The homogenization and multi-scale approaches (which are closely related) lead to a better understanding of large inhomogeneous structures and to the development of new methods for the analysis of such structures. These approaches lead to predictions of a number of new physical effects such as materials with negative Poisson ratio and negative thermal expansion coefficient (Kolpakov, 1985; Almgren, 1985; Milton, 2002; Kolpakov, 1987), and the memory effect (Bensoussan *et al.*, 1975). Homogenization methods stimulated the development of some new theories, for example, topological design by Bendsøe and Kikuchi (1988); Bendsøe and Sigmund (2004); Diaz and Kikuchi (1992).

Several mathematical approaches for the analysis of periodic structures were developed in the framework of homogenization theory, e.g., Bakhvalov and Panasenko (1989); Bensoussan *et al.* (1975, 1978); Del Maso (1993); Jikov *et al.* (1994); Sanchez-Palencia (1974). The homogenization procedure starts from solving the following problem on the periodicity cell P of the structure:

$$\begin{cases} \dfrac{\partial}{\partial x_i}\left(a_{ij}(\mathbf{y})\dfrac{\partial N^k}{\partial x_j} + a_{ik}(\mathbf{y})\right) = 0 \text{ in } P, \\ N^k(\mathbf{y}) \text{ is periodic on } P. \end{cases}$$

The homogenized transport coefficients A_{ij} (thermal or electrical conductivity, coefficient of diffusion, permittivity, etc.) are determined by integration of the cell problem solutions:

$$A_{ij} = \frac{1}{|P|} \int_P \left(a_{ij}(\mathbf{y}) + a_{ik}(\mathbf{y})\frac{\partial N^j}{\partial x_k}\right) d\mathbf{y}. \tag{2.5.1}$$

The key observation here is that in order to find A_{ij} the cell problem must be solved. In the case of densely-packed particles, this problem contains two characteristic length-scales: the size of the particles and the width of the very thin gaps between neighboring particles. This, along with the high contrast in material properties, leads to a high concentration – so-called localization (Kolpakov and Kolpakov, 2010; Gaudiello and Kolpakov, 2011) of physical fields inside the gaps. Thus, we arrive at a multi-scale problem. Using a large number of finite elements and the fine mesh adopted for the geometry of the problem, we can solve this problem for several closely spaced particles. Examples of such solutions are presented below. These examples show that for a large number of closely spaced particles the problem becomes numerically untractable (at least with standard numerical methods).

The problem of the computation of transport properties of disordered materials has attracted the attention of numerous investigators. Various methods for computing the transport properties of inhomogeneous random materials were developed, see, e.g., Cheng and Greengard (1997); Chinh (1997); Garboczi and Douglas (1996); Kellomaki *et al.* (1996); Novikov and Friedrich (2005); Pesetskaya (2005); Karal Jr. and Keller (1966); Thovert and Acrivos (1989); Torquato (2002); Willis (2002), and a rigorous mathematical theory of random homogenization emerged due to the efforts of many researches, see, e.g., Bourgeat and Piatnitski (2004); Bourgeat *et al.* (1994); Jikov *et al.* (1994); Kozlov (1989); Molchanov (1994); Sab (1992); Papanicolaou and Varadhan (1981); Papanicolaou (1995); Torquato (2002); Yurinski (1986, 1980). However, at present random homogenization is much less efficient than periodic homogenization. In particular, there is no efficient method (except direct numerical computations) for the calculation of the effective properties of a random composite with large (0.2–0.5) volume fraction of particles. Most of the rigorous mathematical work addresses the issue of the existence of the homogenized limit, but it does not provide the means of computing this limit. For example, the approach developed by Kozlov (1978, 1980) and Papanicolaou-Varadhan (1981) is based on the construction of a special random field similar to the field $N^k(\mathbf{y})$ introduced in (2.5.1). While the existence of this field was proved, in general there is no efficient method for its computation (except some special examples such as a random checkerboard (Kozlov, 1989)). At present the problem of random homogenization is much less well understood than the periodic case despite the efforts of many researchers.

2.5.3 Keller's analysis of the conductivity of a medium containing a periodic array of perfectly conducting spheres or cylinders

When the particles are placed relatively far from each other, the approach based on the series (harmonic) expansion works well, and it is necessary to save only a few harmonics in the series to obtain accurate formulas. When the particles are closely spaced, it is necessary to keep all harmonics in the series. The numerical results presented below, in Section 2.7, explain this fact. As an example, consider

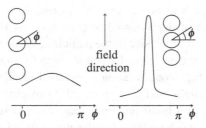

Figure 2.8 Typical distribution of energy around a disk as a function of polar angle.

a periodic array of disks. When the disks are placed relatively far from each other, the electrostatic energy density as a function of polar angle is a smooth function close to the distribution of energy around a single disk, see Figure 2.8 (left). When the disks are closely spaced, the energy density as a function of the polar angle looks like a singular function, see Figure 2.8 (right). Since the energy density is localized, the electric field will also be localized. It is known that a function like that shown in Figure 2.8 (left) usually can be well approximated with a small number of harmonics, whereas for the approximation of a function like that shown in Figure 2.8 (right) a large number of harmonics has to be used. Then the method based on the series (harmonic) expansion is not effective for closely placed bodies.

Keller suggested the use of techniques that are different from series (harmonics) expansions when dealing with functions like that shown in Figure 2.8 (right). In his paper Keller (1963) noted that the previous results of Maxwell and Rayleigh are not valid near the singularity in the effective conductivity (when the particles are close to touching, as $\delta_{ij} \approx 0$) and he derived new formulas for transport properties of periodic arrays of circular disks and spheres different from the Maxwell formula. Keller (1963) also observed that the new formulas agreed well with the numerical results (Keller and Sachs, 1964).

Keller's analysis was based on a hypothesis on the form of the flux between two closely placed disks (spheres). We employ Keller's method to derive an approximate formula for the flux between two disks (the i-th and the j-th) of radii R placed at distance δ_{ij}, see Figure 2.9.

We approximate the disks by the tangential parabolas

$$y = \frac{\delta_{ij}}{2} + \rho\frac{x^2}{2},$$

and

$$y = -\left(\frac{\delta_{ij}}{2} + \rho\frac{x^2}{2}\right),$$

where

$$\rho = \frac{1}{R} \tag{2.5.2}$$

is the curvatures of the disks.

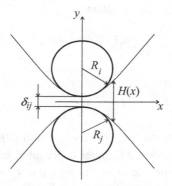

Figure 2.9 Two neighboring disks.

The distance between the parabolas is equal to

$$H_{ij}(x) = \delta_{ij} + \rho x^2. \tag{2.5.3}$$

For simplicity and without loss of generality we assume that the material constant of the matrix is $a(\mathbf{x}) = 1$. The key ingredient of Keller's approach is a "good" guess of the form of the local flux between the disks (we use the notation $\mathbf{x} = (x, y)$):

$$\mathbf{J}(\mathbf{x}) = \left(0, \frac{t_i - t_j}{H_{ij}(x)}\right), \tag{2.5.4}$$

so that it is proportional to the difference in temperature or potential and inversely proportional to the distance between the disks. Alternatively take the potential $\phi(\mathbf{x})$ in the form

$$\phi(\mathbf{x}) = \frac{(t_i - t_j)y}{H_{ij}(x)}. \tag{2.5.5}$$

If we accept the approximation (2.5.4), then the total (integral) flux between the disks can be calculated as follows:

$$J_{ij} = (t_i - t_j) \int_{-\infty}^{\infty} \frac{dx}{\delta_{ij} + \rho x^2} = \frac{t_i - t_j}{(\rho\delta_{ij})^{1/2}} \arctan\left[\left(\frac{\rho}{\delta_{ij}}\right)^{1/2} x\right]\Bigg|_{-\infty}^{\infty}. \tag{2.5.6}$$

In (2.5.6) the limits of integration are $-\infty$ and ∞. However, the flux (2.5.4) is determined only between the neighboring disks. Keller (1963) proposed changing the limits of integration from $-\infty$ and ∞ to $-R$ and R, where R is the radius of the disks. We note that since the integral (2.5.6) converges, the leading term in its asymptotics as $\delta \to 0$ remains the same. A problem arises for two closely spaced spheres, for which the corresponding integral diverges.

If the curvature ρ is not small, and the distance δ between the disks is small, then the value of the arctangent on the right-hand side of (2.5.6) is approximately

π. Taking into account (2.5.2), we obtain

$$J_{ij} = g_{ij}(t_i - t_j), \tag{2.5.7}$$

where

$$g_{ij} = \pi\sqrt{\frac{R}{\delta_{ij}}}. \tag{2.5.8}$$

Note that g_{ij} is the specific flux (i.e., the flux per unit difference of potential).

Formula (2.5.8) can be derived from the exact solution of the Laplace equation for two disks as the leading term in the asymptotic in δ. This exact solution (capacitance of two cylinders) can be found in Smythe (1950).

Also Keller's approach is based on the hypothesis that the flux between two closely placed disks has the form (2.5.4). Unfortunately, the function (2.5.4) (as well as a similar function for spheres) does not satisfy (2.4.2). The exact solution of the problem can be obtained in the framework of the dual variational approach described in Chapter 3. For the network approximation approach for the high-contrast Kozlov model this approach was first developed by Berryman, Borcea and Papanicolaou. We note that the analysis of the Maxwell–Keller model does not require an extensive analysis of the dual problem. Two upper- and lower-sided estimates, well known as Voigt (1910) and Reuss (1929) bounds, are the starting point for the analysis of the Maxwell–Keller model. However, these bounds are not tight enough, and for the complete analysis of the Maxwell–Keller model one needs to construct better trial functions so that the dual estimates match to the leading order. This can be carried out by improving by ordinary estimates the choice of suitable trial functions (see Chapter 3).

2.6 Berryman–Borcea–Papanicolaou analysis of the Kozlov model

As $\epsilon \to 0$, equation (2.4.3) describes a high-contrast composite material. In Borcea (1998) formula (2.4.3) is used to describe the local resistivity of a periodic medium. It was assumed that the function $S(\mathbf{x})$ has a specific form: periodic, differentiable, and has two maxima and two minima inside the periodicity cell. These assumptions guarantee the existence of a *saddle-point* $\mathbf{x}_s$ between the points of maxima and minima. Saddle-points of the function $S(x)$ (see Figure 2.5) play a key role in the construction of the discrete network. Roughly speaking, the network which approximates the electric or heat flow in the medium with conductivity $a(\mathbf{x})$ can be described as follows. The nodes of the network are maxima of $a(\mathbf{x})$ and the branches (edges) of the network connect adjacent maxima via saddle-points; see Figure 2.10. These branches correspond to paths (channels) of maximal conductivity.

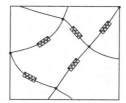

Figure 2.10 The network of paths of maximal conductivity. Resistors placed at saddle-points connect the neighbor cells.

2.6.1 Direct and dual problems

A two-dimensional boundary-value problem corresponding to the Kozlov model was analyzed by Borcea (1998); Borcea and Papanicolaou (1998); Borcea *et al.* (1999). In these works the problem of the rigorous derivation of a network approximation was first formulated for this model.

In Borcea *et al.* (1999) the following problem was considered:

$$\operatorname{div}\!\left(c(\mathbf{x})e^{-S(\mathbf{x})/\varepsilon}\nabla\phi\right) = 0 \text{ in } Q, \tag{2.6.1}$$

$$c(\mathbf{x})e^{-S(\mathbf{x})/\varepsilon}\frac{\partial\phi}{\partial\mathbf{n}}(\mathbf{x}) = I(\mathbf{x}) \text{ on } \partial Q,$$

where $I(\mathbf{x})$ satisfies the condition

$$\int_{\partial Q} I(\mathbf{x})d\mathbf{x} = 0,$$

($\mathbf{n}$ is the unit normal to ∂Q).

The dual problem has the form (Borcea *et al.*, 1999):

$$\nabla \times \left[\frac{e^{S(\mathbf{x})/\varepsilon}}{c(\mathbf{x})}\mathbf{j}\right] = 0 \text{ in } Q,$$

$$\operatorname{div}\mathbf{j} = 0 \text{ in } Q,$$

$$-\mathbf{j}(\mathbf{x})\mathbf{n} = I(\mathbf{x}) \text{ on } \partial Q.$$

2.6.2 Trial functions

In order to construct the trial functions, the authors, Borcea *et al.* (1999), introduced flow lines through the saddle-points of the function $S(\mathbf{x})$, and the local coordinates ξ and η along and perpendicular to the flow lines. In Figure 2.11, the vectors $\hat{\xi}$ and $\hat{\eta}$ are the orthonormal vectors corresponding to the local coordinates ξ and η. Then

$$S(\mathbf{x}) = S(\xi, \eta) = S(\xi, 0) + \frac{k(\xi)}{2}\eta^2 + \dots,$$

where $k(\xi)$ is the curvature in the direction perpendicular to the flow line.

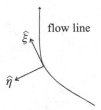

Figure 2.11 Basis vectors of the local coordinate system (ξ, η) related to the flow line.

The flow lines divide the domain into "cells", see Figure 2.10, which form the topology of the Berryman–Borcea–Papanicolaou network. The trial function in the primal problem is constricted as $\phi(\mathbf{x}) = const$ in the cells and

$$\phi(\mathbf{x}) = -\frac{f}{2}\mathrm{erf}\left(\frac{\eta}{\sqrt{\dfrac{2\varepsilon}{k(\xi)}}}\right) + const, \qquad (2.6.2)$$

near the flow lines. In (2.6.2) f is the change of $\phi(\mathbf{x})$ across the flow line.

The trial function in the dual problem is constructed as

$$\mathbf{j}(\mathbf{x}) = \frac{f}{2\sqrt{\pi\varepsilon}}e^{-\frac{k(\xi)\eta^2}{2\varepsilon}}\left[\sqrt{k(\xi)}\hat{\xi} - \frac{\eta}{2\sqrt{k(\xi)}}\frac{dk(\xi)}{d\xi}(1 + O(\delta))\hat{\eta}\right],$$

where $\hat{\xi}$ and $\hat{\eta}$ are the unit tangential and normal vectors of the local coordinate system (ξ, η).

Using these trial functions, the authors (Borcea *et al.*, 1999) found that the effective resistance R between the adjacent cells of the composite material is expressed in the form

$$R = c(\mathbf{x}_s)\sqrt{\frac{k^+}{k^-}}, \qquad (2.6.3)$$

where k^+ and k^- are the principal curvatures at the saddle-point $\mathbf{x}_s$ of the function $S(\mathbf{x})$.

It follows from (2.6.3) that the curvatures k^+ and k^- determine the conductivity of the high-contrast material with continuously distributed properties. Writing the balance equation with the conductivities (2.6.3), Borcea (1998), Borcea and Papanicolaou (1998) and Borcea *et al.* (1999) arrive at the finite-dimensional (network) model for the problem (2.6.1).

2.7 Numerical analysis of the Maxwell–Keller model

Theoretically, the interaction of closely placed bodies leads to an energy and flux concentration in the neck between them (Keller, 1963). One would expect to observe such an important phenomenon in experiments. This is not easy, because

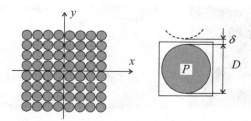

Figure 2.12 A periodic system of densely-packed disks (left) and
periodicity cell P (right).

flux concentration is not easy to measure experimentally. Further, one can only
measure fluxes or strains, although even such measurements are difficult in gaps
between closely spaced particles. Moreover, energy cannot be directly measured
experimentally. In this section we present numerical results (Kolpakov, 2007) that
illustrate energy localization.

2.7.1 Numerical verification of energy localization in the periodic case

Consider a periodic system of disks D_i, $i \in \mathbb{Z}^2$, which are non-overlapping and
not touching (see Figure 2.12). The domain $Q_p = \mathbb{R}^2 \setminus \bigcup_{i \in \mathbb{Z}^2} D_i$ is the matrix.

Let

$$\mathrm{div}(a(\mathbf{x})\nabla\phi) = 0 \text{ in } \mathbb{R}^2, \tag{2.7.1}$$

where the coefficient $a(\mathbf{x})$ has the form

$$a(\mathbf{x}) = \begin{cases} a_m & \text{in matrix } Q_p, \\ a_i & \text{in disks } D_i. \end{cases} \tag{2.7.2}$$

Equation (2.7.2) describes a mixture of two homogeneous components with finite
conductivities.

Let δ be the absolute value of the distance between particles and let $D = 2R$ be
the diameter of the particles (see Figure 2.12). The problem under consideration
involves two parameters: the relative interparticle distance

$$d = \frac{\delta}{D}, \tag{2.7.3}$$

and the contrast between the components of the composite material

$$c = \frac{a_i}{a_m}.$$

For a composite with high-contrast densely-packed particles the value of the
contrast c is large and the interparticle distance d is small:

$$c \gg 1, \ d \ll 1. \tag{2.7.4}$$

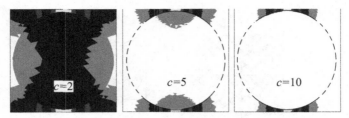

Figure 2.13 Development of the "energy neck" as a geometrically identifiable object.

We assume that the overall potential field with overall gradient vector $(0, E_0)$ (along the Oy-axis) is applied to the composite. For this case, the solution of the problem (2.7.1) has the form

$$\phi(\mathbf{x}) = E_0(y + \phi_p(\mathbf{x})), \quad \mathbf{x} = (x, y)$$

where $\phi_p(\mathbf{x})$ is a periodic function with the periodicity cell $P = [-1, 1]^2$ displayed in Figure 2.12.

Thus (2.7.1) becomes:

$$\text{div}\left(a(\mathbf{x})\nabla\phi^{\pm 1}\right) = 0 \text{ in } P, \tag{2.7.5}$$

$$\phi^{\pm 1}(x, -1) = -1, \, \phi^{\pm 1}(x, 1) = 1,$$

$$\frac{\partial\phi^{\pm 1}}{\partial x}(-1, y) = \frac{\partial\phi^{\pm 1}}{\partial x}(1, y) = 0.$$

Here, $\phi(\mathbf{x}) = E_0\phi^{\pm 1}(\mathbf{x})$.

Problem (2.7.5) was solved numerically for values of the contrast ranging from 2 to 10 000 and values of the relative interparticle distance δ ranging from 0.05 to 0.005. In the numerical computations the transport coefficient was $a_m = 1$, and $D = 2$.

In Figure 2.13 the distribution of the quantity $E = a(\mathbf{x})|\nabla\phi^{\pm 1}(\mathbf{x})|^2$ (the density of the local energy multiplied by two) is shown.

In Figure 2.13, one can observe the development of an "energy neck" as a geometrically identifiable object, starting with the contrast equal to 5. However, for small values of the contrast (from 5 to 500), the value of the energy density in the neck significantly depends on the value of the contrast c (where the relative interparticle distance δ is fixed). As the contrast grows, the value of the energy density in the neck becomes stable, that is it tends to a limit value (see Figure 2.14, which displays the distribution of energy for $c = 1000$ and $c = 10\,000$, respectively). The necks displayed in Figure 2.14 (left) and Figure 2.14 (right) have not only very similar (practically identical) shapes, but the values of the energy densities in the necks are also very similar (see the scale lines under the pictures in Figure 2.14 (left) and Figure 2.14 (right), which show the value of the energy density).

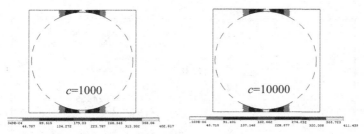

Figure 2.14 The distribution of the local energy in a periodicity cell for the
contrast $c = 1000$.

The value of the contrast at which the value of the energy density in the
neck stabilizes can be estimated to be 1000. This is a direct confirmation of
energy localization between neighboring particles in a high-contrast densely-
packed particle-filled composite proposed in Keller (1963).

2.7.2 Numerics for the non-periodic problem

The numerics for the periodic problem presented in the previous section verify the
existence of "energy necks" and estimate the values of the contrast and relative
interparticle distance at which the effect of energy concentration in high-contrast
densely-packed composite arises.

We present here the numerics for a non-periodic system of disks. The positions
of the disks $\{D_i, \; i = 1, \ldots, N\}$ are displayed in Figure 2.15. The problems were
solved for contrasts c from 10^3 to 10^6 and relative interparticle distances δ from
0.05 to 0.001.

The basic hypothesis of the modeling approach is still the phenomenon of energy
concentration in the domains between inclusions in high-contrast densely-packed
composite materials. In order to observe this phenomenon Figure 2.16 displays
the distribution of the energy in the neck between disks 3 and 4 (the numbers of
the disks are shown in Figure 2.15 in bold figures). One can see from the picture
(see also the detailed comment regarding this picture in Kolpakov (2007)) that the
energy indeed concentrates in the necks between particles.

We mention that standard finite-element (and similar) methods are not directly
applicable to the numerical analysis of the problem with many inclusions. In order
to obtain an accurate solution of the two-dimensional periodic problem with one
disk, it would be necessary use about 5000–10 000 finite elements. Most of them
would be concentrated in the neck. Indeed, the use of a few hundred finite elements
does not guarantee good accuracy.

The numerical method is very informative. As was noted, the experimental
observation of the energy localization effect is a difficult task which has not yet
been performed. The numerical method provides us with convincing evidence
of the existence of the energy localization effect and values of the contrast and
interparticle distance required to observe such energy necks.

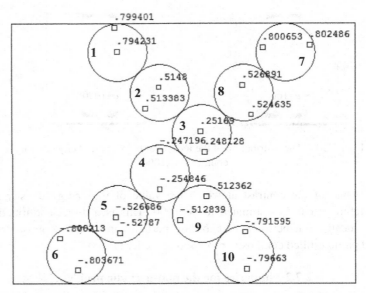

Figure 2.15 The system of disks with values of potentials determined from the numerical solution of the continuum problem. The disks are numbered with bold figures.

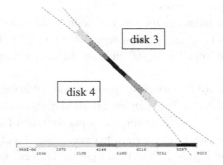

Figure 2.16 Distribution of the density of the local energy in the neck between disks. The numbering of the disks in accordance with Figure 2.15

2.7.3 Difference between localization effect and concentration effect

The results of the numerical computations presented in Section 2.7.1, suggest that the local energy takes large values (as large as 400 for the case presented in Figure 2.14 (right)) in the energy channels. At the same time, it takes relatively small values (as small as 3–10) outside the energy channels. Such an effect is usually referred as the concentration effect. For us the difference between the concentration and localization of fields is as follows. For the concentration phenomenon we assume that the field takes a relatively large value in one region but it is not small outside this region. Elasticity theory (see, e.g., Love (1929); Timoshemko and

Goodier (1951)) provides us with classical examples of the concentrations effects for stress–strains. The strain–stress field in elasticity theory can tend to infinity in the tip of a crack, but the elastic energy is not localized, it is distributed over the whole region. In other words, the energies stored in various regions of the same characteristic dimension (even in the region containing the tip of the crack) have the similar values. So, the defining property of the effect of the concentration of a field is that the absolute value of the field is large in a specific region, but integral values of the field in the region and outside the region have similar values.

A stronger phenomenon of concentration occurs when a field not only takes large values in a specific region but most of the field is accumulated in this specific region. In other words, integral values of the field outside this specific region are small. It is natural to refer to the accumulation of most of the field in some specific regions as localization. Usually, such localization is an asymptotic phenomenon. Let a problem depend on a parameter $\delta \rightarrow 0$.

Definition 2.1 If an effect of localization of the field takes place under the condition $\delta \rightarrow 0$, it is referred to as the effect of asymptotic localization of the field.

In a medium filled by highly-conducting disks, one can observe the transformation of the concentration phenomenon to the localization phenomenon when the interparticle distance $\delta \rightarrow 0$. As demonstrated in Figures 2.13–2.14, the local energy exhibits the localization property more strongly as compared with other fields (local flux and local gradient).

2.8 Percolation in disordered systems

The results presented below demonstrate that transport properties in densely-packed composites occur via neighbor-to-neighbor interactions. If the system of inclusions is random, then these interactions lead to percolation effects. Moreover, for non-periodic arrays one has to use continuum percolation rather than lattice percolation to model these phenomena.

As noted by Sahimi (2003), percolation processes were first developed in papers devoted to polymerization (Flory, 1941; Stockmayer, 1943) and branching processes (Good, 1949). In its present form, percolation theory first appeared in the mathematical literature in Brodbent and Hammerslay (1957). Note that the problem of continuum percolation is not yet completely understood (Meester and Roy, 1992).

Below, we will use standard terminology from percolation theory. Two vertices of a graph are called *connected* if there exists at least one path between them. The set of connected vertices is called a *cluster*.

The transport properties of a graph (network) depend on the transport properties of its edges. For discrete percolation, such transport characteristics are described by the probability p that the edge conducts ($1 - p$ is the probability that the edge does not conduct). For continuum percolation, the transport properties are related to the volume fraction p of inclusions.

It has been found (see Grimet (1992); Kesten (1992); Stauffer and Aharony (1992); Balberg (1987); Sahimi (2003)) that for many types of percolation, both discrete and continuum, the overall transport characteristic $\hat{a}$ of the graph (network) is related to p by

$$\hat{a} = \begin{cases} 0 & \text{if } p < p_c, \\ (p - p_c)^\mu & \text{if } p \geq p_c \end{cases} \tag{2.8.1}$$

(it is assumed that p is "close" to p_c).

In (2.8.1), p_c is the percolation threshold (the volume fraction of conducting edges or inclusions at which the network starts to demonstrate non-zero overall conductivity) and μ is referred to as the critical exponent.

2.9 Summary

The review of results presented above devoted to the problem of determining the conductivity of densely-packed high-contrast composite materials demonstrates that the problem is of both theoretical and practical importance. Although great progress has been made in the field, the previous theories cannot be considered complete. The results described in this Chapter have provided a direction for the work described in this book.

3

Network approximation for boundary-value problems with discontinuous coefficients and a finite number of inclusions

This chapter follows closely the work of Berlyand and Kolpakov (2001). The approach presented here was applied to the modeling of particle-filled composite materials. It is based on dual variational bounds and has been applied to both two- and three-dimensional problems (Berlyand *et al.*, 2005). Further development of this approach allowed us to obtain error estimates for the network approximation (Berlyand and Novikov, 2002). It also provides answers to several unsettled physical questions, such as polydispersity at high concentration (Berlyand and Kolpakov, 2001; Berlyand and Mityushev, 2005), weak and strong blow up of the effective viscosity of disordered suspensions (Berlyand and Panchenko, 2007), and it establishes a connection between the notion of capacitance and the network approximation (Kolpakov, 2005, 2006a). Subsequently this approach was generalized for fluids. Next a new "fictitious fluid" approach was introduced in Berlyand *et al.* (2005). This approach led to a complete description of all singular terms in the asymptotics of the viscous dissipation rate of such suspensions and provided a comprehensive picture of microflows in highly packed suspensions. Note that previous works addressed only certain singularities and therefore provided a partial analysis of such microflows. It also allowed us to predict an anomalous singularity in two-dimensional problems (thin films) which has no analog in three-dimensions (Berlyand and Panchenko, 2007).

This book systematically presents some of the results obtained by Berlyand *et al.* (2005, 2009); Berlyand and Kolpakov (2001); Berlyand and Novikov (2002); Novikov (2009). As mentioned earlier, we simplify the presentation by considering a two-dimensional scalar model. The construction and justification of the network approximations in the framework of our approach for a large variety of problems can be implemented using the following four-step scheme:

1. Formulation of the direct and dual variational principles and derivation of two-sided bounds for the effective properties of materials.

2. Formal construction of the network model.
3. Construction of the trial functions, which account for the concentration of the physical fields in the gaps between the particles. Here it is necessary to satisfy a balance condition for the fluxes of physical fields (electric current, fluid flow, etc.). In order to do this, the solution of the formal network model is employed. As a result, tight two-sided bounds are obtained.
4. Proof that the improved two-sided bounds coincide to leading order (possibly to second order) in the asymptotic limit as the interparticle distance goes to zero. These bounds are called asymptotically exact.

The essence of the above asymptotic approach amounts to addressing two issues:

(i) construction of a formal asymptotic expansion (asymptotic representation) for the effective properties;
(ii) rigorous justification of these asymptotic expansions.

Note that the issue (i) can be considered as a separate self-contained mathematical problem. A classical example of a formal asymptotic construction is given by the so-called asymptotic series introduced by Poincaré (1886). While these series are not necessarily convergent, they have been successfully used to solve a large number of physical problems.

3.1 Variational principles and duality. Two-sided bounds

We now describe a rigorous mathematical model of a medium with a large number of "perfectly conducting" circular particles. Throughout the book "*particles*" mean circular disks embedded into a matrix (solid or liquid).

We begin with the variational formulation of the problem and the corresponding boundary-value problem. Then we discuss the dual variational problem.

3.1.1 The variational formulation of the problem

Consider a two-dimensional mathematical model of a particle-filled composite material. Introduce the domain $\Pi = [-L, L] \times [-1, 1]$ occupied by the composite material, depicted in Figure 3.1. Denote by $\{D_i, i = 1, \ldots, N\}$ the disks which model the particles, where N is the total number of disks. The disks do not necessarily form any periodic array. The only restriction we impose is that the disks do not overlap. Then

$$Q_m = \Pi \setminus \bigcup_{i=1}^{N} D_i$$

is the matrix (the perforated domain Π).

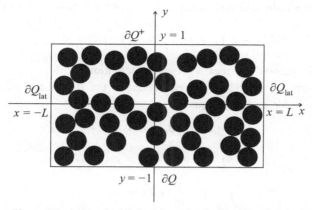

Figure 3.1 A domain filled with randomly distributed disks.

Denote by $\partial Q^+ = \{\mathbf{x} \in \mathbb{R}^2 \,|\, y = 1\}$ and $\partial Q^- = \{\mathbf{x} \in \mathbb{R}^2 \,|\, y = -1\}$ the top and bottom boundary of the domain Q_m and

$$\partial Q_{\text{lat}} = \partial Q_m \setminus (\partial Q^+ \bigcup \partial Q^-)$$

the lateral (left and right in the case under consideration) boundaries of the domain Q_m. We use the notation $\mathbf{x} = (x, y)$; correspondingly, in integrals $d\mathbf{x}$ means $dxdy$.

Introduce the functional space which corresponds to the condition that particles are ideally conducting:

$$V_p = \Big\{ \phi \in H^1(\Pi) \,|\, \phi(\mathbf{x}) = t_i^\phi \in \mathbb{R} \text{ on } D_i, \ i = 1, \dots, N; \qquad (3.1.1)$$

$$\phi(\mathbf{x}) = 1 \text{ on } \partial Q^+, \phi(\mathbf{x}) = -1 \text{ on } \partial Q^- \Big\}.$$

In the definition (3.1.1) each function $\phi(\mathbf{x})$ takes the constant values t_i^ϕ on the disk D_i. Note that it is not necessarily true that for $\phi_1, \phi_2 \in V_p, t_i^{\phi_1} = t_i^{\phi_2}$ $(i = 1, \dots, N)$.

The conditions $\phi(\mathbf{x}) = 1$ on ∂Q^+, and $\phi(\mathbf{x}) = -1$ on ∂Q^- corresponds to applying a potential of ± 1 to the top and the bottom boundaries ∂Q^+ and ∂Q^-, respectively.

Remark. The thickness of the domain Π in the Oy-direction is equal to two. The difference of the potential applied to the top and bottom boundaries ∂Q^+ and ∂Q^- of the domain is also equal to two. Thus, the average difference of potential per unit length in the Oy-direction is equal to one. In other words, with our choice of the domain Π and boundary conditions on the boundaries ∂Q^+ and ∂Q^- we consider a composite material subject to an applied external field of unit value (strength of electric field, temperature gradient, and so on).

Consider the following variational problem for the Dirichlet integral:

$$I(\phi) := \frac{1}{2} \int_{Q_m} a(\mathbf{x})|\nabla\phi(\mathbf{x})|^2 dx \to \min, \, \phi \in V_p \qquad (3.1.2)$$

(as usual, the symbol ":=" means "equal by definition").

The problem (3.1.1), (3.1.2) is equivalent to the following boundary-value problem (the Euler–Lagrange equations):

$$\text{div}(a(\mathbf{x})\nabla\phi) = 0 \text{ in } Q_m; \qquad (3.1.3)$$

$$\phi(\mathbf{x}) = t_i^\phi \text{ on } \partial D_i, \, i = 1, \ldots, N; \qquad (3.1.4)$$

$$\int_{\partial D_i} a(\mathbf{x})\frac{\partial\phi}{\partial\mathbf{n}}(\mathbf{x})dx = 0; \, i \in I; \qquad (3.1.5)$$

$$\phi(\mathbf{x}) = 1 \text{ on } \partial Q^+, \, \phi(\mathbf{x}) = -1 \text{ on } \partial Q^-; \qquad (3.1.6)$$

$$a(\mathbf{x})\frac{\partial\phi}{\partial\mathbf{n}}(\mathbf{x}) = 0 \text{ on } \partial Q_{\text{lat}}. \qquad (3.1.7)$$

Here $\mathbf{n}$ is the outward unit normal to the boundaries of the corresponding domains (Q_m or ∂D_i), and I is the set of interior disk (disks which do not touch the top and bottom boundaries of the domain Π).

The problem (3.1.1), (3.1.2), as well as the equivalent problem (3.1.3)–(3.1.7), describes the thermal flux in an inhomogeneous matrix of thermal conductivity $a(\mathbf{x})$ filled with perfectly conducting disks.

Note the following features of equations (3.1.3)–(3.1.7). The equation (3.1.4) corresponds to $\phi(\mathbf{x}) = t_i^\phi$ on D_i in (3.1.1), which represents the physical condition that the electric potential $\phi(\mathbf{x})$ is constant inside each ideally conducting particle. We emphasize that these constants are unknown (except for the particles which intersect the external boundaries ∂Q^+ or ∂Q^-), and they are in general different for different particles.

Condition (3.1.5) is obtained by integration by parts in the Euler–Lagrange equation, which corresponds to the functional (3.1.2), where the condition (3.1.4) was taken into account. This condition means that the total flux of the current entering each particle D_i is zero. Condition (3.1.7) means that the lateral boundaries of the domain Π are insulated.

3.1.2 Formulas for the effective conductivity

The effective conductivity can be defined in several equivalent forms. We begin from a definition which is natural from the physical point of view, where the effective conductivity is defined as the total flux through a cross-section of the domain (for example, through the upper boundary ∂Q^+). We will need an expression for this flux in terms of the value of the energy integral $I(\phi)$, which is commonly used in homogenization theory as a definition of the effective conductivity (Bensoussan *et al.*, 1978). For the sake of completeness, we show here the equivalence of these two definitions.

Multiply equation (3.1.3) by $\phi(\mathbf{x})$ and integrate by parts. Then

$$0 = -\int_{Q_m} a(\mathbf{x})|\nabla\phi(\mathbf{x})|^2 d\mathbf{x} + \int_{\partial Q^+} a(\mathbf{x})\frac{\partial\phi}{\partial\mathbf{n}}(\mathbf{x})\phi(\mathbf{x})d\mathbf{x} \qquad (3.1.8)$$

$$+ \int_{\partial Q^-} a(\mathbf{x})\frac{\partial\phi}{\partial\mathbf{n}}(\mathbf{x})\phi(\mathbf{x})d\mathbf{x} + \sum_{i=1}^{N} \int_{\partial D_i} a(\mathbf{x})\frac{\partial\phi}{\partial\mathbf{n}}(\mathbf{x})\phi(\mathbf{x})d\mathbf{x}.$$

For the integrals over the boundaries of the particles, we use (3.1.4) and (3.1.5) to obtain

$$\int_{\partial D_i} a(\mathbf{x})\frac{\partial\phi}{\partial\mathbf{n}}(\mathbf{x})\phi(\mathbf{x})d\mathbf{x} = t_i^\phi \int_{\partial D_i} a(\mathbf{x})\frac{\partial\phi}{\partial\mathbf{n}}(\mathbf{x})d\mathbf{x} = 0. \qquad (3.1.9)$$

Furthermore,

$$\int_{\partial Q^+} a(\mathbf{x})\frac{\partial\phi}{\partial\mathbf{n}}(\mathbf{x})\phi(\mathbf{x})d\mathbf{x} + \int_{\partial Q^-} a(\mathbf{x})\frac{\partial\phi}{\partial\mathbf{n}}(\mathbf{x})\phi(\mathbf{x})d\mathbf{x} \qquad (3.1.10)$$

$$= \int_{\partial Q^+} a(\mathbf{x})\frac{\partial\phi}{\partial\mathbf{n}}(\mathbf{x})d\mathbf{x} - \int_{\partial Q^-} a(\mathbf{x})\frac{\partial\phi}{\partial\mathbf{n}}(\mathbf{x})d\mathbf{x}$$

due to (3.1.6).

Multiply equation (3.1.3) by 1 and integrate the result by parts to get

$$0 = \int_{\partial Q^+} a(\mathbf{x})\frac{\partial\phi}{\partial\mathbf{n}}(\mathbf{x})d\mathbf{x} + \int_{\partial Q^-} a(\mathbf{x})\frac{\partial\phi}{\partial\mathbf{n}}(\mathbf{x})d\mathbf{x} - \sum_{i=1}^{N} \int_{\partial D_i} a(\mathbf{x})\frac{\partial\phi}{\partial\mathbf{n}}(\mathbf{x})d\mathbf{x}.$$

The last integral is equal to zero due to (3.1.5). Then

$$\int_{\partial Q^+} a(\mathbf{x})\frac{\partial\phi}{\partial\mathbf{n}}(\mathbf{x})d\mathbf{x} + \int_{\partial Q^-} a(\mathbf{x})\frac{\partial\phi}{\partial\mathbf{n}}(\mathbf{x})d\mathbf{x} = 0. \qquad (3.1.11)$$

The physical meaning of (3.1.11) is that the total flux through the lower boundary ∂Q^- is equal to the total flux through the upper boundary ∂Q^+.

From (3.1.10) and (3.1.11) we obtain

$$\int_{\partial Q^+} a(\mathbf{x})\frac{\partial\phi}{\partial\mathbf{n}}(\mathbf{x})\phi(\mathbf{x})d\mathbf{x} + \int_{\partial Q^-} a(\mathbf{x})\frac{\partial\phi}{\partial\mathbf{n}}(\mathbf{x})\phi(\mathbf{x})d\mathbf{x} = 2\int_{\partial Q^+} a(\mathbf{x})\frac{\partial\phi}{\partial\mathbf{n}}(\mathbf{x})d\mathbf{x}.$$

$$(3.1.12)$$

Combining (3.1.8), (3.1.9) and (3.1.12), we finally obtain

$$\int_{\partial Q^+} a(\mathbf{x})\frac{\partial\phi}{\partial\mathbf{n}}(\mathbf{x})d\mathbf{x} = \frac{1}{2}\int_{Q_m} a(\mathbf{x})|\nabla\phi(\mathbf{x})|^2 d\mathbf{x}, \qquad (3.1.13)$$

where $\phi(\mathbf{x})$ is the solution of the problem (3.1.3)–(3.1.7) (or equivalently, it solves the problem (3.1.1), (3.1.2)).

3.1.3 The effective conductivity of the composite material

In this chapter we introduce a network approximation for the problem of the effective conductivity. When we say that one problem approximates another one, we mean that the value of a specific characteristic determined from one problem approximates (in a rigorous mathematical sense) the value of the same characteristic determined from the other problem. Here such a characteristic is the effective conductivity of the composite material. Note that the characteristic called here the effective conductivity is also referred to in the literature as the overall, average or macroscopic conductivity of the composite material (Beran, 1968; Christensen, 1979; Aboudi, 1991; Nemat-Nasser and Hori, 1993; Shermergor, 1977).

We first define it for the continuum problem and then construct its approximation in the corresponding discrete problem.

Definition 3.1 The flux through the boundary ∂Q^+ per unit length defined by

$$\hat{a} = \frac{1}{|\partial Q^+|} \int_{\partial Q^+} a(\mathbf{x}) \frac{\partial \phi}{\partial \mathbf{n}}(\mathbf{x}) d\mathbf{x} \tag{3.1.14}$$

is called the effective (overall) conductivity of the composite material.

Here $|\partial Q^+| = 2L$ means the length of the boundary segment ∂Q^+.

For our purposes, it is more convenient to use the total flux

$$A = \hat{a}|\partial Q^+| = 2L\hat{a} \tag{3.1.15}$$

across the boundary segment ∂Q^+.

Due to (3.1.13) we have the following formulas for A:

$$A = \frac{1}{2} \int_{Q_m} a(\mathbf{x})|\nabla\phi(\mathbf{x})|^2 d\mathbf{x}, \tag{3.1.16}$$

$$A = \int_{\partial Q^+} a(\mathbf{x}) \frac{\partial \phi}{\partial \mathbf{n}}(\mathbf{x}) d\mathbf{x}, \tag{3.1.17}$$

where $\phi(\mathbf{x})$ is the solution of the problem (3.1.3)–(3.1.7) (or the equivalent variational problem (3.1.1), (3.1.2)).

Since $\phi(\mathbf{x})$ also solves the minimization problem (3.1.2), we can write

$$A = \frac{1}{2} \min_{\phi \in V_p} \int_{Q_m} a(\mathbf{x})|\nabla\phi(\mathbf{x})|^2 d\mathbf{x}. \tag{3.1.18}$$

The equality (3.1.18) can be understood as an extremal principle for the effective conductivity. Extremal principles have been derived for the homogenized characteristics of elastic solids, see Reuss (1929); Voigt (1910); Hill (1963), and plates and beams, see Kolpakov (2004). Now the theory of extremal principles forms a specific direction in the theory of composite materials, see, e.g., Milton (2002).

3.2 Composite material with homogeneous matrix

If the material of the matrix is homogeneous, then

$$a(\mathbf{x}) = a = const. \tag{3.2.1}$$

3.2.1 The dimensionless problem

Observing equations (3.1.3)–(3.1.7), we find that all the equations involving a are homogeneous (i.e., the right-hand sides are zero). Under condition (3.2.1) we can divide these equations by a and, as a result, eliminate a from the boundary problem (3.1.3)–(3.1.7) (but not from the formulas (3.1.16)–(3.1.18) for the computation of the effective conductivity A). Formulas (3.1.16)–(3.1.18) involve a as a multiplier. Thus, without loss of generality, the material constant of the matrix can be taken to be $a = 1$, and therefore, the relationship (2.4.1) between the local flux (e.g., the current) $\mathbf{J}$ and the driving force (e.g., the gradient of the electric field) $\nabla\phi$ in the matrix is $\mathbf{J} = \nabla\phi$.

Thus there is no need to distinguish between the flux and the driving force in the problem under consideration.

3.2.2 Extremal form of the problem. The direct problem

Under condition (3.2.1) with $a = 1$, the extremal problem (3.1.2) can be written as follows:

$$I(\phi) := \frac{1}{2} \int_{Q_m} |\nabla\phi(\mathbf{x})|^2 dx \to \min, \ \phi \in V_p. \tag{3.2.2}$$

The boundary-value problem (3.1.3)–(3.1.7) under condition (3.2.1) takes the form

$$\Delta\phi = 0 \text{ in } Q_m; \tag{3.2.3}$$

$$\phi(\mathbf{x}) = t_i^\phi \text{ on } \partial D_i, \ i = 1, \ldots, N; \tag{3.2.4}$$

$$\int_{\partial D_i} \frac{\partial\phi}{\partial\mathbf{n}}(\mathbf{x})dx = 0; \tag{3.2.5}$$

$$\phi(\mathbf{x}) = 1 \text{ on } \partial Q^+, \quad \phi(\mathbf{x}) = -1 \text{ on } \partial Q^-; \tag{3.2.6}$$

$$\frac{\partial\phi}{\partial\mathbf{n}}(\mathbf{x}) = 0 \text{ on } \partial Q_{\text{lat}}. \tag{3.2.7}$$

This problem will be the main object of our investigation.

The formulas for the effective conductivity (3.1.16) and (3.1.17) take the following form:

$$A = \frac{1}{2} \int_{Q_m} |\nabla\phi(\mathbf{x})|^2 dx, \tag{3.2.8}$$

$$A = \int_{\partial Q^+} \frac{\partial\phi}{\partial\mathbf{n}}(\mathbf{x})dx, \tag{3.2.9}$$

where $\phi(\mathbf{x})$ is the solution of the problem (3.2.3)–(3.2.7) (or equivalently of the extremal problem (3.2.2)), and the formula (3.1.18) takes the form

$$A = \frac{1}{2} \min_{\phi \in V_p} \int_{Q_m} |\nabla \phi(\mathbf{x})|^2 dx. \tag{3.2.10}$$

From (3.2.10) we obtain the simplest well-known upper bound

$$A \leq \frac{1}{2} \int_{Q_m} |\nabla \phi(\mathbf{x})|^2 dx \tag{3.2.11}$$

for any $\phi \in V_p$.

3.2.3 Existence and uniqueness of solution of the problem (3.2.3)–(3.2.7)

The problem (3.2.3)–(3.2.7) is similar to a mixed boundary-value problem (Courant and Hilbert, 1953) (Dirichlet boundary condition on the top and bottom sides of Π and on particles and Neuman boundary condition on the left and right sides of Π). But it is not identical to the classical mixed problem because the constants $\{t_i^\phi, i = 1, \ldots, N\}$ are unknown.

The problem (3.2.3)–(3.2.7) is equivalent to the minimization problem (3.2.2). We introduce the auxiliary functional space

$$V_0 = \left\{ \phi \in H^1(\Pi) \mid \phi(\mathbf{x}) = t_i^\phi \text{ on } D_i, i = 1, \ldots, N; \tag{3.2.12} \right.$$

$$\left. \phi(\mathbf{x}) = 0 \text{ on } \partial Q^+, \phi(\mathbf{x}) = 0 \text{ on } \partial Q^- \right\}$$

corresponding to the functional space (3.1.1) and consider the variational problem corresponding to the minimization problem (3.2.12), which is obtained as follows. Suppose $\phi \in V_p$ is the minimizer of (3.2.2). Then $(\phi + \tau v) \in V_p$, for any $v \in V_0$ and any $\tau \in R$. Thus the variation of the functional $I(\phi)$ around the minimizer $\phi(\mathbf{x})$ in the direction of $v(\mathbf{x})$ is zero for any $v \in V_0$ (or else, smaller values can be achieved):

$$0 = \frac{d}{d\tau} I(\phi + \tau v)\Big|_{\tau=0} = \int_{Q_m} \nabla \phi(\mathbf{x}) \nabla v(\mathbf{x}) dx.$$

That is

$$\int_{Q_m} \nabla \phi(\mathbf{x}) \nabla v(\mathbf{x}) dx = 0 \tag{3.2.13}$$

for any $v \in V_0$.

Writing $\phi(\mathbf{x}) = u(\mathbf{x}) + g(\mathbf{x})$, where $g \in V_p$ is fixed, and $u(\mathbf{x}) = \phi(\mathbf{x}) - g(\mathbf{x}) \in V_0$, we can rewrite the variational problem (3.2.13) in terms of u: Find $u \in V_0$, such that

$$\int_{Q_m} \nabla u(\mathbf{x}) \nabla v(\mathbf{x}) dx = - \int_{Q_m} \nabla g(\mathbf{x}) \nabla v(\mathbf{x}) dx \tag{3.2.14}$$

for any $v \in V_0$.

Define the bilinear form

$$B(u, v) := \int_{Q_m} \nabla u(\mathbf{x}) \nabla v(\mathbf{x}) dx$$

and the linear functional $F_g \in V_0^*$ as

$$F_g(v) := - \int_{Q_m} \nabla g(\mathbf{x}) \nabla v(\mathbf{x}) \, dx.$$

Then the variational form (3.2.14) takes the form: Find $u \in V_0$, such that $B(u, v) = F_g(v)$, for any $v \in V_0$.

Note that $B[u, v]$ defines a scalar product $(u, v)_B := B(u, v)$ on the space V_0. The resulting norm $\|u\|_B^2 := (u, u)_B$ is equivalent to the $H^1(Q_m)$ norm due to Friedrichs's inequality (Zeidler, 1995). Hence, V_0 is a Hilbert space under the scalar product $(u, v)_B$. Therefore, due to the Riesz representation lemma (Zeidler, 1995), for the linear functional $F_g \in V_0^*$ there exists a unique $u \in V_0$ such that $(u, v)_B = F_g(v)$ for any $v \in V_0$. This u is a solution of equation (3.2.14).

3.2.4 The dual problem

The extremal problem, which is dual to the original problem (3.2.2), can be introduced in a standard way (see, e.g., Ekeland and Temam (1976)). The dual problem corresponds physically to the extremal principle for the current. Some specific features are due to the fact that the trial functions of the direct problem take constant values on the particles.

Introduce the following functional space

$$W_p = \left\{ \mathbf{v}(\mathbf{x}) = (v_1(\mathbf{x}), v_2(\mathbf{x})) \in L_2(Q_m) \mid \mathrm{div}\,\mathbf{v} \in L_2(Q_m), \right. \tag{3.2.15}$$

$$\left. \int_{\partial D_i} \mathbf{v}(\mathbf{x})\mathbf{n}dx = 0, \ i = 1, \ldots, N; \ \mathbf{v}(\mathbf{x})\mathbf{n} = 0 \text{ on } \partial Q_{\mathrm{lat}} \right\}.$$

The second (integral) condition in (3.2.15) corresponds to condition (3.2.5). The third condition corresponds to the boundary condition (3.2.7). The divergence of a function $\mathbf{v} \in L_2(Q_m)$ should be understood in the distributional sense (see Section 1.2.4):

$$\int_{Q_m} \mathrm{div}\,\mathbf{v}(\mathbf{x})\phi(\mathbf{x})dx = - \int_{Q_m} \mathbf{v}(\mathbf{x})\nabla\phi(\mathbf{x})dx$$

for any $\phi \in \mathcal{D}(Q_m)$.

We note that conditions $\mathbf{v}(\mathbf{x}) \in L_2(Q_m)$, $\mathrm{div}\,\mathbf{v} \in L_2(Q_m)$ guarantee the existence of the trace of the function $\mathbf{v}(\mathbf{x})\mathbf{n}$ on the boundary $\partial Q_{\mathrm{lat}}$ of the domain Q_m. The trace $\mathbf{v}(\mathbf{x})\mathbf{n}$ is a function from the functional space $H^{-1/2}(\partial Q_{\mathrm{lat}})$, see Section 1.2.3.

We next use the Legendre transform in R^2 (see Section 1.3) which establishes the duality between the quadratic forms $\frac{1}{2}\mathbf{x}^2$ and $\frac{1}{2}\mathbf{v}^2$. By a standard procedure (see Section 1.3.1), we get that

$$\frac{1}{2}\mathbf{x}^2 = \max_{\mathbf{v}\in\mathbb{R}^2}\left(\mathbf{v}\mathbf{x} - \frac{1}{2}\mathbf{v}^2\right), \quad \mathbf{x} \in \mathbb{R}^2. \tag{3.2.16}$$

We apply (3.2.16) to the integrand in (3.2.10). Then (3.2.10) can be written in the form

$$A = \min_{\phi\in V_p}\int_{Q_m}\frac{1}{2}|\nabla\phi(\mathbf{x})|^2 d\mathbf{x} \tag{3.2.17}$$

$$= \min_{\phi\in V_p}\int_{Q_m}\max_{\mathbf{v}\in W_p}\left(\mathbf{v}(\mathbf{x})\nabla\phi(\mathbf{x}) - \frac{1}{2}\mathbf{v}(\mathbf{x})^2\right)d\mathbf{x}$$

$$= \min_{\phi\in V_p}\max_{\mathbf{v}\in W_p}\int_{Q_m}\left(\nabla\phi(\mathbf{x})\mathbf{v}(\mathbf{x}) - \frac{1}{2}\mathbf{v}^2\right)d\mathbf{x}$$

$$= \max_{\mathbf{v}\in W_p}\min_{\phi\in V_p}\int_{Q_m}\left(\nabla\phi(\mathbf{x})\mathbf{v}(\mathbf{x}) - \frac{1}{2}\mathbf{v}(\mathbf{x})^2\right)d\mathbf{x}.$$

We arrive at the problem of interchanging max and min in (3.2.17). This is a well-known minimax problem and the conditions for interchangeability are presented in Ekeland and Temam (1976) (see Section 1.3.2). Since the Lagrangian function

$$L(\mathbf{q}, \mathbf{v}) = \int_{Q_m}\left(\mathbf{q}(\mathbf{x})\mathbf{v}(\mathbf{x}) - \frac{1}{2}\mathbf{v}(\mathbf{x})^2\right)d\mathbf{x}$$

corresponding to the quadratic functional $I(\phi)$ (3.2.2) satisfies the conditions of Proposition 2.2, Chapter 24 from Ekeland and Temam (1976), we can interchange max and min in (3.2.17).

Since the term $\frac{1}{2}\mathbf{v}^2(\mathbf{x})$ in (3.2.17) does not contain $\phi(\mathbf{x})$, it is possible to write (3.2.17) in the form

$$A = \max_{\mathbf{v}\in W_p}\left(\int_{Q_m}-\frac{1}{2}\mathbf{v}(\mathbf{x})^2 d\mathbf{x} + \min_{\phi\in V_p}\int_{Q_m}\nabla\phi(\mathbf{x})\mathbf{v}(\mathbf{x})d\mathbf{x}\right).$$

We integrate by parts in the last integral taking into account that the boundary of Q_m consists of the external boundaries of the rectangle ∂Q^+, ∂Q^-, ∂Q_{lat} and the boundaries ∂D_i of the domains D_i ($i = 1, \ldots, N$), see Figure 3.1.

Then we get

$$A = \max_{\mathbf{v}\in W_p}\left(\int_{Q_m}-\frac{1}{2}\mathbf{v}(\mathbf{x})^2 d\mathbf{x} + \min_{\phi\in V_p}\left[-\int_{Q_m}\phi(\mathbf{x})\,\text{div}\mathbf{v}(\mathbf{x})d\mathbf{x}\right.\right. \tag{3.2.18}$$

$$+ \int_{\partial Q_{\text{lat}}}\phi(\mathbf{x})\mathbf{v}(\mathbf{x})\mathbf{n}d\mathbf{x} + \int_{\partial Q^+}\phi(\mathbf{x})\mathbf{v}(\mathbf{x})\mathbf{n}d\mathbf{x}$$

$$\left.\left.+ \int_{\partial Q^-}\phi(\mathbf{x})\mathbf{v}(\mathbf{x})\mathbf{n}d\mathbf{x} + \sum_{i=1}^N\int_{\partial D_i}\phi(\mathbf{x})\mathbf{v}(\mathbf{x})\mathbf{n}d\mathbf{x}\right]\right).$$

Due to the definition of the functional space W_p (in particular, using the fact that $\mathbf{v}(\mathbf{x})\mathbf{n} = 0$ on ∂Q_{lat}) and since $\phi(\mathbf{x}) = const$ on ∂D_i in accordance with (3.2.4), we observe that the third and the sixth integrals in (3.2.18) are equal to zero.

On the horizontal boundaries $\partial Q^+ \bigcup \partial Q^-$ the function $\phi(\mathbf{x})$ takes the values ± 1, respectively, since it is in V_p. To simplify the notation we introduce a function of the variable y, such that

$$\phi^0(1) = 1, \quad \phi^0(-1) = -1, \qquad (3.2.19)$$

i.e., $\phi^0(y) = 1$ and -1 on the surfaces ∂Q^+ and ∂Q^-, correspondingly.

Using the function $\phi^0(y)$, we rewrite (3.2.18) as follows:

$$A = \max_{\mathbf{v} \in W_p} \left(-\frac{1}{2} \int_{Q_m} \mathbf{v}(\mathbf{x})^2 d\mathbf{x} + \int_{\partial Q^+ \cup \partial Q^-} \phi^0(y)\mathbf{v}(\mathbf{x})\mathbf{n}d\mathbf{x} \right. \qquad (3.2.20)$$

$$\left. + \min_{\phi \in V_p} \left[-\int_{Q_m} \phi(\mathbf{x}) \operatorname{div}\mathbf{v}(\mathbf{x})d\mathbf{x} \right] \right).$$

The last term in the right-hand side of (3.2.20) is linear in ϕ and it is equal to $-\infty$ if $\operatorname{div}\mathbf{v} \neq 0$. To understand this, it is sufficient to take a function $\tilde{\phi} \in V_p$, such that

$$\int_{Q_m} [\tilde{\phi}(\mathbf{x}) - \phi^0(y)] \operatorname{div}\mathbf{v}(\mathbf{x})d\mathbf{x} \neq 0$$

(the function $\tilde{\phi}(\mathbf{x})$ exists if $\operatorname{div}\mathbf{v} \neq 0$) and then take

$$\phi(\mathbf{x}) = t[\tilde{\phi}(\mathbf{x}) - \phi^0(y)] + \phi^0(y),$$

where t is an arbitrary number, as a trial function. Note that $\phi \in V_p$ for any $t \in R$ if $\tilde{\phi} \in V_p$. Thus, we can assume that $\phi(\mathbf{x}) - \phi^0(y)$ is an arbitrary function from $H^{01}(Q_m)$. For this choice of trial function

$$\int_{Q_m} \phi(\mathbf{x}) \operatorname{div}(\mathbf{v})d\mathbf{x} = t \int_{Q_m} [\tilde{\phi}(\mathbf{x}) - \phi^0(y)] \operatorname{div}\mathbf{v}(\mathbf{x})d\mathbf{x} + \int_{Q_m} \phi^0(y) \operatorname{div}\mathbf{v}(\mathbf{x})d\mathbf{x}.$$

Since t is arbitrary, this expression can take any value from $-\infty$ to $+\infty$ and the minimal value is $-\infty$.

Thus, the maximum in (3.2.20) is attained on the functions satisfying the condition

$$\operatorname{div}\mathbf{v} = 0 \text{ in } Q_m. \qquad (3.2.21)$$

As a result, we obtain

$$A = \max_{\substack{\mathbf{v} \in W_p, \\ \operatorname{div} \mathbf{v}=0}} \left(-\frac{1}{2} \int_{Q_m} \mathbf{v}(\mathbf{x})^2 d\mathbf{x} + \int_{\partial Q^+ \cup \partial Q^-} \phi^0(y)\mathbf{v}(\mathbf{x})\mathbf{n}d\mathbf{x} \right). \qquad (3.2.22)$$

In formulas, $\operatorname{div}\mathbf{v} = 0$ means (3.2.21).

From (3.2.22) we get an ordinary lower-sided bound for the effective constant:

$$A \geq J(\mathbf{v}) := \left(-\frac{1}{2} \int_{Q_m} \mathbf{v}(\mathbf{x})^2 d\mathbf{x} + \int_{\partial Q^+ \cup \partial Q^-} \phi^0(y)\mathbf{v}(\mathbf{x})\mathbf{n}d\mathbf{x} \right) \qquad (3.2.23)$$

for any $\mathbf{v} \in W_p$ satisfying the condition (3.2.21).

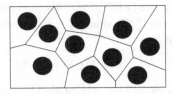

Figure 3.2 Voronoi tessellation in a bounded domain.

The solution $\phi(\mathbf{x})$ of the problem (3.1.1), (3.2.2) satisfies:

$$J(\mathbf{v}) = I(\phi), \tag{3.2.24}$$

where $\mathbf{v}(\mathbf{x}) = \nabla\phi(\mathbf{x})$.

Equality (3.2.24) follows from the definition of the functionals $I(\phi)$ and $J(\mathbf{v})$ (3.2.2) and (3.2.23) and Green's formula. Using the boundary conditions (3.2.4)–(3.2.6) this identity can be written as follows:

$$\int_{Q_m} |\nabla\phi(\mathbf{x})|^2 d\mathbf{x} = 2 \int_{\partial Q^+} \frac{\partial\phi}{\partial\mathbf{n}}(\mathbf{x})\phi^0(y)d\mathbf{x}.$$

3.2.5 Modeling of particle-filled composite materials using the Delaunay–Voronoi method. The notion of pseudo-particles

The general idea of modeling disordered materials using the Delaunay–Voronoi method was presented in Chapter 1. A particle-filled composite material with randomly distributed particles is a typical disordered composite material and its geometry can be effectively described by the Delaunay–Voronoi method (Voronoi, 1908). An additional difficulty in our consideration arises due to the interaction between near-boundary particles and the external boundary (particle–wall interactions).

In our case the Voronoi tessellation is constructed in a bounded domain, see Figure 3.2.

The Delaunay–Voronoi method also works for such domains. In fact, it partitions the boundary of the domain. Namely, consider a particle, which lies near the boundary of a rectangular domain described above. Then a face of its Voronoi cell coincides with part of the boundary of this domain. We call this part of the boundary a *"pseudo-disk"* (*"pseudo-particle"* in the general case (Kolpakov, 2005, 2006a)). Complications arise in the construction of the Delaunay graph for the problem (3.2.3)–(3.2.7) due to the boundary conditions (3.2.6). The particle–wall interaction at the boundaries $\partial Q^+ \bigcup \partial Q^-$ is taken into account by modifying the standard construction of the Delaunay graph. We incorporate such interactions using edges of the Delaunay graph, which connect centers of near-boundary particles with corresponding pseudo-disks. These edges are chosen to be perpendicular to the boundary, see Figure 3.3, which reflects the physics of the particle–wall

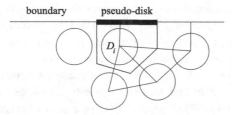

Figure 3.3 Pseudo-disk corresponding to the near-boundary disk D_i, Voronoi cells and edges of the Delaunay graph.

interactions. These edges are the shortest possible, reflecting the concentration of physical fields between the wall and a particle in its vicinity.

This construction allows us not to distinguish the disks and pseudo-disks, which is a very convenient way to incorporate the boundary effects. Also, since a pseudo-disk has no curvature, when local fluxes are computed one can treat any pseudo-disk as a disk of infinite radius.

Let K denote the total number of disks and pseudo-disks. Then obviously $K \geq N$ (N is the number of actual disks).

3.3 Trial functions and the accuracy of two-sided bounds. Construction of trial functions for high-contrast densely-packed composite materials

The key technical point of our approach (as well as the other approach mentioned above) is the construction of two-sided bounds for the variational definition of the effective conductivity. These bounds are obtained from (3.2.11) and (3.2.23) taking into account that $\phi^0(\pm 1) = \pm 1$:

$$\frac{1}{2} \int_{Q_m} |\nabla \phi(\mathbf{x})|^2 dx \geq A \geq \left(-\frac{1}{2} \int_{Q_m} \mathbf{v}(\mathbf{x})^2 dx + \int_{\partial Q^+} \mathbf{v}(\mathbf{x})\mathbf{n}dx - \int_{\partial Q^-} \mathbf{v}(\mathbf{x})\mathbf{n}dx \right).$$
(3.3.1)

These hold for any trial functions $\phi \in V_p$ and $\mathbf{v} \in W_p$, such that div$\mathbf{v} = 0$.

3.3.1 Trial functions and the accuracy of two-sided bounds

A successful choice of trial functions leads to tight (matching to some asymptotic order) bounds.

The problem of determining variational bounds for effective material properties is well-known in homogenization theory and the theory of composite materials. It dates back to the work of Vought and Reuss (Reuss, 1929; Voigt, 1910), which paved the way for a new research area in the mechanics of composite materials. A large number of researchers contributed to this area. Extremal principles were obtained for the homogenized properties of linear and nonlinear solid (two- and three-dimensional) composite materials (Cherkaev, 2000; Jikov *et al.*, 1994; Kolpakov, 2004; Milton, 2002; Prager, 1963; Walpole, 1966; Yeh, 1970b,a). In Kolpakov (2004), the extremal principles were obtained for the stiffnesses of composite plates and beams (inhomogeneous structures occupying regions of small thickness or small diameter). The variational bounds were derived in Avellaneda (1987); Beran and Molyneux (1966); Francfort and Murat (1986); Hill (1963); Kozlov (1992); Lipton (1994) (see also Cherkaev (2000); Milton (2002) for a comprehensive list of references). We remark here that progress in this area relied essentially on the development of improved variational principles and two-sided bounds analogous to (3.3.1).

The key issue in obtaining such two-sided bounds is the construction of trial functions for the direct and dual variational principles. In relation to high-contrast composite materials, we mention here the papers by Bhattacharya *et al.* (1999); Borcea and Papanicolaou (1998); Bruno (1991) and Kolpakov (1988, 1992), where various novel types of trial functions were constructed. For instance, in Bhattacharya *et al.* (1999), the reader can find an interesting example, which shows how physical intuition "prompts" a sophisticated mathematical construction of a "spiral" trial function. We emphasize that the construction of the trial function is a highly non-trivial and creative task. There is no general recipe for the construction of such functions. Such a construction depends on the specific features of the problem, both physical and geometrical. For example, the physics and geometry in each of the above mentioned works (Bhattacharya *et al.*, 1999; Borcea and Papanicolaou, 1998; Bruno, 1991) are different and, as a result, trial functions from one paper are not useful in others.

The authors use the trial function (2.5.4) and a function similar to the Keller function (2.5.5) in the neck. Since in physical problems the fields are concentrated inside the necks, our trial functions must be asymptotically exact. We also pay significant attention to the construction of a trial function outside the necks, where the trial function may be less accurate. Since the variational problem is defined in the entire domain, this construction gives formal mathematical completeness to our analysis.

3.3.2 Construction of trial functions for high-contrast densely-packed composite materials

The specific feature of high-contrast densely-packed composite materials is the localization of fluxes in the thin gaps (so-called "necks") between neighboring particles. In the model with continuously distributed properties (Borcea, 1998; Borcea and Papanicolaou, 1998; Borcea *et al.*, 1999), described by the Kozlov function (2.4.3) the construction of trial functions is specific to this form of the coefficients $a(\mathbf{x})$ (see Section 2.4.3).

We consider particle-filled composite materials whose local material properties are described by piecewise constant functions (for example, the resistivity of

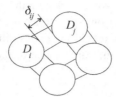

Figure 3.4 Distance between the disks.

a composite material is 1 in the matrix and 0 in the particles) that reflect the geometry of the problem (the locations of particles and their shapes). This results in essential differences in the construction of trial functions if compared with Borcea (1998). Also our trial functions take into account the long-range interactions between non-neighboring particles as well as the concentration of fields in the necks between neighboring particles, so these functions capture the physics of the entire disordered array of particles and the matrix. In order to capture the concentration of the fields, we construct a trial function, which resembles the Keller function (see Keller (1963)) in the necks.

3.4 Construction of a heuristic network model. Two-dimensional transport problem for a high-contrast composite material filled with densely packed particles

We now introduce the discrete network which corresponds to the continuum model (3.2.3)–(3.2.7). We consider a non-periodic array of circular disks $\mathbf{X} = \{D_i;\ i = 1, \ldots, N\}$ inside the domain $\Pi = [-L, L] \times [-1, 1]$ with variable distances between the neighboring disks which are of the same order $\delta \ll 1$. Unlike in classical periodic homogenization, we only assume that the disks are non-overlapping. For simplicity, we first assume that the disks are identical of radius R. The polydispersity case when the radii of the disks are different will be considered later.

We denote by $\delta_{ij} = |\mathbf{x}_i - \mathbf{x}_i| - 2R$ the distance between the i-th and the j-th disks, see Figure 3.4. More precisely, this means that

$$\delta_{ij} = \delta d_{ij}. \tag{3.4.1}$$

Here d_{ij} are called the rescaled interparticle distances and the small parameter δ is called the *characteristic interparticle distance parameter*.

We take into account the fluxes between the neighboring disks and neglect the fluxes between the non-neighboring disks. The latter is motivated by physical intuition. More precisely, we account for the value of the flux in a gap (neck) Π_{ij} which connects the disks D_i and D_j, see Figure 3.4. Note that the fluxes between the disks and the neighboring pseudo-disks can be taken into account in the same manner. This leads to a formal asymptotic network model, referred to below as the "heuristic network model". We remark here that this formal asymptotic model will be rigorously justified later on. An interesting feature of this proof is that it relies on the solution of the discrete problem described by this heuristic network.

According to Keller (1963) the capacitance of the i-th and j-th disks is given asymptotically (for $\delta_{ij} \ll 1$) by

$$g_{ij} = \pi \sqrt{\frac{R}{\delta_{ij}}} \qquad (3.4.2)$$

and the capacitance of the i-th and j-th pseudo-disks is given asymptotically by

$$g_{ij} = \pi \sqrt{\frac{2R}{\delta_{ij}}} \qquad (3.4.3)$$

This represents the asymptotic of the specific flux (the flux corresponding to a difference of potential equal to unity) between the disks when δ_{ij} is small. Note that the flux g_{ij} corresponds to unit potential drop. If the values of the potential on the disks are equal to t_i and t_j, respectively, then the flux in the neck Π_{ij} is equal to $g_{ij}(t_i - t_j)$ and the corresponding energy in the neck Π_{ij} is

$$\frac{1}{2} g_{ij}(t_i - t_j)^2. \qquad (3.4.4)$$

On the other hand, for a given potential $\phi(\mathbf{x})$ the energy in the neck Π_{ij} is given by the Dirichlet integral

$$\frac{1}{2} \int_{\Pi_{ij}} |\nabla \phi(\mathbf{x})|^2 dx \qquad (3.4.5)$$

and thus our assumption is equivalent to the following formal asymptotic formula relating the exact (3.4.5) and approximated (3.4.4) values of energy in the neck:

$$\frac{1}{2} \int_{\Pi_{ij}} |\nabla \phi(\mathbf{x})|^2 dx \approx \frac{1}{2} g_{ij}(t_i - t_j)^2. \qquad (3.4.6)$$

Thus the heuristic network model is described by the following weighted graph:

$$\mathcal{G} = \{\mathbf{x}_i, g_{ij}; \ i, j = 1, \ldots, K\}$$

(K is the number of disks and pseudo-disks, see Section 3.2.5).

The centers of the disks $\mathbf{x}_i$ are the vertices of this graph, and the weights g_{ij} are assigned to the corresponding edges. If N_i stands for the set of neighboring vertices to the vertex $\mathbf{x}_i$, then $g_{ij} \neq 0$ when $j \in N_i$. The sizes of the disks and the distances between them enter the model through the set $\{g_{ij}\}$.

Next we observe that (3.4.6) implies the discrete representation of the Dirichlet energy integral $I(\phi)$ (3.2.2)

$$\frac{1}{4} \sum_{i,j=1}^{K} g_{ij}(t_i - t_j)^2,$$

where the factor $\frac{1}{4}$ appears due to the fact that we count each neck twice (a more detailed explanation will be given later). Thus the discrete version of the minimization problem (3.2.16) is given by

$$\frac{1}{4} \sum_{i,j=1}^{K} g_{ij}(t_i - t_j)^2 \to \min, \qquad (3.4.7)$$

which corresponds to (3.2.2), where the minimum is taken over all t_i which satisfy the conditions

$$t_i = 1 \text{ for } i \in S^+, \ t_i = -1 \quad \text{for } i \in S^-, \qquad (3.4.8)$$

where S^+ and S^- are the set of vertices corresponding to the pseudo-disks and disks which touch the boundaries ∂Q^+ and ∂Q^-, respectively. The equality (3.4.8) is a discrete version of the boundary condition (3.2.4). Note that all pseudo-disks belong to the set $S^+ \bigcup S^-$. The set of remaining vertices, whose indices are denoted by

$$I = \{i = 1, \ldots, K\} \setminus (S^+ \bigcup S^-), \qquad (3.4.9)$$

is called the set of internal vertices of the network.

It is clear that the minimization problem (3.4.7), (3.4.8) is equivalent to solving the following algebraic system

$$\sum_{j \in N_i} g_{ij}(t_i - t_j) = 0, \qquad (3.4.10)$$

where $i \in I$. We rewrite this as

$$\sum_{j \in n_i} g_{ij}(t_i - t_j) + \sum_{j \in (S^+ \bigcup S^-) \cap N_i} g_{ij}(t_i - t_j) = 0,$$

where $n_i = N_i \bigcap I$ are the interior neighbors of the vertex $\mathbf{x}_i$.

Taking into account (3.4.8), we obtain the following Kirchhoff's type equations

$$\sum_{j \in n_i} g_{ij}(t_i - t_j) + \sum_{j \in (S^+ \bigcup S^-) \cap N_i} g_{ij}t_i = \sum_{j \in S^+ \cap N_i} g_{ij} - \sum_{j \in S^- \cap N_i} g_{ij}, \ i \in I. \quad (3.4.11)$$

The first sum on the left-hand side of (3.4.11) corresponds to fluxes between neighboring pairs of interior particles; the second sum corresponds to fluxes between the near-boundary particles and the boundary (the particle–wall interaction). The right-hand side in (3.4.11) corresponds to the boundary conditions (3.4.8).

Finally, we write a discrete formula for the effective conductivity, which is a discrete analog of (3.2.9):

$$A_d = \sum_{i \in S^+} \sum_{j \in N_i} g_{ij}(1 - t_j). \qquad (3.4.12)$$

Here $\{t_i, i \in I\}$ is a solution of the minimization problem (3.4.7), (3.4.8) or, equivalently, equations (3.4.11). Analogously to the equivalence of (3.2.9) and (3.2.10) it can be shown that A_d is the minimum value of the quadratic form (3.4.7).

Hereafter the subscript "d" stands for the discrete analog of a quantity defined in the continuum problem.

The discrete analog of (3.2.10) has the form

$$A_d = \frac{1}{4} \sum_{i \in I} \sum_{j \in N_i} g_{ij}(t_i - t_j)^2, \tag{3.4.13}$$

where $\{t_i, i \in I\}$ is the solution to (3.4.7), (3.4.8) or (3.4.11). A factor $\dfrac{1}{4}$ appears instead of the usual $\dfrac{1}{2}$ in (3.4.13) because we count each neck Π_{ij} twice.

A rigorous justification of the network model would require estimates of the gradients of trial functions. As mentioned above, we construct the trial functions using the solution of the heuristic discrete problem (3.4.11). That is why we need estimates of the solution of the discrete problem given by the following lemma.

Lemma 3.2 (Discrete maximum principle). *The solution of the problem (3.4.11) satisfies the inequalities* $-1 \le t_i \le 1$ *for all* $i = 1, \ldots, N$.

Proof Recall from equation (3.4.10) that

$$\sum_{j \in N_i} g_{ij}(t_i - t_j) = 0 \tag{3.4.14}$$

for $i \in I$.

From (3.4.14) we have the following analog of the mean value theorem for harmonic functions:

$$t_i \sum_{j \in N_i} g_{ij} = \sum_{j \in N_i} g_{ij}t_j, \quad i \in I, \tag{3.4.15}$$

where the summation is taken over all adjacent (neighboring) vertices.

We call (3.4.15) an analog of the mean value theorem because it relates the value t_i of the potential at a given vertex $\mathbf{x}_i$ to the values of the potential t_j at the adjacent vertices $\mathbf{x}_j$, $j \in N_i$.

We will show that the maximum value cannot be achieved at an interior site. We prove this by contradiction. Suppose that the maximum value M is achieved at some interior vertices $i \in I$. Then $t_j \le M$ for all vertices adjacent to the i-th vertex (i.e., for all $j \in N_i$). Because $g_{ij} \ge 0$, we have for the right-hand side of (3.4.15) the following non-strict inequality:

$$\sum_{j \in N_i} g_{ij}t_j \le M \sum_{j \in N_i} g_{ij} \tag{3.4.16}$$

in which equality is possible only if $t_j = M$ for all $j \in N_i$.

Since $t_i = M$ by assumption, we obtain from (3.4.16) that $M \leq M$. If at least one t_j, $j \in N_i$, in (3.4.16) is smaller than M, then (3.4.16) leads to the impossible inequality $M < M$. Thus, $t_j = M$ for all $j \in N_i$. Since the graph $\mathcal{G}$ is connected (since it is a Delaunay graph), it follows that $t_j = M$ for all vertices of the graph. But the graph $\mathcal{G}$ includes the vertices on the upper and lower boundaries S^+ and S^-, where $t_j = 1$ and $t_j = -1$, respectively. This leads us to a contradiction, which means that the maximum value can be achieved only at the boundaries and this maximum value is 1.

The same argument works for the minimum value -1. $\qquad\qquad\square$

The proposition provides an analog of the maximum principle, and it provides a bound on the solution which does not depend on the coefficients g_{ij}. Finally, we remark that an analog of the maximum principle also holds in three or more dimensions.

3.5 Asymptotically matching bounds

The main point of our analysis throughout the book is to show that for our specific problems we can construct the upper- and lower-sided bounds for effective material constants in the form of an asymptotic expansion in the interparticle distance δ so that they coincide to leading order. The idea of using tight variational bounds is not new and it has been used by many authors; see, for example, Kozlov (1989); Borcea (1998); Bhattacharya *et al.* (1999). In each problem, the main question is the actual construction of the trial functions which guarantees asymptotic matching of the lower- and upper-sided bounds. This is usually a highly non-trivial task and there is no generic approach for such a construction since it relies on the specific features (both physical and mathematical) of the problem. For example, in our problem we were not able to utilize the trial functions from Kozlov (1989); Borcea (1998); Bhattacharya *et al.* (1999). We emphasize that the actual construction of such trial functions is the part of the mathematical approach presented in this book.

We now define the asymptotically matching bounds.

Definition 3.3 Let I_δ and J_δ be lower- and upper-sided bounds, respectively, for a quantity A which depends on a parameter δ: $I_\delta \leq A \leq J_\delta$ for every δ. We call these bounds asymptotically matching to leading order as $\delta \to 0$ if

$$\frac{I_\delta}{J_\delta} \to 1 \text{ as } \delta \to 0.$$

Let I_δ and J_δ be variational estimates, i.e., there exist functionals I and J and functions ϕ_δ and $\mathbf{v}_\delta$ such that $I_\delta = I(\phi_\delta)$, $J_\delta = J(\mathbf{v}_\delta)$, and

$$A = \min_{\phi \in V_p} I(\phi), \qquad\qquad (3.5.1)$$

$$A = \max_{\mathbf{v} \in W_p} J(\mathbf{v}).$$

We assume that the functional $I(\phi)$ is differentiable in the sense of Gateaux (see Chapter 1). Since its derivative $I'(\phi)$ is a linear coercive operator, there exists $\gamma > 0$ such that we can write

$$\langle I'(\phi) - I'(\psi), \phi - \psi \rangle \geq \gamma ||\phi - \psi||^2 \text{ for every } \phi, \psi \in V_p, \quad (3.5.2)$$

where $\langle \, , \, \rangle$ stands for the dual coupling.

Under the conditions formulated above, the maximization problem

$$I(\phi) \to \max, \phi \in V_p$$

has a unique solution ϕ_0.

Let us consider the expression $I(\phi_0 + \lambda(\psi - \phi_0))$, where λ is a real variable. We can write

$$\frac{d}{d\lambda} I(\phi_0 + \lambda(\psi - \phi_0)) = \langle I'(\phi_0 + \lambda(\psi - \phi_0)), \psi - \phi_0 \rangle$$

and transform it into the form

$$\frac{d}{d\lambda} I(\phi_0 + \lambda(\psi - \phi_0)) \quad (3.5.3)$$

$$= \langle I'(\phi_0 + \lambda(\psi - \phi_0)), \psi - \phi_0 \rangle - \langle I'(\phi_0), \psi - \phi_0 \rangle + \langle I'(\phi_0), \psi - \phi_0 \rangle.$$

For the solution ϕ_0 of the minimization problem (3.5.1) we have Ekeland and Temam (1976)

$$I'(\phi_0) = 0. \quad (3.5.4)$$

The first two terms on the right-hand side of (3.5.3) can be written as

$$\langle I'(\phi_0 + \lambda(\psi - \phi_0)) - I'(\phi_0), \psi - \phi_0 \rangle$$

and due to the coercivity condition (3.5.2)

$$\langle I'(\phi_0 + \lambda(\psi - \phi_0)) - I'(\phi_0), \psi - \phi_0 \rangle \geq \lambda \gamma ||\psi - \phi_0||^2. \quad (3.5.5)$$

We have from (3.5.3)–(3.5.5) the inequality

$$\frac{dI}{d\lambda} \geq \lambda \gamma ||\psi - \phi_0||^2. \quad (3.5.6)$$

Integrating (3.5.6) with respect to λ from 0 to 1, we obtain

$$I(\psi) - I(\phi_0) \geq \frac{\gamma}{2} ||\psi - \phi_0||^2.$$

Obviously $I(\phi_\delta) - I(\phi_0) \leq J_\delta - I_\delta$. Then

$$\frac{\gamma}{2} ||\phi_\delta - \phi_0||^2 \leq J_\delta - I_\delta. \quad (3.5.7)$$

This means that the closeness of the variational estimates leads to the closeness of the solution ϕ_0 of the minimization problem and the function ϕ_δ. Note that this closeness usually takes place in an integral (energy) norm.

3.6 Proof of the network approximation theorem

Here we present in detail a justification of the heuristic network model for the two-dimensional transport problem developed in Section 2.4.

3.6.1 The refined lower-sided bound

Due to (3.2.23) the effective conductivity can be bounded from below by

$$A \geq -\frac{1}{2} \int_{Q_m} \mathbf{v}(\mathbf{x})^2 dx + \int_{\partial Q^+ \cup \partial Q^-} \phi^0(y) \mathbf{v}(\mathbf{x}) \mathbf{n} dx, \qquad (3.6.1)$$

where the trial function $\mathbf{v}(\mathbf{x})$ satisfies the following conditions:

$$\text{div}\mathbf{v} = 0 \quad \text{in } Q_m, \qquad (3.6.2)$$

$$\int_{\partial D_i} \mathbf{v}(\mathbf{x}) \mathbf{n} dx = 0, \quad i = 1, \dots, N. \qquad (3.6.3)$$

From the physical point of view, these conditions mean that there are no sources and sinks. Condition (3.6.2) holds for the matrix (perforated domain) Q_m and it is a local (differential) condition. The integral condition (3.6.3) holds for each particle D_i.

Also,

$$\mathbf{v}(\mathbf{x})\mathbf{n} = 0 \text{ on } \partial Q_{\text{lat}} \qquad (3.6.4)$$

since there is no flux through the vertical boundaries ∂Q_{lat} (insulating conditions, see (3.2.7)). As noted above (see the comment after (3.2.15)) the trace $\mathbf{v}\mathbf{n}$ is defined in the sense of the trace theorem (Bensoussan *et al.*, 1978; Temam, 1979), see Section 1.2.3.

3.6.2 Construction of a trial function for the dual problem

We now present a construction of the trial function for the functional (3.6.1). But first we make a general observation about the construction of trial functions for our problem. If a trial function can be constructed based on analysis of the field $\phi(\mathbf{x})$ for a neighboring pair of particles, then we say that such a trial function is "local". This will be the case for the trial functions for the upper-sided bound from the functional space (3.1.1), where we use a Keller type of asymptotic solution for the flux between two closely spaced disks. For the lower-sided bound, the Keller-type solution is not sufficient. Indeed, the functional space W_p (3.2.15) contains the condition (3.6.3). This condition means that for a given disk the total balance of fluxes from *all* neighboring disks is zero. Moreover, these neighbors have their own neighbors different from the initial disk. Thus, the condition (3.6.3) is "non-local" (we call it "global") since it cannot be satisfied for an individual disk. We note that this problem does not arise in periodic systems of bodies. In other words (3.6.3) is a system of N coupled conditions. In the construction of trial functions from W_p,

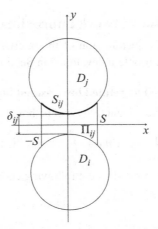

Figure 3.5 The "neck" between two neighboring disks.

this issue will be resolved by using the solution of the heuristic network model (3.4.11). Note that the trial function must satisfy the divergence-free condition (3.6.2). It is well known that construction of a divergence-free function inside a domain requires satisfying integral conditions on the boundary of the domain. This implies the non-local nature of the condition (3.6.2) when constructing the trial functions for the lower-sided bound, while in the scalar problem under consideration we found a technical trick which allows us to resolve this issue relatively easily.

Consider two disks D_i and D_j as shown in Figure 3.5 with a neck between these disks (usually a disk has some necks attached; here we analyze one specific neck).

The choice of the coordinate system as shown in Figure 3.5 does not lead to a loss of generality because all quantities in the formulas (3.6.1)–(3.6.4) are invariant under translation and rotation of the coordinate system. We begin from the construction of the trial function in the neck Π_{ij}. This neck is chosen so that it does not overlap with any other necks attached to the same disk and does not intersect any other disk. More specific conditions on the width S of the neck are presented below.

We choose a function for this pair of disks of the form

$$\mathbf{v}(\mathbf{x}) = (v_1(\mathbf{x}), v_2(\mathbf{x})) = \begin{cases} (0, \zeta_{ij}(x)) & \text{in the neck } \Pi_{ij}, \\ (0, 0) & \text{in } Q_m \setminus \Pi_{ij}. \end{cases} \qquad (3.6.5)$$

Here $\zeta_{ij}(x)$ is a smooth function of a single variable x.

We need to make sure that it is in the class of admissible functions. To this end we note that it satisfies (3.6.2) since for $\mathbf{v}(\mathbf{x})$ given by (3.6.5)

$$\text{div}\mathbf{v} = \frac{\partial v_1}{\partial x} + \frac{\partial v_2}{\partial y} = 0 + \frac{\partial \zeta_{ij}(x)}{\partial y} = 0. \qquad (3.6.6)$$

In general, equation (3.6.6) in the space W_p should be understood in the weak sense; however, for the chosen trial functions one can check directly that the divergence-free condition is satisfied in the classical sense as well. Indeed, our trial function (v_1, v_2) is vectorial and moreover, $v_1 = 0$ everywhere, and v_2 is discontinuous. However, v_2 depends on x only, which is why the partial derivative of v_2 with respect to y is

$$\frac{\partial v_2}{\partial y} = 0$$

and it does not lead to the δ-function at the lateral boundary of the neck.

Substitute (3.6.5) into the integral (3.6.3). For this function we integrate only over the arc $S_{ij} \in \partial D_i$ (the bold line in Figure 3.5). Then the flux of the vector $\mathbf{v}(\mathbf{x})$ through the curve $S_{ij} \in \partial D_i$ is equal to

$$\int_{S_{ij}} \mathbf{v}(\mathbf{x})\mathbf{n}dx = \int_{-S}^{S} \mathbf{v}(\mathbf{x})\mathbf{n}dx = \int_{-S}^{S} \zeta_{ij}(\mathbf{x})dx. \tag{3.6.7}$$

The first equality in (3.6.7) follows from the divergence theorem, equality (3.6.6) and the fact that $\mathbf{v}(\mathbf{x})\mathbf{n} = 0$ on the lateral sides of the neck for functions of the form (3.6.5).

We have constructed here a function for a single neck connecting a given pair of disks. The trial function for the entire collection of disks $\{D_i; i = 1, \ldots, K\}$ is obtained by summing up such functions over all necks (in the original coordinate system). In general a disk D_i has several neighbors and due to (3.6.3)

$$\int_{\partial D_i} \mathbf{v}(\mathbf{x})\mathbf{n}dx = \sum_{j \in N_i} \int_{S_{ij}} \mathbf{v}(\mathbf{x})\mathbf{n}dx = 0 \tag{3.6.8}$$

since the trial function is of the form (3.6.5). Here the summation in j is taken over all neighbors N_i of the disk D_i.

Therefore, equations (3.6.8) are some kind of balance equations (the sum of all integral fluxes entering a given disk from all neighbors is equal to zero). The trial function (3.6.5) is constructed so that all fluxes can flow through the necks (channels) only. The flux through each neck is given by (3.6.7), and we need to consider all necks between neighbors. Therefore, we need to choose our trial function so that for each disk the balance equation (3.6.3) holds. In other words, we need a set of numbers, which can be chosen as the integrals of fluxes over all the necks. Where do we get such a set of numbers? In short, the answer is: from the heuristic network model (3.4.11), whose solution satisfies the required balance condition.

Indeed, let us revisit the network problem (3.4.11). Consider the set of numbers (the fluxes of current)

$$p_{ij} = g_{ij}(t_i - t_j), \tag{3.6.9}$$

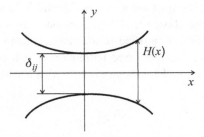

Figure 3.6 The local geometry of the "neck" between two neighboring disks.

where $\{t_i; \ i \in I\}$ are the solutions of the algebraic system (3.4.11), and the numbers $\{g_{ij}\}$ are defined by (3.4.2) and (3.4.3). Then equation (3.4.11) has the form of Kirchhoff's law

$$\sum_{j \in N_i} p_{ij} = 0, \ i \in I \tag{3.6.10}$$

for an interior vertex I. Therefore, in order to satisfy the balance equation (3.6.3) it is sufficient to impose the following condition on the trial function $\zeta(x)$:

$$\int_{-S}^{S} \zeta_{ij}(x)dx = p_{ij}. \tag{3.6.11}$$

3.6.3 Construction of the function $\zeta_{ij}(x)$, which determines the trial function

Recall that we are analyzing the lower-sided bound (3.6.1), and our goal is to make it as tight as possible, which means that we wish to increase the right-hand side of (3.6.1). The latter can be achieved by decreasing the value of the integral $\int_{\Pi_{ij}} |\mathbf{v}(\mathbf{x})|^2 d\mathbf{x}$, which for the trial function (3.6.5) is written as follows

$$\int_{-S}^{S} \zeta_{ij}^2(x)H_{ij}(x)dx, \tag{3.6.12}$$

where $H_{ij}(x) = \delta_{ij} + 2R - 2\sqrt{R^2 - x^2}$ is the distance between the i-th and j-th disks, see Figure 3.6. Note that equation (3.6.12), as well as subsequent equations with repeated indices, should be interpreted as the formula for one value of (i, j) (no summation in repeating indices here).

Minimization of the integral (3.6.12) is to be carried out under the constraints (3.6.11). Therefore, we need to use the Lagrange multiplier $2\lambda_{ij}$ in the corresponding Euler–Lagrange equation:

$$2\zeta_{ij}(x)H_{ij}(x) - 2\lambda_{ij} = 0. \tag{3.6.13}$$

From (3.6.13) we obtain

$$\zeta_{ij}(x) = \frac{\lambda_{ij}}{H_{ij}(x)}, \tag{3.6.14}$$

where λ_{ij} is an unknown multiplier to be determined.

The equality (3.6.14) implies that the desired trial function (which determines the local flux) is inversely proportional to the distance between the disks. The latter agrees with the physical assumptions about the flux between two closely spaced disks formulated in Keller (1963).

Substitute (3.6.14) into (3.6.11) to obtain

$$\int_{-S}^{S} \lambda_{ij} \frac{dx}{H_{ij}(x)} = p_{ij} = g_{ij}(t_i - t_j), \tag{3.6.15}$$

where the numbers p_{ij} are defined in (3.6.9).

Introduce the following quantities:

$$g_{ij}^s = \int_{-S}^{S} \frac{dx}{H_{ij}(x)}. \tag{3.6.16}$$

From (3.6.15) we obtain $\lambda_{ij} g_{ij}^s = p_{ij}$ or

$$\lambda_{ij} = \frac{p_{ij}}{g_{ij}^s} = \frac{g_{ij}(t_i - t_j)}{g_{ij}^s}. \tag{3.6.17}$$

If we choose λ_{ij} according to (3.6.17), then the trial function $\mathbf{v}(\mathbf{x})$ defined by (3.6.5) and (3.6.14) satisfies all the conditions (3.6.2)–(3.6.4).

Substitute (3.6.5), (3.6.14), (3.6.17) into the integral

$$\int_{\Pi_{ij}} \frac{1}{2} |\mathbf{v}(\mathbf{x})|^2 d\mathbf{x} \tag{3.6.18}$$

over the neck Π_{ij}. Since the integrand does not depend on the variable y, we evaluate the latter integral as follows:

$$\int_{-S}^{S} \zeta_{ij}^2(x) H_{ij}(x) dx = \int_{-S}^{S} \left[\frac{\lambda_{ij}}{H_{ij}(x)} \right]^2 H_{ij}(x) dx \tag{3.6.19}$$

$$= \lambda_{ij}^2 \int_{-S}^{S} \frac{dx}{H_{ij}(x)} = \lambda_{ij}^2 g_{ij}^s = \frac{g_{ij}^2 (t_i - t_j)^2}{g_{ij}^s}.$$

This calculation is done for a pair of disks. Consider now all the disks in the domain Π. Recall that the set also includes the pseudo-disks, therefore, the set of all the disks D_i can be decomposed into two disjoint sets: the set of internal disks denoted by I and the set of the boundary disks, which is the union of all disks touching the boundaries $\partial Q^+ \bigcup \partial Q^-$ and the pseudo-disks (those two sets are denoted by S^+ and S^-, respectively). The pseudo-disks incorporate information about the interaction between the boundary of the domain Π and the internal disks, and they can be treated as disks of radius $R = \infty$, see Figure 3.3. Thus, all calculations for the disks and the "pseudo-disks" can be carried out using the same formulas, and from the mathematical point of view there is no reason to distinguish the disks and the pseudo-disks. But it is important to take into account the difference in the enumeration of the disks and the pseudo-disks. In particular, the "pseudo-disks" belong to the sets $S^+ \bigcup S^-$ (corresponding to the horizontal boundaries only). We also remark that the disks which touch the boundary are treated as regular disks with the potential ± 1.

We now evaluate the functional (3.6.1). It consists of the double and boundary integrals.

3.6.4 Calculation of the double integral (3.6.18) from (3.6.1)

The integral (3.6.18) was evaluated above. Since in our construction different necks do not overlap, this evaluation amounts to the summation of the integrals over all necks Π_{ij} (for both the disks and the pseudo-disks). The integral over one neck is given by (3.6.19) and therefore we get

$$\int_{Q_m} -\frac{1}{2}\mathbf{v}^2 dx = -\frac{1}{2}\sum_{\Pi_{ij}} \frac{g_{ij}^2 (t_i - t_j)^2}{g_{ij}^s}. \tag{3.6.20}$$

The index Π_{ij} in the sum means that we sum up over the necks. When we change to the summation over the indices $i, j = 1, \ldots, K$ (K is the number of disks including pseudo-disks) we will count each neck Π_{ij} twice. Therefore, a factor $\frac{1}{4}$ appears in the sum over i, j instead of $\frac{1}{2}$ in (3.6.20). Then we write (3.6.20) in the form

$$\int_{Q_m} -\frac{1}{2}|\mathbf{v}(\mathbf{x})|^2 dx = -\frac{1}{4}\sum_{i,j=1}^{K} \frac{g_{ij}^2 (t_i - t_j)^2}{g_{ij}^s}. \tag{3.6.21}$$

Alternatively, one can sum over $i \le j = 1, \ldots, K$. In this case, the factor $\frac{1}{4}$ will become a factor of $\frac{1}{2}$, because we will no longer be double-counting the necks. We pay attention to these technical issues because our goal is to make sure that the leading terms of the direct and dual functionals are actually the same.

3.6.5 Calculation of the boundary integral in (3.6.1)

The flux $\mathbf{v}(\mathbf{x})\mathbf{n}$ through the boundary ∂Q^+, where the trial function is defined by (3.6.5), (3.6.14), is given by

$$\mathbf{v}(\mathbf{x})\mathbf{n} = \frac{\lambda_{ij}}{H_{ij}(x)}. \tag{3.6.22}$$

Using (3.6.16), (3.6.17) and the definition of g_{ij}^s (3.6.16), we compute the sum

$$\int_{\partial Q^+} \phi^0(y)\mathbf{v}(\mathbf{x})\mathbf{n}dx + \int_{\partial Q^-} \phi^0(y)\mathbf{v}(\mathbf{x})\mathbf{n}dx \tag{3.6.23}$$

for the function $\mathbf{v}(\mathbf{x})$ given by (3.6.22) and the function $\phi^0(y)$ given by (3.2.19). We denote by $\Pi_{ij}^{\pm}$ the necks adjacent to the boundaries ∂Q^+ and ∂Q^-. Then (3.6.23) is equal to

$$\sum_{\Pi_{ij}^{\pm}} \int_{-S}^{S} \frac{\lambda_{ij}dx}{H_{ij}(x)} = \sum_{\Pi_{ij}^{\pm}} \lambda_{ij} \int_{-S}^{S} \frac{dx}{H_{ij}(x)} \tag{3.6.24}$$

$$= \sum_{\Pi_{ij}^{\pm}} \frac{g_{ij}(t_i - t_j)g_{ij}^s}{g_{ij}^s} = \sum_{\Pi_{ij}^{\pm}} g_{ij}(t_i - t_j).$$

If $i \in S^+$ (S^+ is the set of pseudo-disks at the boundary $y = +1$ and disks touching this boundary) or $i \in S^-$ (S^- is the set of pseudo-disks at the boundary $y = -1$ and disks touching this boundary), then we do not count any neck twice as was the case in the double integral.

Now the integral over the boundary $y = +1$ in (3.6.1) can be evaluated as follows:

$$P^+ := \sum_{i \in S^+} \sum_{j \in N_i} g_{ij}(1 - t_j). \tag{3.6.25}$$

The integral over the boundary ∂Q^- in (3.6.1) can be evaluated similarly (recall that $\phi^0(y) = -1$ when $y = -1$) and is given by

$$P^- := \sum_{i \in S^-} \sum_{j \in N_i} g_{ij}(-1 - t_j). \tag{3.6.26}$$

The sums (3.6.25) and (3.6.26) represent fluxes through the upper and the lower boundaries S^+ and S^-, respectively (S^+ and S^- correspond to $\partial Q^+ \bigcup \partial Q^-$ in the continuum model), in the discrete network model (3.4.11). Since the vertical boundaries are insulated, the conservation of flux in the domain Π implies that these two fluxes are equal.

We have evaluated both integrals in (3.6.1) for the trial function (3.6.5) and established the following lemma:

Lemma 3.4 *The following lower-sided bound holds*

$$A \geq J(\mathbf{v}) = -\frac{1}{4} \sum_{i,j=1}^{K} \frac{g_{ij}^2(t_i - t_j)^2}{g_{ij}^s} + (P^+ - P^-), \tag{3.6.27}$$

where g_{ij}^s and g_{ij} are defined by (3.6.16) and by (3.4.2), respectively; P^+ and P^- are given by (3.6.25) and (3.6.26), respectively.

To understand the "$-$" sign in front of P^- on the right-hand side of (3.6.27), one should look at (3.3.1) and the definition of the function $\phi^0(y)$ (3.2.19).

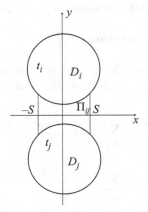

Figure 3.7 Two neighboring disks.

3.6.6 The refined upper-sided bound

Our goal is to construct a trial function $\phi \in V_p$ which mimics the behavior of the exact solution and substitute it into the following inequality, see (3.2.2)

$$A \leq I(\phi) = \frac{1}{2} \int_{Q_m} |\nabla \phi(\mathbf{x})|^2 d\mathbf{x}. \tag{3.6.28}$$

We shall use the knowledge gained in the development of the formal network model above. Roughly speaking, the trial function behaves according to the Keller approximation in a narrow neck between each two closely spaced disks. Outside this neck we make the extension described below. Then the integral on the right-hand side of (3.6.28) will be decomposed into two parts: the first part is due to the contribution from the necks, and the second part is due to the contribution from the extension. It will be shown that the first part behaves like $\delta^{-1/2}$ as $\delta \to 0$ and it is precisely the energy of the heuristic network (3.4.7). The second part is of order $O(1)$ as $\delta \to 0$.

Any trial function $\phi(\mathbf{x})$ should be piecewise-differentiable and satisfy the other conditions in the definition of the functional space V_p (3.1.1). In particular, this means that the trial function takes constant values on the disks D_i and is equal to 1 on the boundary S^+ and -1 on the boundary S^-.

Consider two disks D_i and D_j, see Figure 3.7, with potentials t_i and t_j. The choice of the coordinate system in Figure 3.7 is not important because the integral in (3.6.28) is invariant under any translation and rotation of the coordinate system.

We first describe a sketch of the construction of a trial function. We need to ensure that the trial function takes values t_i and t_j on the disks D_i and D_j. In fact, it is sufficient for this function to take these values on the boundaries ∂D_i and ∂D_j and to be smooth (namely, belong to C^1) outside the disks.

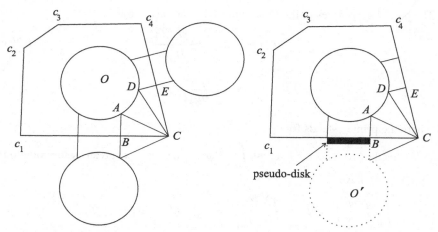

Figure 3.8 Decomposition of a Voronoi cell for the inner disk (left) and near-boundary disks (right). The decomposition for the near-boundary disk may be reduced to decomposition for the inner disk by the introduction of the disk O' mirror-symmetric with respect to the boundary for the near-boundary disk O.

Again consider the neck Π_{ij} connecting the disks D_i and D_j, see Figure 3.7. In this neck the trial function changes linearly from t_i to t_j when moving from disk D_i to disk D_j according to the Keller approximation for two closely spaced disks. Next, we extend the trial function from this neck onto the Voronoi cells of these two particles.

We now describe the construction of the trial function in each Voronoi cell. As shown in Figure 3.8, the Voronoi cell $c_1 c_2 c_3 c_4 C$ of a disk is portioned into the necks connecting this disk to its neighbors and remaining subdomains.

3.6.7 Step A. Construction of the trial function in the neck

From the physical point of view, the potential field in the narrow channel (the neck Π_{ij}) must be linear in y (see Keller (1963)). This motivates us to take the trial function $\phi(\mathbf{x})$ linear in y in the neck Π_{ij} with the values t_i and t_j on the disks ∂D_i and ∂D_j or equivalently, for $y = -H_{ij}(x)/2$ and $y = H_{ij}(x)/2$, respectively. These conditions uniquely determine the function $\phi(\mathbf{x})$ as follows:

$$\phi(\mathbf{x}) = t_i + \frac{(t_j - t_i)\left(y + \dfrac{H_{ij}(x)}{2}\right)}{H_{ij}(x)} \qquad (3.6.29)$$

$$= t_i + (t_j - t_i)\left(\frac{y}{H_{ij}(x)} + \frac{1}{2}\right),$$

$$y \in \left[-\frac{H_{ij}(x)}{2}, \frac{H_{ij}(x)}{2}\right].$$

The partial derivatives of the function (3.6.29) are given by

$$\frac{\partial \phi}{\partial x}(\mathbf{x}) = -(t_j - t_i)\frac{yH'_{ij}(x)}{H^2_{ij}(x)}, \tag{3.6.30}$$

$$\frac{\partial \phi}{\partial y}(\mathbf{x}) = \frac{t_j - t_i}{H_{ij}(x)}, \tag{3.6.31}$$

where the prime ' stands for $\dfrac{d}{dx}$.

We need to evaluate the integral $\displaystyle\int_{Q_m} |\nabla \phi(\mathbf{x})|^2 d\mathbf{x}$ for $\phi(\mathbf{x})$ given by the equation (3.6.31). We have

$$\int_{\Pi_{ij}} \left(\frac{\partial \phi}{\partial y}\right)^2 dxdy = \int_{-S}^{S} \frac{(t_j - t_i)^2}{H^2_{ij}(x)} H_{ij}(x) dx = (t_j - t_i)^2 g^s_{ij}. \tag{3.6.32}$$

In (3.6.32), we switch from the double integral to the iterated integral and integrated in the variable y. Since $\dfrac{\partial \phi}{\partial y}$ in (3.6.31) does not depend on y, we arrive at the integral in the variable x. Recall that

$$\Pi_{ij} = \left\{ \mathbf{x} = (x, y) \,|\, x \in (-S, S); \; \frac{-H_{ij}(x)}{2} \le y \le \frac{H_{ij}(x)}{2} \right\}. \tag{3.6.33}$$

From (3.6.30), we have

$$\int_{\Pi_{ij}} \left(\frac{\partial \phi}{\partial x}\right)^2 dxdy = \int_{\Pi_{ij}} (t_j - t_i)^2 \left[\frac{yH'_{ij}(x)}{H^2_{ij}(x)}\right]^2 dxdy.$$

Since $|y/H_{ij}(x)| \le \dfrac{1}{2}$ due to the condition for y in (3.6.33), the last integral is bounded by the following quantity:

$$(t_j - t_i)^2 \int_{\Pi_{ij}} \left[\frac{y}{H_{ij}(x)}\right]^2 \frac{[H'_{ij}(x)]^2}{[H_{ij}(x)]^2} dxdy \tag{3.6.34}$$

$$\le \frac{(t_j - t_i)^2}{4} \int_{\Pi_{ij}} \frac{[H'_{ij}(x)]^2}{[H_{ij}(x)]^2} dxdy = \frac{(t_j - t_i)^2}{4} \int_{-S}^{S} \frac{[H'_{ij}(x)]^2}{H_{ij}(x)} dx.$$

In the last equality of (3.6.34) we again reduce the double integral to an integral in x, using the fact that the integrand in the second integral in (3.6.34) does not depend on the variable y, which leads to the multiplication of the integrand by $H_{ij}(x)$.

Since $H_{ij}(x) = \delta_{ij} + 2R - 2\sqrt{R^2 - x^2}$, the integrand in (3.6.34) is given by

$$\frac{[H'_{ij}(x)]^2}{H_{ij}(x)} = \frac{4x^2}{R^2 - x^2} \cdot \frac{1}{\delta_{ij} + 2R - 2\sqrt{R^2 - x^2}} \tag{3.6.35}$$

$$\le \frac{2}{R^2 - x^2} \cdot \frac{x^2}{R - \sqrt{R^2 - x^2}} = \frac{2}{R^2 - x^2}\left(R + \sqrt{R^2 - x^2}\right)$$

$$= \frac{2R}{R^2 - x^2} + \frac{2}{\sqrt{R^2 - x^2}}.$$

If we restrict the thickness of the neck with the condition

$$S \leq \frac{R}{2},\tag{3.6.36}$$

then $-R/2 \leq x \leq R/2$ and thus we can bound the last expression in (3.6.35).

The maximum of the last expression on the right-hand side of (3.6.35) in the interval $-R/2 \leq x \leq R/2$ is denoted by C. Then the constant $C < \infty$ does not depend on δ and the integral (3.6.34) is bounded from above by the quantity

$$E_A = \frac{(t_i - t_j)^2}{4} C \cdot 2S \leq 2CS < \infty,\tag{3.6.37}$$

which does not depend on δ. Here we have used the discrete maximum principle (Lemma 3.2), which implies that $|t_i| \leq 1$, $i = 1, \ldots, N$.

3.6.8 Step B. Construction of the trial function in the triangle

In the triangle ABC, see Figure 3.8, we construct a linear function $\phi(\mathbf{x})$ which takes the following values:

(i) t_i at the point A;
(ii) $\dfrac{t_i + t_j}{2}$ at the point B;
(iii) 0 at the point C.

In the coordinate system Oxy oriented so that Ox and Oy are parallel to the segments BC and AB, respectively, centered at the point B, we set

$$\phi(\mathbf{x}) = \alpha x + \beta y + \gamma.$$

It is clear that

$$\gamma = \frac{t_i + t_j}{2}$$

and we satisfy the condition (i) at the point B.

The above conditions (ii) and (iii) at the points A and C imply

$$\beta |AB| + \gamma = t_i,$$

$$\alpha |BC| + \gamma = 0.$$

From these equations, we obtain

$$\alpha = -\frac{\gamma}{|BC|},$$

$$\beta = -\frac{\gamma}{|AB|} + \frac{t_i}{|AB|}.$$

Then

$$\phi(\mathbf{x}) = \frac{t_i + t_j}{2}\left(-\frac{x}{|BC|} + 1\right) - \frac{t_i - t_j}{2} \cdot \frac{y}{|AB|}.\tag{3.6.38}$$

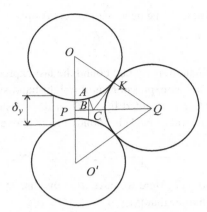

Figure 3.9 A neck for closely placed disks.

For the function $\phi(\mathbf{x})$ given by (3.6.38) we have

$$\nabla\phi(\mathbf{x}) = \left(-\frac{t_i + t_j}{2} \cdot \frac{1}{|BC|}, -\frac{t_i - t_j}{2} \cdot \frac{1}{|AB|} \right)$$

and we evaluate the integral over the triangle ABC:

$$\int_{ABC} \frac{1}{2} |\nabla\phi(\mathbf{x})|^2 d\mathbf{x} = \left[\frac{(t_i + t_j)^2}{8} \cdot \frac{1}{|BC|^2} + \frac{(t_i - t_j)^2}{8} \cdot \frac{1}{|AB|^2} \right] |ABC|, \quad (3.6.39)$$

where $|ABC|$ stands for the area of the triangle ABC.

The next lemma provides lower-sided bounds for $|AB|$ and $|BC|$ which are independent of δ.

Lemma 3.5 *There exists a positive number a_0, independent of δ, such that the quantities $|BC|$ and $|AB|$ are bounded below by a_0 for an arbitrary δ.*

The lower-sided bound on $|BC|$ follows from elementary geometrical considerations. In Figure 3.9 for a neck of width $2S$ connecting two disks centered at O and O', the length $|BC|$ is the smallest if the third disk centered at Q touches the other two. Consider the right triangle $\triangle OPQ$. By the Pythagorean theorem

$$|PQ| = S + |BC| + |CQ| = \sqrt{(2R)^2 - \left(R + \frac{\delta_{ij}}{2} \right)^2},$$

$$S = |PB|.$$

Also,

$$\frac{|OK|}{|CQ|} = \frac{R}{|CQ|} = \cos(\angle OQP) = \frac{|PQ|}{|OQ|} = \frac{\sqrt{(2R)^2 - \left(R + \frac{\delta_{ij}}{2} \right)^2}}{2R}.$$

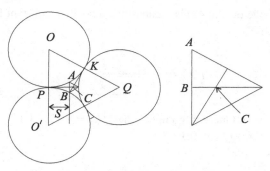

Figure 3.10 Neck for three touching disks.

Therefore

$$|BC| = |PQ| - |CQ| - S = \frac{R^2 - R\delta_{ij} - \dfrac{\delta_{ij}^2}{4}}{\sqrt{3R^2 - R\delta_{ij} - \dfrac{\delta^2}{4}}} - S. \qquad (3.6.40)$$

An upper-sided bound on S independent of δ follows if we consider the same three disks when all of them touch each other and are connected by necks of the same width, see Figure 3.10. Then we have $S + |BC| = |PC|$.

Next consider the right triangle $\triangle ABC$ and observe that $|BC| = \frac{1}{2}|AC|$ and $|PC| = \frac{1}{3}|PQ|$ from the equilateral triangle OQO'. Thus

$$S + \frac{1}{2}|AC| = \frac{1}{3}|PQ|,$$

$$S + \frac{1}{2}(|OC| - R) = \frac{1}{3}\sqrt{3}R,$$

$$S + \frac{1}{2}\left(\frac{2}{3}|PQ| - R\right) = \frac{1}{3}\sqrt{3}R.$$

Since $|PQ| = \sqrt{3}R$, we obtain

$$S = \frac{R}{2}. \qquad (3.6.41)$$

Combining (3.6.40) and (3.6.41), we have

$$|BC| \geq \frac{R^2 - R\delta_{ij} - \dfrac{\delta_{ij}^2}{4}}{\sqrt{3R^2 - R\delta_{ij} - \dfrac{\delta_{ij}^2}{4}}} - \frac{R}{2}. \qquad (3.6.42)$$

The right-hand side of (3.6.42) is greater than a positive constant for small δ_{ij}. In fact, the function

$$f(x) = \frac{1 - x - \dfrac{x^2}{4}}{\sqrt{3 - x - \dfrac{x^2}{4}}} - \frac{1}{2}$$

decreases for $x > 0$. One can compute that $f(x = 0.1) > 0.027$. Then, $f(x) \geq f(0.1) > 0.027$ for $0 \leq x \leq 0.1$.

Then

$$|BC| \geq \frac{R^2 - R\delta_{ij} - \dfrac{\delta_{ij}^2}{4}}{\sqrt{3R^2 - R\delta_{ij} - \dfrac{\delta_{ij}^2}{4}}} - \frac{R}{2} \geq 0.027R$$

for $0 \leq \delta_{ij} \leq 0.1R$.

The lower-sided bound on $|AB|$ can be obtained when the disks centered at O and O' touch each other ($\delta_{ij} = 0$). We have

$$|AB| = |OP| - |OA|\cos(\angle POA) = R - \sqrt{R^2 - S^2}. \tag{3.6.43}$$

Combining (3.6.43) with the upper-sided bound on S given by (3.6.41) we arrive at a lower-sided bound on $|AB|$ which does not depend on δ_{ij}.   □

It follows from Lemma 3.5 and Lemma 3.2 that the quantities on the right-hand side of (3.6.39) are bounded independently of δ.

Then from (3.6.39) we have

$$\int_{ABC} \frac{1}{2}|\nabla\phi(\mathbf{x})|^2 d\mathbf{x} \leq C_1 |ABC|, \tag{3.6.44}$$

where $C_1 < \infty$ does not depend on δ.

Construction of the trial function in the triangle CDE in Figure 3.8 is analogous.

3.6.9 Step C. Construction of the trial function in the curvilinear triangle

According to the construction in Step B, the function $\phi(\mathbf{x})$ is linear on the sides AC and DC, Figure 3.8. Moreover it takes the value t_i at the points A and D (and along the entire arc AD) and the value 0 at the point C. We extend this function into the curvilinear triangle ADC by taking $\phi(\rho, \Theta)$ in the polar coordinates (ρ, Θ) centered at O of the form

$$\phi(\rho, \Theta) = \alpha\rho + \beta \tag{3.6.45}$$

and imposing the following conditions

$$\alpha R + \beta = t_i, \tag{3.6.46}$$
$$\alpha|OC| + \beta = 0,$$

where O is the center of the disk under consideration, R is the radius of the disk and ρ is the polar coordinate (polar radius), which corresponds to the center O.

Solving the system (3.6.46), we obtain

$$\alpha = -\frac{t_i}{\left(1 - \dfrac{R}{|OC|}\right)|OC|}, \tag{3.6.47}$$

$$\beta = \frac{t_i}{1 - \dfrac{R}{|OC|}}. \tag{3.6.48}$$

For the function $\phi(\mathbf{x})$ given by (3.6.45), (3.6.47) and (3.6.48) we have

$$\int_{ACD} \frac{1}{2}|\nabla\phi(\mathbf{x})|^2 d\mathbf{x} = \int_{adc} \frac{1}{2}\left[\left(\frac{\partial\phi}{\partial\rho}\right)^2 + \frac{1}{\rho}\left(\frac{\partial\phi}{\partial\Theta}\right)^2\right]\rho d\rho d\Theta \tag{3.6.49}$$

$$= \frac{1}{2}\int_{adc} \alpha^2 \rho d\rho d\Theta = \frac{1}{2}\alpha^2|ADC|,$$

where ρ, Θ are the polar coordinates, and the domain of integration adc corresponds to the curvilinear triangle ACD in the polar coordinate system.

The length $|OC|$ would have its minimal value when all three disks in Figure 3.8 are touching each other. We denote by L this minimal value of OC. Clearly L does not depend on δ and $L > R$. Then the denominators in formulas (3.6.46) are bounded below by a positive constant which does not depend on δ.

Due to (3.6.48), (3.6.49) and Lemma 3.2 (the discrete maximum principle) we have

$$\int_{ACD} \frac{1}{2}|\nabla\phi(\mathbf{x})|^2 d\mathbf{x} \leq C_2|ACD|, \tag{3.6.50}$$

where $C_2 < \infty$ does not depend on δ.

3.6.10 Step D. Construction of the trial function for the near boundary disks

Above we constructed the extension of the trial function from the neck to the entire Voronoi cell for internal disks whose neighbors are also internal disks. We now consider a disk lying near the boundary so that one of its neighbors is a pseudo-disk, see Figure 3.8.

Recall that we introduced pseudo-disks as the segments obtained by partitioning the boundary ∂Q^+ by Voronoi cells and that these pseudo-disks can be considered as disks of infinite radius for the purpose of calculating the flux according to the Keller formula (3.4.2).

For a boundary disk (or equivalently for disk–pseudo-disk pair) the boundary of the Voronoi cell is also part of the boundary of the domain Π. Therefore we

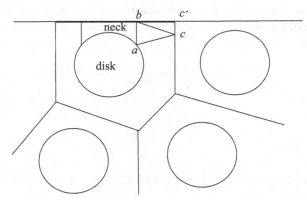

Figure 3.11 Decomposition of a Voronoi cell for a boundary disk.

cannot require $\phi(\mathbf{x}) = 0$ at c', see Figure 3.11, as was done in Step B for the point C.

However, the above construction works after a slight modification. Namely, we introduce a point c at the boundary of the Voronoi cell, see Figure 3.11, at a distance

$$|cc'| = \frac{\sqrt{R^2 - S^2}}{2}$$

$\left(S \geq \dfrac{R}{2}\right.$ is the width of the neck$\left.\right)$ from the boundary ∂Q^+ and require $\phi(\mathbf{x}) = 0$ at c. Then, in the triangle $bc'c$ we construct a linear function with values 1 on bc' and 0 at c. It is given by

$$\phi(\mathbf{x}) = 1 - \frac{y}{|cc'|}.$$

For this function we have

$$\int_{bc'c} \frac{1}{2}|\nabla\phi(\mathbf{x})|^2 dx = \int_{bc'c} \frac{1}{2|cc'|^2} dx = \frac{1}{R^2 - S^2}|bc'c|.$$

The quantity $S > 0$ does not depend δ, and therefore we have

$$\int_{bc'c} \frac{1}{2}|\nabla\phi(\mathbf{x})|^2 dx \leq C_3|bc'c|, \qquad (3.6.51)$$

where $C_3 < \infty$ does not depend on δ_{ij}.

3.6.11 Step E. Evaluation of the Dirichlet integral in the whole domain

We consider now the domain

$$U = \Pi \setminus \bigcup_{i=1}^{N} D_i \setminus \left(\bigcup_{i=1}^{N} \bigcup_{j \in N_i} \Pi_{ij} \right)$$

outside the disks. This domain can be decomposed into the set of necks and the set of triangles, which have been considered in Steps A–D.

We remark here that the construction in Steps A–D can be viewed as the introduction of a special type of finite element, which takes into account localization of large energy within the necks and boundedness of the energy outside the necks.

From (3.6.44), (3.6.50) and (3.6.51) we have

$$\int_U \frac{1}{2}|\nabla\phi(\mathbf{x})|^2 dx \leq \max(C_1, C_2, C_3)|U|, \tag{3.6.52}$$

where $|U|$ stands for the area of the domain U. It is clear that $|U|$ is less than the area of the domain Q_m.

Using (3.6.52), we can write

$$\int_U \frac{1}{2}|\nabla\phi(\mathbf{x})|^2 dx \leq C|\Pi|, \tag{3.6.53}$$

where the constant $C < \infty$ does not depend on δ ($|\Pi|$ is the area of the domain Π).

Summing over all necks and triangles and using equation (3.6.32), we obtain

$$\int_{Q_m} \frac{1}{2}|\nabla\phi(\mathbf{x})|^2 dx = \frac{1}{4}\sum_{i,j=1}^{K} g_{ij}^s(t_i - t_j)^2 \tag{3.6.54}$$

$$+ \sum_{\Pi_{ij}} \int_{\Pi_{ij}} \frac{1}{2}\left(\frac{\partial\phi}{\partial x}\right)^2 dx + \int_U \frac{1}{2}|\nabla\phi(\mathbf{x})|^2 dx.$$

The factor $\frac{1}{4}$ in (3.6.54) appears for the same reason as in (3.6.21): it is the double count of the necks when summing in $i, j = 1, \ldots, K$. In order to keep the usual factor $\frac{1}{2}$ in (3.6.54) one can sum over $i < j$.

Finally, from (3.6.53) and (3.6.54) we obtain

$$A \leq \frac{1}{4}\sum_{i,j=1}^{K} g_{ij}^s(t_i - t_j)^2 + M, \tag{3.6.55}$$

where g_{ij}^s is defined by (3.6.16), and the constant

$$M = \max(C_1, C_2, C_3)|U| + KE_A$$

is uniformly bounded in δ. Here K is the number of disks (including pseudo-disks), and E_A is defined by (3.6.37).

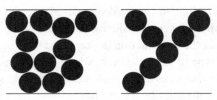

Figure 3.12　Examples of closely-packed structures: quasi-hexagonal (left) and a chain-like (right) packed particles.

3.7　Close-packing systems of bodies

We begin by introducing the following definitions.

3.7.1　Some definitions for close-packing systems of bodies

Definition 3.6 Let δ be the interparticle distance parameter introduced in (3.4.1) and let R be the radius of the disks. The ratio $\dfrac{\delta}{R}$ is called the relative interparticle distance parameter.

In this section, we prove that when the relative interparticle distance $\dfrac{\delta}{R}$ tends to 0, the upper- and lower-sided bounds obtained above become matching to leading (0-th) order in the sense of Definition 3.3. This would imply that the effective conductivities A and A_d defined by (3.2.8) and (3.4.12), respectively, are asymptotically equivalent as $\delta \to 0$.

We consider an array of disks. Introduce the parameter

$$\delta_{\max} = \max_{i,j \in N_i} \delta_{ij}. \tag{3.7.1}$$

Here the maximum is taken over the neighboring pairs of vertices (recall that N_i means vertices connected to the i-th vertex).

Definition 3.7 Consider a family of arrays of disks. For each array from this family consider the value of the parameter $\delta_{\max}$ defined by (3.7.1). If $\delta_{\max} \ll 1$, then we say that such a family satisfies the close-packing condition.

We present here two examples of arrays, which satisfy the close-packing condition (Figure 3.12).

Further, we distinguish the uniform packing condition and non-uniform packing condition.

Definition 3.8 Suppose a family of disks satisfies the close-packing condition. If, additionally, d_{ij} from (3.4.1) is such that $C_1 \leq d_{ij} \leq C_2$, where $0 < C_1 < C_2 < \infty$ are absolute constants and ij vary over all neighboring pairs of vertices, then we say that the family of arrays satisfies the uniform packing condition. Otherwise, we say that this family satisfies the non-uniform packing condition.

Roughly speaking, the uniformity means that all distances between neighbors are of the same order, whereas non-uniformity means that some neighbors may be much closer than others.

For a uniformly packed family of arrays, the quantities δ_{max}, δ and δ_{ij} for neighboring disks have the same order.

Definition 3.9 If a family of disks satisfies conditions 3.7 and 3.8, then we say that the family of arrays satisfies the uniform close-packing condition.

We will use this definition in the main theorem of this chapter. Note that this definition does not cover all closely packed configurations, which are of interest for applications. Later in Chapter 5, we consider problems which require a more general definition of closely packed systems, the so-called δ-**N** *close packing condition*.

3.7.2 Modeling of close-packing systems of bodies

To understand the possible mechanisms of close-packing, we recall the periodic model, see Section 2.7. The limit model is a periodic system of disks touching one another. The near-limit configuration of the disks can be obtained in two different ways:

1. In the limit position let the disks (we assume the disks to be identical) have radii equal to R_0. Decreasing the radius of the i-th disk for $R_i < R_0$, we obtain a system of disks, which form a close-packing system as $R_i \rightarrow R_0$. In this case the interparticle distances between the i-th disk and the j-th disk is $\delta_{ij} = |R_i - R_j| \rightarrow 0$. In this case the radii of the disks are not fixed, but have finite (non-zero) limit values.

2. In the situation described in item 1 above, we can extend the system of disks while retaining the radii of the disks. It this case the interparticle distances δ_{ij} between the i-th disk and the j-th disk become greater then zero and when we compress the system, $\delta_{ij} \rightarrow 0$. In a periodic system this extension can be modeled by a sum of uniform extension and individual movement of the particles (the shaking of particles; for a detailed description of the shaking model see Section 8.4).

For a random system of disks both models described above can be applied. In many cases, for a random model we can use a model with many degrees of freedom. In many cases, for a random system of disks in the limit configuration it is possible to move disks a small distance without overlapping other disks (in this case we say that the disks have degrees of freedom). In other words, we can "strew" the limit random structure. In this case the radii of the disks are preserved.

3.8 Finish of the proof of the network approximation theorem

We now use bounds (3.6.27) and (3.6.55) to justify the heuristic network model. We will need the following formulas obtained above:

$$A \geq -\frac{1}{4} \sum_{i,j=1}^{K} \frac{g_{ij}^2 (t_i - t_j)^2}{g_{ij}^s} + (P^+ - P^-), \qquad (3.8.1)$$

$$A \leq \frac{1}{4} \sum_{i,j=1}^{K} g_{ij}^s (t_i - t_j)^2 + M, \qquad (3.8.2)$$

where M is uniformly bounded in δ,

$$g_{ij}^s = \int_{-S}^{S} \frac{dx}{H_{ij}(x)}, \qquad (3.8.3)$$

$H_{ij}(x)$ is the distance between the neighboring i-th and i-th disks as a function of x, see Figure 3.4,

$$H_{ij}(x) = \delta_{ij} + 2R - 2\sqrt{R^2 - x^2},$$

the quantities

$$g_{ij} = \pi \sqrt{\frac{R}{\delta_{ij}}}$$

are determined by (3.4.2), and S is the half-width of the neck Π_{ij}. As of yet, S has not been defined precisely. In order to do this, we first summarize all the restrictions on S introduced above:

(1) $S > 0$ and does not depend on δ (we assume that R is fixed);
(2) $S \leq \dfrac{R}{2}$ (see (3.6.36));
(3) necks do not overlap.

We choose S so that conditions (1)–(3) hold for a sufficiently small $\dfrac{\delta}{R}$. We next specify what "for a sufficiently small $\dfrac{\delta}{R}$" means.

A neck of width $2S$ is adjacent to the arc L corresponding to the angle ψ, Figure 3.13. We justify item (1) if we show that there exist $\psi_0 > 0$ such that $\psi > \psi_0$ for arbitrary positions of disks if the distances between the disks are sufficiently small. This fact it obvious and we give only brief explanations.

We consider two disks D_2 and D_3 for which $\delta_{23} = R$ and disk D_1, which penetrates between these two disks, Figure 3.13. In this case the thickness of the neck is minimal and the corresponding value of the angle $\angle AO_2O_3$ is $\arctan\left(\dfrac{\sqrt{7}-2}{6}\right)$.

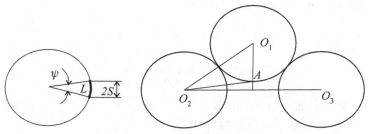

Figure 3.13 The width S of a neck.

When $\delta_{23} < R$, the angle

$$\angle AO_2O_3 > \arctan\left(\frac{\sqrt{7}-2}{6}\right).$$

We take

$$\psi_0 = \arctan\left(\frac{\sqrt{7}-2}{6}\right).$$

Then the choose the thickness S satisfying the condition $S_0 < S < R\sin\psi_0$, where S_0 is an arbitrary positive number such that $S_0 < R\sin\psi_0$, guarantees that the thickness S is separated from zero for arbitrary positions of disks, and necks are not overlapping if the distances between the disks are sufficiently small.

We shall need several auxiliary facts. First we establish a relation between the quantities g_{ij}^s and g_{ij} defined in (3.8.3) and (3.4.2), respectively. The first quantity came from the rigorous bounds on the effective conductivity A and the second appeared in the construction of the heuristic network.

Lemma 3.10 *Let* $\dfrac{\delta_{ij}}{R}$ *be the relative interparticle distance between the i-th and the i-th disks. Then, as* $\delta_{ij} \to 0$

$$g_{ij}^s = g_{ij} + O(1) + \sqrt{\frac{R}{\delta_{ij}}} O\left(\frac{\delta_{ij}}{R}\right), \qquad (3.8.4)$$

where $g_{ij} = \pi\sqrt{\dfrac{R}{\delta_{ij}}}.$

Proof For circular disks (*i*-th and *i*-th) of radius R we have $H_{ij}(x) = \delta_{ij} + 2R - 2\sqrt{R^2 - x^2}$. Then

$$g_{ij}^s = \int_{-S}^{S} \frac{dx}{\delta_{ij} + 2R - 2\sqrt{R^2 - x^2}} = \int_{-S^*}^{S^*} \frac{R\cos\tau\, d\tau}{\delta_{ij} + 2R - 2R\cos\tau}, \qquad (3.8.5)$$

where we made the change of variables

$$x = R \sin \tau,$$

$$S^* = \arcsin\left(\frac{S}{R}\right) > 0.$$

Note that S^* does not depend on δ_{ij}.

The last integral in (3.8.5) is equal to

$$-\tau + \frac{(\delta_{ij} + 2R)}{\sqrt{(\delta_{ij} + 2R)^2 - 4R^2}} \arctan\left(\frac{\tan\left(\frac{\tau}{2}\right)}{\sqrt{\frac{\delta_{ij}}{\delta_{ij} + 4R}}}\right)\Bigg|_{-S^*}^{S^*}. \tag{3.8.6}$$

Consider this integral for small δ_{ij}. If we evaluate the leading terms of order $\frac{R}{\delta_{ij}}$ in explicit form and estimate the lowest order terms, we have

$$\frac{\delta_{ij} + 2R}{\sqrt{(\delta_{ij} + 2R)^2 - 4R^2}} = \frac{\frac{\delta_{ij}}{R} + 2\sqrt{\frac{R}{\delta_{ij}}}}{\sqrt{\frac{\delta_{ij}}{R} + 4R}} = \sqrt{\frac{R}{\delta_{ij}}}\left[1 + O\left(\frac{\delta_{ij}}{R}\right)\right],$$

$$\frac{1}{\sqrt{\frac{\delta_{ij}}{\delta_{ij} + 2R}}} = 2\sqrt{\frac{R}{\delta_{ij}}}\left[1 + O\left(\frac{\delta_{ij}}{R}\right)\right].$$

Then (3.8.6) becomes

$$-S^* + \sqrt{\frac{R}{\delta_{ij}}}\left[1 + O\left(\frac{\delta_{ij}}{R}\right)\right]\left[\arctan\left(2\sqrt{\frac{R}{\delta_{ij}}}\tan\left(\frac{\tau}{2}\right)\right)\Bigg|_{-S^*}^{S^*} + O\left(\frac{\delta_{ij}}{R}\right)\right],$$

as $\frac{\delta_{ij}}{R} \to 0$.

Since S^* does not depend on δ_{ij}, we have

$$\arctan\left(2\sqrt{\frac{R}{\delta_{ij}}}\tan\left(\frac{\tau}{2}\right)\right)\Bigg|_{-S^*}^{S^*} = \pi + O\left(\frac{\delta_{ij}}{R}\right).$$

Finally

$$g_{ij}^s = 2S^* + \pi\sqrt{\frac{R}{\delta_{ij}}}\left[1 + O\left(\frac{\delta_{ij}}{R}\right)\right].$$

It is seen that the term

$$g_{ij} = \pi \sqrt{\frac{R}{\delta_{ij}}},$$

is the leading term in the asymptotic expansion of g_{ij}^s as $\delta_{ij} \to 0$. □

Lemma 3.10 gives a limiting value of the quantity g_{ij}^s for a "disk–disk" pair (i.e., it describes the asymptotic value of the specific flux between two closely spaced disks). In the considered problem, in addition, there exist fluxes between disks and pseudo-disks. The following lemma gives a limiting value of the quantity g_{ij}^s for a disk–pseudo-disk pair.

Lemma 3.11 *Let $\dfrac{\delta_{ij}}{R}$ be the relative interparticle distance between the i-th disk and the j-th pseudo-disk. Then as $\delta_{ij} \to 0$*

$$g_{ij}^s = g_{ij}^\infty + O(1) + \sqrt{2\frac{R}{\delta_{ij}}} O\left(\frac{\delta_{ij}}{R}\right), \qquad (3.8.7)$$

where

$$g_{ij}^\infty = \pi \sqrt{\frac{2R}{\delta_{ij}}}.$$

The proof is similar to the proof of Lemma 3.10. The single distinction is that the distance between the i-th disk and the j-th pseudo-disk (which is a portion of a straight line) is $H_{ij}(x) = \delta_{ij} + R - \sqrt{R^2 - x^2}$.

Introduce the following quantity called the energy of the discrete network:

$$E_d = \frac{1}{4} \sum_{i,j=1}^{K} g_{ij}(t_i - t_j)^2, \qquad (3.8.8)$$

where $\{t_i; \ i \in I\}$ is the solution of the problem (3.4.11) (the factor $\dfrac{1}{4}$ was explained above). We wish to show that the energy E_d grows as $\sqrt{\dfrac{R}{\delta}}$ (here δ means the characteristic value of δ_{ij}). Note that although each coefficient g_{ij} is of order $\sqrt{\dfrac{R}{\delta}}$, the differences $(t_i - t_j)^2$ may become small. The next lemma shows that this is not the case for the scalar problem under consideration. Note that for vectorial problems, such degeneration of the leading term in the asymptotic of the effective coefficients may take place and it leads to interesting physical effects.

Lemma 3.12 *Under the close-packing condition, the energy E_d of the discrete network is of order not less than $\sqrt{\dfrac{R}{\delta_{\max}}}$, where $\delta_{\max}$ is determined by (3.7.1).*

Proof We shall employ the comparison method introduced in Kesten (1992). Define another network with the coefficients

$$g_{ij}^1 = \begin{cases} \pi\sqrt{\dfrac{R}{\delta_{\max}}} & \text{if } g_{ij} > 0, \\ 0 & \text{if } g_{ij} = 0. \end{cases}$$

Since $\delta_{\max} = \max\limits_{ij} \delta_{ij}$, we have $g_{ij} \geq g_{ij}^1$ and therefore

$$\sum_{i,j=1}^{K} g_{ij} x_i x_j \geq \sum_{i,j=1}^{K} g_{ij}^1 x_i x_j.$$

Let $\{t_i; \ i \in I\}$ be the solutions of the problem (3.4.11) and $\{\tilde{t}_i; \ i \in I\}$ be the solutions of the problem (3.4.11) with the coefficients $\tilde{g}_{ij}$. The latter problem can be written as

$$\pi\sqrt{\dfrac{R}{\delta_{\max}}} \sum_{j=1}^{K} \tilde{g}_{ij}(\tilde{t}_i - \tilde{t}_j) = 0 \text{ for } i \in I; \tag{3.8.9}$$

$$\tilde{t}_i = 1 \text{ for } i \in S^+, \ \tilde{t}_i = -1 \text{ for } i \in S^-.$$

Here we sum over i, j such that $g_{ij}^1 > 0$ (i.e., over neighboring disks); I is the set of internal sites of the network, S^+ and S^- are boundary sites, and the coefficients are

$$\tilde{g}_{ij} = \begin{cases} 1 & \text{if } g_{ij} > 0, \\ 0 & \text{if } g_{ij} = 0. \end{cases}$$

The energy of the network $\{\mathbf{x}_i, g_{ij}; \ i, j = 1, \dots, K\}$ is greater than the energy of the network $\{\mathbf{x}_i, \tilde{g}_{ij}; \ i, j = 1, \dots, K\}$. Indeed, the energies in these networks are the minimum value of the functions

$$\sum_{i,j=1}^{K} g_{ij}(t_i - t_j)^2$$

and

$$\sum_{i,j=1}^{K} \tilde{g}_{ij}(t_i - t_j)^2,$$

respectively.

Due to the above relation between the quadratic forms we have

$$\sum_{i,j=1}^{K} g_{ij}(t_i - t_j)^2 \geq \sum_{i,j=1}^{K} g_{ij}^1(t_i - t_j)^2 \geq \pi \sqrt{\frac{R}{\delta_{\max}}} \sum_{i,j=1}^{K} (\tilde{t}_i - \tilde{t}_j)^2. \qquad (3.8.10)$$

The last sum is taken over i, j such that $g_{ij}^1 > 0$.

The solution of the problem (3.8.9) does not depend on $\pi \sqrt{\dfrac{R}{\delta_{\max}}}$ because we can cancel out $\pi \sqrt{\dfrac{R}{\delta_{\max}}}$ in equation (3.8.9) and obtain a problem with respect to $\{\tilde{t}_i\}$, which is equivalent to (3.8.9) but does not involve $\pi \sqrt{\dfrac{R}{\delta_{\max}}}$. The left-hand side of (3.8.10) is $4E$ and its right-hand side is of the form $\pi \sqrt{\dfrac{R}{\delta_{\max}}} const$, where $const < \infty$ does not depend on δ and is positive. Due to the uniform δ-close packing there exists a path $\{(\mathbf{x}_{i_p}, \mathbf{x}_{i_{p+1}}); \ p = 1, \dots, P\}$ (a conducting spanning cluster) such that $\tilde{g}_{i_p, i_{p+1}} = 1$ and the disk $D_{p_1} \in S^+$ and the disk $D_{p_{P+1}} \in S^-$.

It is impossible that all $\tilde{t}_i - \tilde{t}_j = 0$ along this path. Thus the sum

$$\sum_{p=1}^{P} (\tilde{t}_{p+1} - \tilde{t}_p)^2 \qquad (3.8.11)$$

is positive and does not involve $\pi \sqrt{\dfrac{R}{\delta_{\max}}}$. The sum in the right-hand part of (3.8.10) is not less than (3.8.11) multiplied by $\pi \sqrt{\dfrac{R}{\delta_{\max}}}$. $\qquad \square$

Using Lemma 3.10 and Lemma 3.2, we can replace g_{ij}^s by g_{ij} as $\delta \to 0$ in (3.8.1) and (3.8.2). Then (3.8.1) and (3.8.2) provide the following bounds:

$$A \geq -\frac{1}{4} \sum_{i,j=1}^{K} g_{ij}(t_i - t_j)^2 + (P^+ + P^-) + \sqrt{\frac{R}{\delta}} O\left(\frac{\delta}{R}\right), \qquad (3.8.12)$$

$$A \leq \frac{1}{4} \sum_{i,j=1}^{K} g_{ij}(t_i^- - t_j^-)^2 + M + \sqrt{\frac{R}{\delta}} O\left(\frac{\delta}{R}\right), \qquad (3.8.13)$$

where M is uniformly bounded in δ as $\delta \to 0$ and $\{t_i; i \in I\}$ are the solutions of the algebraic system (3.4.11). We recall that δ and $\delta_{\max}$ are of the same order for a uniform closely-packed family of disks.

The first two terms on the right-hand side of (3.8.12) and the first term on the right-hand side (3.8.13) are the discrete analogs of the dual variational principle for the continuum medium. We have explained above how to replace $\frac{1}{4}$ by the usual $\frac{1}{2}$. It is known Ekeland and Temam (1976) that the values of the direct and the dual energy functionals for the continuum problem are the same when the exact minimizers are substituted into the functionals. The next lemma is the discrete analog of this statement.

Lemma 3.13 *The following equality holds*

$$\frac{1}{4} \sum_{i,j=1}^{K} g_{ij}(t_i - t_j)^2 = P^+ - P^-, \qquad (3.8.14)$$

where $\{t_i;\ i \in I\}$ is the solution of the problem (3.4.10).

Formula (3.8.14) differs from the usual Green's formula (for partial differential equations) by the factor $\frac{1}{4}$. It appears for the same reason that $\frac{1}{4}$ appears in the discrete formulas for the energy presented above.

Proof We write (taking into account that $g_{ij} = g_{ji}$)

$$\frac{1}{2} \sum_{i,j=1}^{K} g_{ij}(t_i - t_j)^2 = \frac{1}{2} \sum_{i,j=1}^{K} g_{ij}(t_i - t_j)(t_i - t_j) \qquad (3.8.15)$$

$$= \frac{1}{2} \sum_{i=1}^{K} t_i \left[\sum_{j=1}^{N} g_{ij}(t_i - t_j) \right] + \frac{1}{2} \sum_{j=1}^{K} t_j \left[\sum_{i=1}^{N} g_{ji}(t_j - t_i) \right].$$

The two sums in the square brackets are equal to zero due to (3.4.10) (for $i \in I$ in the first sum and for $j \in I$ in the second sum). Then (3.8.15) becomes:

$$\sum_{i \in S^+ \cup S^-} t_i \sum_{j=1}^{K} g_{ij}(t_i - t_j). \qquad (3.8.16)$$

Only terms which correspond to S^- and S^+ remain. Note that $t_i = 1$ for $i \in S^+$, and $t_i = -1$ for $i \in S^-$ in (3.8.16). Hence (3.8.16) is exactly equal to $P^+ - P^-$. $\qquad\qquad\square$

The following theorem shows that the continuum problem for the effective conductivity is asymptotically equivalent to the discrete network.

Theorem 3.14 *Let the uniform close-packing condition hold. Then the effective coefficient (3.2.8)*

$$A = \frac{1}{2} \int_{Q_m} |\nabla\phi(\mathbf{x})|^2 d\mathbf{x}$$

is given by the following asymptotic formula (as $\delta_{\max} \to 0$):

$$A = \frac{1}{4} \sum_{i,j=1}^{N} g_{ij}(t_i - t_j)^2 + \sqrt{\frac{R}{\delta}} O\left(\frac{\delta}{R}\right) + O(1), \qquad (3.8.17)$$

where

$$g_{ij} = \pi \sqrt{\frac{R}{\delta_{ij}}}$$

and $\{t_i; i \in I\}$ is the solution of the discrete network problem (3.4.10).
    The leading term in (3.8.17)

$$A_d = \frac{1}{4} \sum_{i,j=1}^{N} g_{ij}(t_i - t_j)^2$$

is of order $\sqrt{\dfrac{R}{\delta}}$ and is expressed through the solution $\{t_i; i \in I\}$ of the discrete network problem (3.4.10).

Proof Using (3.8.12), (3.8.13) and (3.8.14) we obtain

$$\frac{1}{4} \sum_{i,j=1}^{K} g_{ij}(t_i - t_j)^2 + \sqrt{\frac{R}{\delta}} O\left(\frac{\delta}{R}\right) + O(1) \leq A \qquad (3.8.18)$$

$$\leq \frac{1}{4} \sum_{i,j=1}^{K} g_{ij}(t_i - t_j)^2 + M + \sqrt{\frac{R}{\delta}} O\left(\frac{\delta}{R}\right) + O(1).$$

By Lemma 3.12 the term $\dfrac{1}{4} \sum_{i,j=1}^{K} g_{ij}(t_i - t_j)^2$ in (3.8.18) is of order $\sqrt{\dfrac{R}{\delta}}$ and, consequently, is the leading term in the asymptotic expansion. We recall that $\delta_{\max}$ and δ are of the same order under the uniform close-packing condition. $\qquad \square$

In terms of Definition 3.3, we say that the bounds (3.8.1) and (3.8.2) asymptotically match to leading order $\dfrac{1}{4} \sum_{i,j=1}^{K} g_{ij}(t_i - t_j)^2$ as $\delta \to 0$.

In Chapter 5, the proof will be modified so that the remainder in the asymptotic formula (3.8.17) can be evaluated in explicit terms. The latter will allow us to obtain an error bound for our network approximation.

Remark. The proof of Theorem 3.14 was given for fixed radius of the disks. The proof works for the case of disks of variable radius $R(\delta)$ under the condition that $R(\delta) \to R > 0$ as the parameter $\delta \to 0$. In fact, in this case $O\left(\dfrac{\delta}{R(\delta)}\right)$ and $O(1)$ in (3.8.1) and (3.8.2) can be replaced by $O\left(\dfrac{\delta}{R}\right)$ and $O(1)$, maybe with a constant not depending on δ. Thus, Theorem 3.14 is valid for all dense-packing models described in Section 3.7.2.

We give one useful interpretation of Theorem 3.14. First, we rewrite (3.8.17) as

$$\frac{|A - A_d|}{\sqrt{\dfrac{R}{\delta}}} = O\left(\frac{\delta}{R}\right) + O(1)\sqrt{\frac{\delta}{R}}. \tag{3.8.19}$$

Now recall that A_d is of the order of $\sqrt{\dfrac{R}{\delta}}$ and replace in the left-hand side of (3.9.1) $\sqrt{\dfrac{R}{\delta}}$ by A_d. In addition, we note that the sum in the right-hand side of the equality (3.9.1) has the order of $O\left(\sqrt{\dfrac{\delta}{R}}\right)$, as $\delta \to 0$. Then, we can rewrite (3.9.1) in the following form:

$$\frac{|A - A_d|}{A_d} = O\left(\sqrt{\frac{\delta}{R}}\right). \tag{3.8.20}$$

Formula (3.9.2) implies that the continuum energy is approximated by the discrete energy in the following sense. The difference between the energies of the original continuum problem and the discrete network problem normalized by the energy of the discrete problem (or by the energy of the continuum problem, since A and A_d are of the same order, as $\delta \to 0$) tends to 0, as $\delta \to 0$. Note that the normalized differences of the energies has the order of $O\left(\sqrt{\dfrac{\delta}{R}}\right)$ (see also Theorem 5.2 in Chapter 5).

3.9 The pseudo-disk method and Robin boundary conditions

The pseudo-disks are artificial objects of a mathematical nature. They do not exist physically. Thus, to have a "physical" network model, we have to remove the pseudo-disk from the final model. It may be done in the following way.

The sum $\dfrac{1}{2} \displaystyle\sum_{i,j=1}^{K} g_{ij}(t_i - t_j)^2$ in Theorem 3.14 is naturally decomposed as

$$\frac{1}{2} \sum_{i \in I} \sum_{j \in N_i} g_{ij}(t_i - t_j)^2 + \frac{1}{2} \sum_{i \in S^+ \bigcup S^-} \sum_{j \in N_i} g_{ij}(t_i - t_j)^2.$$

The first sum

$$\frac{1}{2} \sum_{i \in I} \sum_{j \in N_i} g_{ij}(t_i - t_j)^2 \tag{3.9.1}$$

corresponds to the energy in the necks connecting the original (real) disks.

The second sum

$$\frac{1}{4} \sum_{i \in S^+ \bigcup S^-} \sum_{j \in N_i} g_{ij}(t_i - t_j)^2 \qquad (3.9.2)$$

corresponds to the energy in the necks connecting the original disks and pseudo-disks. Thus, the energy form (3.9.2) describes the particle–wall interaction. The fluxes (the particle–wall fluxes) corresponding to the energy form (3.9.2) are proportional to the difference of potentials of the pseudo-disks (i.e., potential ± 1 of the upper and low boundaries S^+ and S^-) and the potentials of the near-boundary particles.

Minimization of (3.9.2) in the process of minimization of (3.9.1) leads to the sum of the expression of the form $g_{ij}(t_i - t_j)$, $i \in S^+ \bigcup S^-$, which are $g_{ij}(1 - t_j)$ for $i \in S^+$ and $g_{ij}(-1 - t_j)$ for $i \in S^-$.

These terms describe the fluxes from the boundaries S^+ and S^- to the near-boundary particles. This kind of relationship for continuous problems is known as a Robin (or third-type) boundary condition (Courant and Hilbert, 1953). The introduction of the notion of pseudo-disks allowed us to account for the particle–wall interaction in a uniform style.

4

Numerics for percolation and polydispersity via network models

In this section, we present an application of the network model developed in Chapter 3 to the numerical analysis of high-contrast composite materials.

In Chapter 3, we expressed the leading term A of the conductivity of high-contrast composite materials through the solution of the network problem (3.4.11). The dimension of the network problem (3.4.11) is significantly smaller than the dimension of a non-structural (for example, finite elements or finite differences) approximation of the original problem (3.2.3)–(3.2.7). We demonstrate that the network approximation also provides us with an effective tool for the numerical analysis of high-contrast composite materials.

We consider models of a composite material filled with mono- and polydispersed particles (once again, we will model particles by disks). A composite material is called monodispersed if all disks have the same radii. If the radii of the disks vary, then the composite material is called polydispersed.

4.1 Computation of flux between two closely spaced disks of different radii using the Keller method

In order to analyze polydispersed composite materials, we need to know the flux between two disks (from one disk to another) of different radii if the potential on each disk is constant. A simple approximate formula for this flux was obtained in Keller (1987) for identical disks. We employ Keller's method to derive an approximate formula for the flux between two disks (the i-th and the j-th) of arbitrary radii R_i and R_j placed at a distance δ_{ij} from one another (see Figure 4.1). We approximate the disks by their tangential parabolas, namely

$$y = \frac{\delta_{ij}}{2} + \rho_i \frac{x^2}{2} \qquad (4.1.1)$$

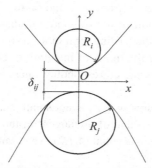

Figure 4.1 Two neighboring disks and their tangential parabolas.

and

$$y = -\left(\frac{\delta_{ij}}{2} + \rho_j \frac{x^2}{2}\right),$$

where

$$\rho_i = \frac{1}{R_i}, \rho_j = \frac{1}{R_j} \qquad (4.1.2)$$

are the curvatures of the disks, equal to the curvatures of the tangential paraboloids (4.1.1) at the points of tangency.

The distance between the parabolas (4.1.1) is equal to

$$H_{ij}(x) = \delta_{ij} + (\rho_i + \rho_j)\frac{x^2}{2}.$$

We assume that the material constant of the matrix is $a(\mathbf{x}) = 1$ and that the local flux $\mathbf{J}$ (the gradient of the potential) between the disks is of the form $(\mathbf{x} = (x, y))$

$$\mathbf{J}(\mathbf{x}) = \left(0, \frac{t_i - t_j}{H_{ij}(x)}\right). \qquad (4.1.3)$$

That is, we assume that it is proportional to the difference in temperature or potential and inversely proportional to the distance between the disks. Then the total flux between the disks can be calculated as follows:

$$J_{ij} = (t_i - t_j)\int_{-\infty}^{\infty} \frac{dx}{\delta_{ij} + (\rho_i + \rho_j)\frac{x^2}{2}}$$

$$= \frac{t_i - t_j}{\left(\frac{\rho_i + \rho_j}{2}\right)^{1/2}\delta_{ij}^{1/2}} \arctan\left[\left(\frac{\rho_i + \rho_j}{2\delta_{ij}}\right)^{1/2}x\right]\Bigg|_{-\infty}^{\infty}. \qquad (4.1.4)$$

In (4.1.4), the limits of integration are $-\infty$ and ∞. However, the flux (4.1.3) is determined only between the neighboring disks. Following Keller (1963), we change the limits of integration from $-\infty$ and ∞ to $-R_i$ and R_i or $-R_j$ and R_j, where R_i and R_j are the radii of the i-th and j-th disks, respectively. Since the integral (4.1.4) converges, the leading term in its asymptotics as $\delta_{ij} \to 0$ remains the same for any choice of the limits of integration (except zero). Note that for two closely spaced spheres the corresponding integral diverges (Kolpakov and Kolpakov, 2010).

If the sum of the curvatures $\rho_i + \rho_j$ is not small, and the distance δ_{ij} between the disks is small, then the value of arctangent in the right-hand side (4.1.4) is approximately π. Taking into account (4.1.2), we obtain

$$J_{ij} = g_{ij}(t_i - t_j), \qquad (4.1.5)$$

where

$$g_{ij} = \frac{\pi \sqrt{\dfrac{2R_i R_j}{R_i + R_j}}}{\sqrt{\delta_{ij}}}. \qquad (4.1.6)$$

Note that g_{ij} is the specific flux (i.e., the flux per unit difference of the potential).

Formula (4.1.5) can be derived from the exact solution of the Laplace equation for two disks by taking the leading term asymptotic in δ_{ij}. This exact solution (capacitance of two cylinders) can be found in Smythe (1950).

4.2 Concept of neighbors using characteristic distances

It is possible to extend the results of Chapter 3, including the network approximation Theorem 3.14, to the case of a composite material filled with disks of various radii.

In Chapter 3, we introduced the concept of neighbors using the Voronoi method. Next we introduce the concept of neighbors using the concept of characteristic distances. The Voronoi method is a universal method for introducing the concept of neighbors. The concept of characteristic distances is based on empirical results collected on the basis of statistical analysis of random distributions of particles. In numerical computations, the method of characteristic distances is a useful alternative to the Voronoi method due to its simplicity. The method of characteristic distances is in agreement with the localization of large fluxes between closely spaced particles in the composite material noted in Chapter 3. In order to estimate the characteristic values of the flux for the different characteristic distances between the disks, we will introduce two characteristic distances between a given particle and all other particles:

(i) δ^* represents the maximum distance between two particles such that they will be considered the nearest neighbors.

(ii) A second distance, corresponding to the distance between particles that are not neighbors with a given particle yet are still close. For concreteness, we choose this distance to be equal to the particle radius R in the case of monodispersed particles and $R = \min_i R_i$ in the case of polydispersed particles. Particles in the range (δ^*, R) form a "second belt" around a given particle. Note that $\delta^* \ll R$ for densely-packed particles.

In order to choose the cut-off distance δ^* in numerical simulations, we use a visual control. This is performed by examining many randomly generated pictures. Two disks are considered neighbors when they share a common edge in the Voronoi tessellation. Neighbors in a random array of disks are not necessarily closely spaced. The heuristic idea behind our approach is that we take into account only the fluxes between closely spaced particles. Precisely, we choose the cut-off distance δ^* in the interval $0.3R$–$0.5R$ and obtain a connected graph (network), which consists of the centers of closely spaced disks. The graph is a discrete object controlled by the continuous real variable δ^*. Since edges can form or disappear as δ^* changes, the graph does not deform continuously with changes in δ^*.

The choice of the interval $0.3R$–$0.5R$ is motivated by the fact that numerical simulations show that even if the disks are not quite of identical sizes, but the variability in the sizes is not large (a factor of 2 or 3), then we obtain stable, consistent pictures. More precisely, in this range for δ^*, we observe that all closely spaced neighbors are connected and the disks which are not closely spaced neighbors are not connected. We have also used another numerical criterion to support this choice of the cut-off distance δ^*. Namely, if δ^* is increased, then the effective conductivity practically does not change.

The characteristic values of the fluxes corresponding to these distances are given by

$$J(\delta_{ij} \le \delta^*) \sim \sqrt{\frac{R}{\delta^*}}, \tag{4.2.1}$$

$$J(\delta^* < \delta_{ij} \le R) \sim 1.$$

The first equation (4.2.1) is obtained from (4.1.5).
Then, we get

$$\frac{J(\delta_{ij} \le \delta^*)}{J(\delta^* < \delta_{ij} \le R)} \sim \sqrt{\frac{R}{\delta^*}}.$$

Thus, by results obtained in Chapter 3, we can neglect the fluxes of order 1 and keep the fluxes of order $\sqrt{\frac{R}{\delta^*}}$, where R is the characteristic radius of the particles and δ^* is the characteristic distance between the particles.

4.3 Numerical implementation of the discrete network approximation and fluxes in the network

We now define the discrete network corresponding to the original continuum model and formulate the discrete problem.

Numerically, we create a distribution of disks of a given radius (radii for the polydispersed case) in the rectangle $\Pi = [-L, L] \times [-1, 1]$. The center $\mathbf{x}$ of each disk is generated as a uniformly distributed random variable in Π. If a disk with center $\mathbf{x}$ and radius R overlaps with any disk which is already present, then it is rejected. Otherwise, it is accepted and added to the list of disks

$$X = \{\mathbf{x}_i, \ i = 1, \ldots, N\}. \tag{4.3.1}$$

In (4.3.1), we identify disks with those centered at $\mathbf{x}_i$. If the disks have different radii, the radii are included in the list of disks: $X = \{(\mathbf{x}_i, R_i), \ i = 1, \ldots, N\}$.

For the monodispersed case, we stop adding disks when the prescribed volume fraction of the disks

$$V = \frac{1}{|\Pi|} \sum_{i=1}^{N} \pi R^2$$

has been achieved. Here, N is the total number of disks and $|\Pi|$ is the volume of Π. This procedure works well for disks with relatively small radii, when the fraction of the boundary disks is not too high. After finishing the procedure of the generation of random disks, we construct pseudo-disks corresponding to the near-boundary disks and denote the total number of the disks by K.

We compute the distances δ_{ij} between the i-th and the j-th disks for the obtained distribution of disks. We then determine the flux between neighboring disks and neglect the fluxes between non-neighboring disks. Thus, the flux between the i-th and j-th disks is described by

$$g_{ij} = \begin{cases} g_{ij} \text{ calculated with formula (4.1.6)} & \text{if } \delta_{ij} \leq \delta^*, \\ 0 & \text{if } \delta_{ij} \geq \delta^* \end{cases} \tag{4.3.2}$$

(for a pseudo-disk in (4.3.2) the corresponding radius tends to infinity).

Thus we have defined a discrete network model (a weighted graph),

$$\mathcal{G} = \{\mathbf{x}_i, g_{ij}; \ i, j = 1, \ldots, K\}, \tag{4.3.3}$$

which consists of points or sites $\mathbf{x}_i$ (centers of the disks) and edges with assigned numbers g_{ij}. The discrete model (4.3.3) does not explicitly contain the sizes of the disks and the distances between them. This information is incorporated via the set g_{ij}. Note that if the concentration is small, g_{ij} will be zero (see equation (4.3.2)) and thus the effective conductivity will also be zero. On the other hand, the conductivity of the matrix is 1 and therefore the effective conductivity should approach 1. Nevertheless, this is not a contradiction. This is because our methods are tailored for densely-packed high-contrast composites, where the effective conductivity is a very large number. In comparison with such a large number, 1 and 0 are numerically indistinguishable.

For the purposes of these numerics, we now redefine the characteristic distances. If one looks at the pictures obtained as a result of the numerical procedure for placing random disks on the plane, it is easy to see that with respect to each given disk all other disks fall into one of the two categories: closely spaced neighbors (or simply "neighbors") and remote disks ("aliens"). More precisely, a typical disk has $5 - 6$ "neighbors", and the distance between the given disk and its neighbors can be small ($\delta_{ij} \ll R$), whereas the distance between the given disks and the "aliens" can not be small (at least of order of R). The "neighbors", which can be almost touching the disk in question, form a belt which separates the given disk from the rest ("aliens"). The fluxes between non-neighbors (i.e., a disk and the "aliens") are negligible and so the network approximation should ignore them. We thus ignore interactions with elements of the "second belt" discussed above and introduce the cut-off distance $\delta^* = R$ based on the analysis of the arrays obtained numerically in the procedure described above.

The equations that determine the values t_i of the potential on the sites $\mathbf{x}_i$ of the network (4.3.3) are (see (3.4.11))

$$\sum_{j \in n_i} g_{ij}(t_i - t_j) = \sum_{j \in S^+} g_{ij} - \sum_{j \in S^-} g_{ij}, \ i = 1, \dots, K. \tag{4.3.4}$$

The discrete formula for the effective conductivity is (see (3.4.12)):

$$A_d = \sum_{i \in S^+} \sum_{j \in N_i} g_{ij}(1 - t_j). \tag{4.3.5}$$

4.4 Property of the self-similarity problem (3.2.4)–(3.2.7)

The boundary-value problem (3.2.4)–(3.2.7) is invariant with respect to the transformation

$$\mathbf{x} \to \mathbf{z} = t\mathbf{x}, \tag{4.4.1}$$

where $t > 0$. This is because the equalities in problem (3.2.4)–(3.2.7), which involve the derivative $\dfrac{\partial \phi}{\partial \mathbf{n}}$, are homogeneous (have zero right-hand side).

Now, we consider the total flux through the boundary ∂Q^+ per unit length defined by formula (3.1.14)

$$\hat{a} = \frac{1}{|Q^+|} \int_{\partial Q^+} \frac{\partial \phi}{\partial \mathbf{n}}(\mathbf{x}) d\mathbf{x} = \frac{1}{2L} \int_{-L}^{L} \frac{\partial \phi}{\partial \mathbf{n}}(\mathbf{x}) d\mathbf{x}, \tag{4.4.2}$$

which also involves $\dfrac{\partial \phi}{\partial \mathbf{n}}(\mathbf{x})$. Recall that $2L$ is the length of the upper boundary ∂Q^+ of the domain Π.

In (4.4.2) ∂Q^+ is the top boundary of the domain $\Pi = [-L, L] \times [-1, 1]$. The transformation (4.4.1) transforms the interval $[-L, L]$ into the interval $[-tL, tL]$.

Subjected to the action of the transformation (4.4.1), the quantity (4.4.2) becomes

$$\hat{a} = \frac{1}{2tL} \int_{-tL}^{tL} \frac{\partial \phi}{\partial \mathbf{n}} (\mathbf{z}) dz. \tag{4.4.3}$$

We have several characteristic dimensions in the problem under consideration. The most important is, probably, δ/R. We take R as a unity. In other words, we make a transformation, which transform disks of radius R into disks equal to unity. To do this, we take $t = \frac{1}{R}$.

For this choice, the formula (4.4.3) becomes

$$\hat{a} = \frac{1}{2\frac{L}{R}} \int_{-\frac{L}{R}}^{\frac{L}{R}} \frac{\partial \phi}{\partial \mathbf{n}} (\mathbf{z}) dz. \tag{4.4.4}$$

Here $\frac{2L}{R}$ is the length of the top boundary ∂Q^+ "measured" in the radius R of the disks.

When the distribution of the disks is random and uniform (in the probability sense), we wait for an ergodic property of the problem, which means that

$$\frac{1}{2L} \int_{-L}^{L} \frac{\partial \phi}{\partial \mathbf{n}} (\mathbf{z}) d\mathbf{z} \tag{4.4.5}$$

does not depend on L as $L \to \infty$ and it is equal to the expected value of $\hat{a}$ with respect to random configurations of the disks. In this case, the quantity (4.4.5) has the physical meaning of average flux through unit length of the boundary. The quantity (4.4.4) has the same meaning, but it can demonstrate variations of value for a random configuration of disks if the ratio L/R is finite.

4.5 Numerical simulations for monodispersed composite materials. The percolation phenomenon

A composite material filled with particles of the same size is called monodispersed (see Figure 4.2).

We have implemented a numerical method to

(i) generate a configuration ω as a system of random disks.
(ii) compute the coefficients g_{ij} for this system of disks.
(iii) solve the linear system (4.3.4).
(iiii) compute the effective conductivity $A(V, \omega)$ using formula (4.3.5) for this configuration.

The control parameters are the diameter of the disks R and the total volume fraction V of the disks.

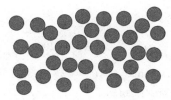

Figure 4.2 A monodispersed composite material.

As a result of each run of all four codes, we compute the value of the effective conductivity denoted by $A(\omega)$ for a given configuration ω.

In order to collect the statistical information, we run the program repeatedly and collect the data, consisting of $\{A(V, \omega), \omega \in \Omega\}$ (Ω means the set of all numerical experiments). Then we compute

(i) the expected value of the effective conductivity $A = A(V) = \mathbb{E}A(V, \omega)$.
(ii) the mean deviation $DA = \mathbb{E}|A(V, \omega) - A(V)|$.
(iii) the maximal conductivity over all collected data $m = \max_{\omega \in \Omega} A(V, \omega)$.

The expected value of the conductivity is then defined as

$$\mathbb{E}A(V, \omega) = \frac{1}{|\Omega|} \sum_{\omega \in \Omega} A(V, \omega).$$

where $|\Omega|$ is the number of experiments.

All the above quantities have been computed for given values of the quantities R and V, which is why $A, DA, m(R, V)$ are functions of R and V.

In order to test our numerical algorithms based on the network approximation, we perform our numerics for monodispersed composite materials. In this manner, we will make sure that we can consistently recover (in a very efficient way) known results, in particular, percolation phenomena.

4.5.1 Description of the numerical procedure

In our numerical simulations we have used an integer lattice $\Pi_0 = 550 \times 400$. That is, $L = 550/400 = 1.375$. This choice is made for convenience of visual control on the computer screen.

The centers of the disks have been generated in the form

$$\mathbf{x}_i = (random(550) + random([0, 1]), random(400) + random([0, 1]))$$

where $random(n)$ is a pseudo-random number uniformly generated from the set $\{0, 1, \ldots, n\}$ and $random([0, 1])$ is a pseudo-random number uniformly generated from the interval $[0, 1]$. Disks near the boundary were retained if their centers were inside Π_0. The boundary values 1 and 0 were prescribed at the upper and lower edges of Π_0, respectively.

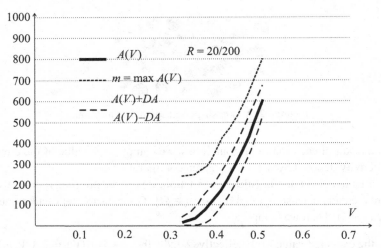

Figure 4.3 Effective conductivity as a function of volume fraction. A monodispersed composite material with $R = 20/200$.

In the numerical procedure we have used disks of radii 35, 25, 20, 15, and 10 in undimensional coordinates (in dimensional coordinates $35/200$, $25/200$, $20/200$, $15/200$, $10/200$, since $[-1, 1]$ corresponds to $[0, 400]$) and there were from 5 to 20 layers of disks on top of each other in the vertical direction. Statistical data were collected mainly for disks of radii 25, 20, and 15. The average number of disks in the rectangle Π_0 was between 50 and 150.

We computed the specific flux as

$$g_{ij} = \pi \sqrt{\frac{R}{\delta_{ij}}}$$

for a "disk–disk" pair (formula (4.1.6) with $R_1 = R_2 = R$) and

$$g_{ij} = \pi \sqrt{\frac{2R}{\delta_{ij}}}$$

for a "disk–pseudo-disk" pair (formula (4.1.6) with $R_1 = R$ and $R_2 = \infty$).

In the numerical experiments for the monodispersed composites, the expected value of A and the deviation DA was calculated over 320 configurations for several values of the volume fraction V, which are indicated in the plots on the horizontal axis. The values of A at these points are also indicated in Figures 4.3, 4.4, and 4.5.

We observe that the graphs, which corresponds to different volume fractions, are similar both qualitatively and quantitatively. Also, one can see that $m(V)$ and $\mathbb{E}A(V)$ have the same qualitative behavior. Thus we conclude that our numerical procedure is stable.

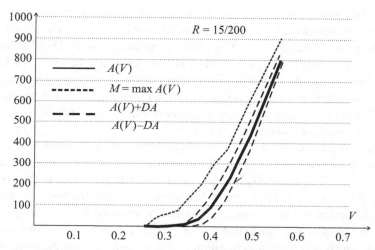

Figure 4.4 Effective conductivity as a function of volume fraction. A monodispersed composite material with $R = 15/200$.

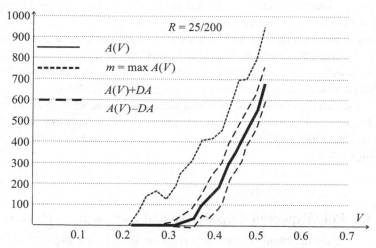

Figure 4.5 Effective conductivity as a function of volume fraction. A monodispersed composite material with $R = 25/200$.

4.5.2 Verification of the computer program

Recall that we developed computer programs in order to perform simulations of random distributions of disks and compute the effective conductivity. In order to test the computer programs, we computed $A(\omega)$ for a monodispersed composite for a range of volume fractions separated by the step size δV.

For each fixed V, the quantities A, DA and $m(\omega)$ were computed repeatedly for different random configurations (100–300 configurations for each value of V). The sizes of the disks used in these computations were (15/200, 20/200 and 25/200). Indeed, since the problem is scale invariant, the choice of radius should not change the effective conductivity. This scale invariance was used to test the numerical procedure. As can be seen in Figures 4.3, 4.4, and 4.5, there is no difference in the qualitative and quantitative dependence of A on the volume fraction V when the sizes were changed.

A typical plot of the function $A = A(\omega)$ is shown in Figure 4.3. The plot consists of two parts: for $V \leq 0.2$, the total flux through the composite material layer is zero. More precisely, it is very small and it is very close to the Ox-axis on the plot. For $V \geq 0.35$, the total flux is always positive (taking into account the mean deviation). In the range $V = 0.2$–0.35, the total flux is never positive. Thus, the plots of the function $A = A(V)$ reveal a dependence on the volume fraction which is typical in percolation processes (Stauffer and Aharony, 1992; Sahimi, 2003). The percolation threshold in all computations can be estimated as $V_0 \approx 0.3$.

Above, the function $A(V)$ behaves like $A(V) = const(V - V_0)^p$ with $p \geq 1$, which is consistent with the conventional percolation picture observed in experiments and explained by percolation theory scaling arguments.

We have done this plot for several sizes of identical particles to make sure that there is no size effect, which, in principle, can appear as a numerical artifact. Thus we conclude that our numerical analysis captures basic features of the real physical phenomenon and provides realistic quantitative predictions.

We have used the graph of the maximal value $m(\omega)$ as an additional test for our code. Indeed, from the physical point of view it should behave (qualitatively) in the same way as the graph for the average conductivity, which is what one can see in Figures 4.3–4.5.

4.6 Polydispersed densely-packed composite materials

The effect of the distribution of particle sizes on the effective properties of composite materials arises in numerous areas of science and engineering (see, e.g., Chang and Powell (1994); Goto and Kuno (1984); Robinson and Friedman (2001); Torquato (2002)). For a long time, various opinions (different and sometimes opposite) were expressed about the influence of polydispersity on the overall properties of composite materials. The key question was if polydispersity increases or decreases the transport properties of composite materials.

In this section, we apply the network model to the analysis of this problem for densely-packed composite material and demonstrate that the network model is an effective tool for the analysis of the influence of polydispersity on the effective transport properties of composite materials.

4.6.1 The influence of polydispersity on the effective transport properties of a composite material

There is considerable interest in the effect of polydispersity on both the effective conductivity and the effective viscosity of suspension of rigid inclusions in a solid and fluid, respectively (see Chang and Powell (1994); Hill (1996); Hill and Supancic (2002); Goto and Kuno (1984); Jeffrey and Acrivos (1976); Poslinski *et al.* (1988)). The main question here is whether the presence of polydispersity results in an increase or decrease of the effective conductivity and viscosity. Theoretical results on this subject, reviewed below, point to an increase (see formula (4.6.1) below). This question is of significant practical interest; when polymer/ceramic composites are prepared for capacitors or thermal insulting packages (see examples in Section 2.1), should the ceramic powder be monodispersed or polydispersed to achieve the desired (e.g., high) dielectric or thermal properties? How significant is the effect of polydispersity? Limited experimental data (Chang and Powell, 1994; Hill, 1996; Hill and Supancic, 2002; Goto and Kuno, 1984; Jeffrey and Acrivos, 1976; Poslinski *et al.*, 1988; Robinson and Friedman, 2001) are inconclusive and point to an increase (Robinson and Friedman, 2001) when the concentration of inclusions is not high.

Most theoretical works on this subject study polydispersity in the dilute limit. In Thovert and Acrivos (1989), Thovert *et al.* (1990) and Torquato (2002), the technique of variational bounds (Milton, 2002) was used to estimate the effective characteristic (effective conductivity or effective viscosity) A of polydispersed composite materials in the low-concentration regime in both two and three dimensions. They performed a rigorous asymptotic analysis in V and evaluated the terms of order $O(V)$ and $O(V^2)$ in the expansion of A in powers of V, where V is the total volume fraction of the inclusions. Moreover, they performed a partial analysis of the cubic term. The analysis of Torquato (2002), Thovert and Acrivos (1989) and Thovert *et al.* (1990) shows that if the conductivity of inclusions is greater than the conductivity of the matrix $(a_i > a_m)$, then

$$A > A_{\text{mono}}, \tag{4.6.1}$$

where A_{mono} is the effective characteristic of the monodispersed composite material with the same volume fraction of inclusions.

The key ingredient of the analysis of Torquato (2002), Thovert and Acrivos (1989) and Thovert *et al.* (1990) is the observation that the so-called three- and four-point lower-sided bounds on A and A_{mono} provide a good approximation of the effective conductivity when the contrast parameter (Bergman and Dunn, 1992)

$$\rho = \frac{a_i - a_m}{a_i + a_m},$$

is not large. Note that the formulas for A derived in Torquato (2002), Thovert and Acrivos (1989) and Thovert *et al.* (1990) provide good numerical results for $V < V_0$, where V_0 is the percolation threshold.

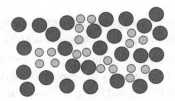

Figure 4.6 A polydispersed composite material.

In Robinson and Friedman (2001), the problem of calculating the effective conductivity was considered for spherical inclusions of two different sizes in three dimensions. Their derivation of the effective conductivity is based on the Maxwell–Garnett formula. Note that the Maxwell–Garnett formula is an approximation in V for small V (Milton, 2002). This formula was derived as an approximation for the dilute limit case when interactions between inclusions are small (see Milton (2002)). While it is known that for special geometric arrays the Maxwell–Garnett formula holds for fairly large concentrations (e.g., for periodic structures; Berdichevskij (2009)) and other well-separated geometries, in general it only holds to second order in V for macroscopically isotropic composites (see Milton (2002)). In particular, the Maxwell–Garnett approximation may not hold for high volume fractions in the presence of pronounced percolation effects, when the conductivity patterns dominate the behavior of the effective properties.

Considering three-dimensional non-dilute composites, we meet additional difficulties, which have not yet been completely worked out.

4.6.2 Numerical analysis of effective transport properties of a polydispersed composite material

We call a composite material polydispersed, if it is filled with particles of different sizes (see Figure 4.6).

In our model, this amounts to considering disks of different radii. The distribution of sizes can be continuous or discrete. We consider the discrete case – that is, we consider disks of several fixed sizes. We describe a polydispersed composite material with total volume fraction of disks V and relative volume fraction of disks of the i-th sort p_i, where $i = 1, ..., K$. The volume fraction of disks of the i-th sort in the mixture is then Vp_i. We have $p_1 + \cdots + p_K = 1$, and $Vp_1 + \cdots + Vp_K = V$.

For the sake of simplicity, we consider the bimodal case. In this case, we introduce a polydispersity parameter p as

$$p = p_2, 0 < p < 1,$$

that characterizes the relative volume fraction of small disks ($p = 0$ means all disks are large and $p = 1$ all disks are small). Alternatively, the polydispersity parameter $q = 1 - p = p_1$ characterizes the relative volume fraction of large disks.

For a fixed total volume fraction V, Vp is the total volume fraction of small disks and Vq is the total volume fraction of large disks. Our objective is to obtain the dependence of the effective conductivity A_{poly} on the polydispersity parameters. We have analyzed the following two-component mixtures:

Mixture 1. The radii of the disks are $R_1 = 25/200$, $R_2 = 15/200$ and the relative fractions are $p_1 = q = 33\%$, $p_2 = p = 67\%$, respectively.

Mixture 2. The radii of the disks are $R_1 = 25/200$, $R_2 = 15/200$ and the relative fractions are $p_1 = q = 67\%$, $p_2 = p = 33\%$, respectively.

Mixture 3. The radii of the disks are $R_1 = 35/200$, $R_2 = 15/200$ and the relative fractions are $p_1 = q = 33\%$, $p_2 = p = 67\%$.

Mixture 4. The radii of the disks are $R_1 = 35/200$, $R_2 = 15/200$ and the relative fractions are $p_1 = q = 67\%$, $p_2 = p = 33\%$.

The total volume fraction of the disks varies in the interval $[0.4, 0.55]$. The total volume fraction 0.55 is practically the maximal volume fraction which can be achieved in our procedure (we have assumed the disks do not overlap each other). We used the following two step procedure for generating the polydispersed (bimodal) random configurations of the disks.

First we used the algorithm described at the beginning of this section for the disks of larger size until we reach the volume fraction Vq. Next, we used the same algorithm for the small disks until we reached the volume fraction Vp, where Vp is the total volume fraction of disks (kept fixed). In the second step, we take into account the large disks generated in the first step – that is, if a small disk overlaps any other disk (small or large), it is rejected. The cut-off distance for both steps is chosen such that $\delta^* \in [0.3R_2, 0.5R_2]$.

The specific flux g_{ij} between two disks is given by (4.1.6) and by

$$ g_{ij} = \pi \sqrt{\frac{2R_i}{\delta_{ij}}} $$

for the flux between a disk and a pseudo-disk (the disk is indexed by "i", the pseudo-disk is indexed by "j"); only R_i enters the above formula because $R_j = \infty$.

The graphs of the effective constant A as a function of the total volume fraction V are presented in Figure 4.7. One can see that the plots corresponding to mixtures 1, 2 and 4 in Figure 4.7 are very similar. This means that in these cases, we observe no influence of polydispersity on the effective constant.

At the same time, the graph corresponding to mixture 3 differs from the others. This leads us to assume that the polydispersity may or may not have an essential influence on the effective conductivity, depending on the relative volume fractions of the large and the small disks. Note that mixture 3 has a large relative volume fraction of small disks.

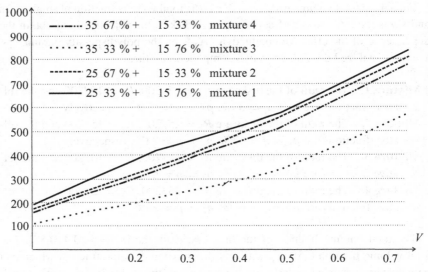

Figure 4.7 Graphs of the effective conductivity A as a function of the total volume fraction V.

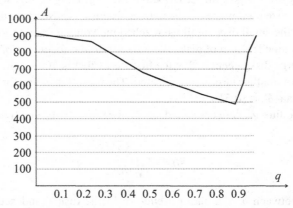

Figure 4.8 Graph of the effective constant A as a function of the relative volume fraction $q = p_1$ of large disks.

By the latter observation, we decided to study the influence of polydispersity in more detail. For this purpose we have studied a mixture with disks of radii $R_1 = 25/200$, $R_2 = 10/200$ and total volume fraction $V = 0.55$.

We then vary the relative volume fraction of the disks $p_1 = q$ and $p_2 = p$ $(p + q = 1)$ from 0 to 1. In Figure 4.8, the effective conductivity is presented as a function of p, which is the relative volume fraction of the disks of radius $R_1 = 25/200$.

This graph shows the same values of A for $p = 0$ and $p = 1$. This means that there is no size effect, which agrees with the general theory, since the problem is scale invariant and it *plunges down sharply* in the interval $[0.5; 0.9]$. The latter agrees with the above result for mixture 4. Thus, we observe that for this mixture, polydispersity can decrease the effective conductivity by a factor of approximately 2. This sharp drop occurs when the relative volume fraction p of the large particles is quite high. If we compare the minimum value $A(p = 0.9) = 484$ with the corresponding value in a monodispersed composite material, $A(0.9V \approx 0.5) = 680$, then we conclude that the small particles practically do not contribute to the effective conductivity.

5

The network approximation theorem for an infinite number of bodies

In this chapter, we present a method that allows one to obtain an a priori error estimate for the discrete network approximation independent of the total number of filling particles. Such estimates are referred to as homogenization estimates. These estimates can be derived under the natural δ-**N** close-packing condition (Berlyand and Novikov, 2002), which, loosely speaking, allows for "holes" (regions containing no particles) to be present in the medium of order $\mathbf{N}R$ (see Figure 5.1). Here, R is the radius of the particles and $\mathbf{N}$ is the number of particles in the perimeter of the largest hole in the conducting cluster (see Figure 5.2). We demonstrate that the error of the network approximation is determined not by the total number of particles in the composite material but by the perimeter of these "holes". The explicit dependence of the network approximation and its error on the irregular geometry of the particle array is explicitly evaluated.

5.1 Formulation of the mathematical model

We consider here the composite material described in Section 3.1.1. It is a two-dimensional rectangular specimen of a two-phase composite material that consists of a matrix filled by a large number of ideally conducting disks that do not intersect. In this chapter, we do not assume any restriction on the number of particles and prove the network approximation theorem independent of the total number of particles.

As above, we denote the domain occupied by the composite material by $\Pi = [-L, L] \times [-1, 1]$ (see Figure 5.3), a system of disks (inclusions) $\{D_i, i = 1, \ldots, N\}$, where N is the total number of disks and domain $Q_m = \Pi \backslash \bigcup_{i=1}^{N} D_i$ (matrix). Unlike in Chapter 3, we will allow disks to intersect the boundary $\partial \Pi$ as long as their centers lie inside Π.

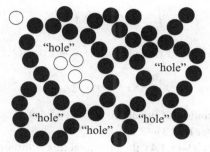

Figure 5.1 A conducting cluster in a composite material with "holes". Black disks form the conducting cluster.

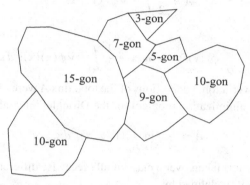

Figure 5.2 The graph that corresponds to the conducting cluster in Figure 5.1.

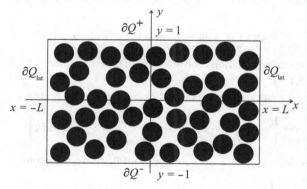

Figure 5.3 A composite material.

The potential $\phi(\mathbf{x})$ as above satisfies

$$\Delta\phi = 0 \text{ in } Q_m; \tag{5.1.1}$$

$$\phi(\mathbf{x}) = t_i \text{ on } \partial D_i, i = 1, \ldots N; \tag{5.1.2}$$

$$\int_{\partial D_i} \frac{\partial \phi}{\partial \mathbf{n}} d\mathbf{x} = 0, \, i \in I; \tag{5.1.3}$$

$$\phi(\mathbf{x}) = 1 \text{ on } \partial Q^+, \quad \phi(\mathbf{x}) = -1 \text{ on } \partial Q^-; \tag{5.1.4}$$

$$\frac{\partial \phi}{\partial \mathbf{n}}(\mathbf{x}) = 0 \text{ on } \partial Q_{\text{lat}}. \tag{5.1.5}$$

The physical interpretation of equations (5.1.1)–(5.1.4) was given in Chapter 3 (see the comments after (3.2.3)–(3.2.7)).

If $\phi(\mathbf{x})$ satisfies (5.1.1)–(5.1.4), then the total flux A through the specimen Π is defined by (see Chapter 3)

$$A = \frac{1}{2} \int_{Q_m} |\nabla \phi(\mathbf{x})|^2 d\mathbf{x}, \tag{5.1.6}$$

or

$$A = \int_{\partial Q^+ \cup \partial Q^-} \phi^0(y) \nabla \phi(\mathbf{x}) \mathbf{n} d\mathbf{x} - \frac{1}{2} \int_{Q_m} |\nabla \phi(\mathbf{x})(\mathbf{x})|^2 d\mathbf{x}. \tag{5.1.7}$$

In Chapter 3, two variational definitions of the total flux A were obtained. The first one is given by a minimization problem for the Dirichlet integral

$$A = \frac{1}{2} \min_{\phi \in V_p} \int_{Q_m} |\nabla \phi(\mathbf{x})|^2 d\mathbf{x}, \tag{5.1.8}$$

where the minimum is taken over a class of all piecewise differentiable potentials $\phi(\mathbf{x}) \in V_p$, where V_p is defined by

$$V_p = \left\{ \phi \in H^1(\Pi) \mid \phi(\mathbf{x}) = t_i^\phi \in \mathbb{R} \text{ on } D_i, \, i = 1, \ldots, N; \tag{5.1.9} \right.$$

$$\left. \phi(\mathbf{x}) = 1 \text{ on } \partial Q^+, \phi(\mathbf{x}) = -1 \text{ on } \partial Q^- \right\}.$$

The Euler–Lagrange equations for the minimization problem (5.1.8) and (5.1.9) are (5.1.1)–(5.1.4).

The second variational formulation is given by the dual problem

$$A = \max_{\mathbf{v} \in W_p^0} \left\{ -\frac{1}{2} \int_{Q_m} \mathbf{v}(\mathbf{x})^2 d\mathbf{x} + \int_{\partial Q^+ \cup \partial Q^-} \phi^0(y) \mathbf{v}(\mathbf{x}) \mathbf{n} d\mathbf{x} \right\}, \tag{5.1.10}$$

where the minimum is taken over a class of all fluxes (see Chapter 3 for details)

$$W_p^0 = \left\{ \mathbf{v} = (v_1(\mathbf{x}), v_2(\mathbf{x})) \in L_2(Q_m) \mid \text{div} \mathbf{v} = 0 \text{ in } Q_m, \tag{5.1.11} \right.$$

$$\left. \mathbf{v}(\mathbf{x}) \mathbf{n} = 0 \text{ on } \partial Q_{\text{lat}}; \int_{\partial D_i} \mathbf{v}(\mathbf{x}) \mathbf{n} d\mathbf{x} = 0, \, i = 1, \ldots, N \right\}.$$

The functional space W_p^0 can also be described as $\{\mathbf{v} \in W_p \mid \text{div} \mathbf{v} = 0\}$ (W_p is defined by (3.2.15)).

Hence for any $\phi \in V_p$ and $\mathbf{v} \in W_p^0$ we have the bounds

$$-\frac{1}{2}\int_{Q_m}\mathbf{v}^2 dx + \int_{\partial Q^+\cup\partial Q^-}\phi^0(y)\mathbf{v}(\mathbf{x})\mathbf{n}dx \leq A \leq \frac{1}{2}\int_{Q_m}|\nabla\phi(\mathbf{x})|^2 dx. \quad (5.1.12)$$

Moreover, if $\mathbf{v}(\mathbf{x})\mathbf{n} = \nabla\phi(\mathbf{x})\mathbf{n}$, then the upper-sided bound equals the lower-sided bound in (5.1.12).

As in Chapter 3, we can write the network problem corresponding to the problem (5.1.1)–(5.1.4). Both problems – the original and network problems – depend on the total number of particles N. In this chapter, we develop a network model, which approximates the original problem independent of the total number of particles.

5.2 Triangle–neck partition and discrete network

As in Chapter 3, we construct the discrete network model corresponding to problem (5.1.1)–(5.1.4) using the notion of the Voronoi tessellation. We partition the matrix Q_m into simple non-overlapping geometric figures – necks and triangles. This triangle–neck partition is an auxiliary construction, which is used below as a convenient and efficient tool for the construction of the trial functions for the error estimates.

5.2.1 Triangle–neck partition

Let $\mathbf{x}_i$ represent the center of D_i. We begin by constructing the Voronoi tessellation with vertices $\{\mathbf{x}_i, \ i = 1, 2, \ldots, N\}$. Denote by V_i the Voronoi cell associated with $\mathbf{x}_i$. Let e_{ij} be the set of edges in the Delunay graph dual to the Voronoi tessellation. Specifically, the edge e_{ij} connects vertices $\mathbf{x}_i$ and $\mathbf{x}_j$ if their respective cells share a common edge in the Voronoi tessellation (see Figure 5.4). Any two disks D_i and D_j are said to be neighbors if their centers $\mathbf{x}_i$ and $\mathbf{x}_j$ are connected by the edge e_{ij}.

Let D_i and D_j be two neighbors with centers $\mathbf{x}_i$ and $\mathbf{x}_j$, respectively (see Figure 5.5). Denote by O_n and O_p (in fact, O_n and O_p should be indexed by the pair (i, j), but we will suppress this dependence for simplicity of presentation) the endpoints of the common edge of their Voronoi cells (V_i and V_j). In order to construct the necks, we first connect the center $\mathbf{x}_j$ with all the vertices of its Voronoi cell V_j by auxiliary line segments (dotted lines on Figure 5.5). Let A_{jn} denote the intersection of the line segment $\mathbf{x}_j O_n$ with the circumference of the disk D_j. Finally define similarly points A_{in} and A_{kn} and connect the points A_{in}, A_{jn} and A_{kn}.

Definition 5.1 The neck Π_{ij} between the neighbors D_i and D_j is the curvilinear quadrangle $A_{in}A_{ip}A_{jp}A_{jn}$, bounded by the two line segments $A_{in}A_{jn}$ and $A_{ip}A_{jp}$ and the two arcs $A_{in}A_{ip}$ and $A_{jn}A_{jp}$.

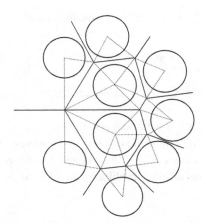

Figure 5.4 Voronoi tessellation and Delaunay graph.

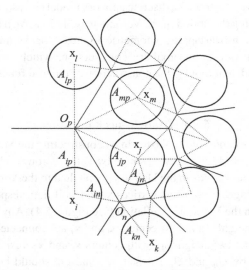

Figure 5.5 Decomposition of a Voronoi cell. Voronoi tessellation.

When we apply this algorithm to all sets of neighbors, in general, all these line segments $A_{in}A_{jn}$ partition the domain Q_m into necks Π_{ij} between neighboring disks and triangles Δ_{ijk}. In exceptional cases, instead of triangles, we obtain polygons (for example the quadrangle $A_{ip}A_{lp}A_{mp}A_{kp}$ on Figure 5.6) which can be further partitioned into triangles by drawing auxiliary diagonal lines (an arbitrary diagonal can be used).

In order to define the partition of Q_m near the boundary $\partial\Pi$, we reflect all disks D_i such that $\partial V_i \cap \partial\Pi \neq \emptyset$ about $\partial\Pi$. Since the centers $\mathbf{x}_i$ lie inside Π, the centers

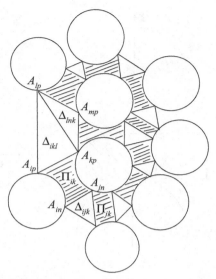

Figure 5.6 Triangle–neck partition. Necks are hatched.

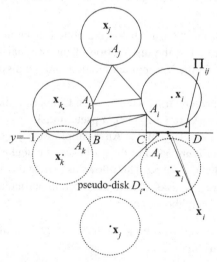

Figure 5.7 Horizontal boundary and pseudo-disk $D_{i''}$.

of the reflected disks will lie outside Π (however, the case where the centers of the disks are allowed to lie outside the boundary can also be treated by the model by adding simple, but cumbersome modifications). This algorithm is illustrated in Figure 5.7 for the boundary $y = -1$. Here, we reflect the disks, including those that intersect the boundary, symmetrically about the line $y = -1$. The latter disks

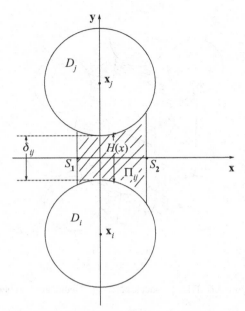

Figure 5.8   The hatched region is the neck between two neighbors.

partially overlap with the "ghost" disks (the dotted disks in Figure 5.7), which are their mirror images. For the distribution of the original disks and the ghost disks, we can still apply the Voronoi tessellation and the algorithm proposed for the interior disks.

For uniformity of presentation, we use, as in Chapter 3, the notion of pseudo-disks on the boundary. As in Chapter 3, a pseudo-disk $D_{i''}$, is the part of a neck $\Pi_{ii'}$ that lies on the boundary of Π. An example of a pseudo-disk is the line segment $CD = D_{i''}$ in Figure 5.7. For the purposes of calculations, the radius of a pseudo-disk is considered to be ∞. This allows us to treat the pseudo-disks and the original disks uniformly as disks of different radii. In particular, for a pseudo-disk the same definition of neighbors applies.

Definition 5.2   The triangle–neck partition $\mathbb{P} = \mathbb{P}(Q_m)$ of the domain Q_m is the set of necks Π_{ij} and triangles Δ_{ijk}.

The triangle–neck partition is unique up to partitioning of the degenerate (exceptional) polygons into triangles. Typically a neck Π_{ij} is not symmetric with respect to the line connecting the centers of the disks D_i and D_j. An example of such a neck is given in Figure 5.8, where we have used the local coordinate system where the centers of both disks lie on the Oy-axis. In this coordinate system, the width of the left half-neck is $|S_1|$, $S_1 < 0$, and the width of the right half-neck is $|S_2|$, $S_2 > 0$. Note that the inequalities $S_1 < 0$ and $S_2 > 0$ are not true in general, but

$S_1 \leq S_2$ always holds by our construction. For uniformity of presentation, we view the auxiliary diagonals as necks with zero width (i.e., $S_1 = S_2$). For example, the line segment A_iB in Figure 5.7 corresponds to such a neck.

Definition 5.3 The maximal and the minimal relative half-neck widths are defined by

$$\beta_{ij}^{max} = \max\left(\frac{|S_1|}{R}, \frac{|S_2|}{R}\right), \quad \beta_{ij}^{min} = \min\left(\frac{|S_1|}{R}, \frac{|S_2|}{R}\right),$$

$$0 \leq \beta_{ij}^{min} \leq \beta_{ij}^{max} < 1.$$

We use the relative half-neck widths β_{ij}^{max} and β_{ij}^{min} in the error estimates below.

Using the triangle–neck partition we decompose the Dirichlet integral (5.1.6) into integrals over necks and triangles:

$$A = \frac{1}{2}\int_{Q_m} |\nabla\phi(\mathbf{x})|^2 dx$$

$$= \frac{1}{2}\sum_{\Pi_{ij}\in\mathbb{P}}\int_{\Pi_{ij}} |\nabla\phi(\mathbf{x})|^2 dx + \frac{1}{2}\sum_{\Delta_{ijk}}\int_{\Delta_{ijk}} |\nabla\phi(\mathbf{x})|^2 dx.$$

A subscript like Π_{ij} or Δ_{ijk} in the summation symbol means that the summation is carried out over Π_{ij} or Δ_{ijk} (or other objects indicated as the subscript in the summation symbol).

In Chapter 3 it was observed that for a high concentration of disks, the fluxes $\nabla\phi$ are significant only in the necks Π_{ij} between closely spaced disks. Thus,

$$\sum_{\Delta_{ijk}}\int_{\Delta_{ijk}} |\nabla\phi(\mathbf{x})|^2 dx \ll \sum_{\Pi_{ij}\in\mathbb{P}}\int_{\Pi_{ij}} |\nabla\phi(\mathbf{x})|^2 dx. \tag{5.2.1}$$

Moreover, in Chapter 3 it was observed that the fluxes in the neck Π_{ij} between two neighbors D_i and D_j with the vertical direction as indicated on Figure 5.8, can be approximated by $((\mathbf{x}) = (x, y))$

$$\mathbf{J}(\mathbf{x}) = \left(0, \frac{t_i - t_j}{H_{ij}(x)}\right), \tag{5.2.2}$$

where

$$H_{ij}(x) = \begin{cases} \delta_{ij} + 2R - 2\sqrt{R^2 - x^2}, & \text{for disks,} \\ \delta_{ij} + R - \sqrt{R^2 - x^2}, & \text{for pseudo-disks.} \end{cases} \tag{5.2.3}$$

where δ_{ij} is given by the following.

Definition 5.4 The length δ_{ij} of a neck Π_{ij} is

$$\delta_{ij} = \begin{cases} d_{ij} - 2R, & \text{if } D_i \text{ and } D_j \text{ are disks,} \\ d_{ij} - R, & \text{if one of } D_i \text{ is a pseudo-disk,} \end{cases} \tag{5.2.4}$$

where $d_{ij} = |\mathbf{x}_i - \mathbf{x}_j|$ is the (Euclidean) distance between $\mathbf{x}_i$ and $\mathbf{x}_j$.

Note the function $\mathbf{J}(\mathbf{x})$ (5.2.2) is not a potential function and only an approximation:

$$\nabla \phi(\mathbf{x}) \approx \mathbf{J}(\mathbf{x}). \tag{5.2.5}$$

Using (5.2.2) and (5.2.5), we can write

$$\int_{\Pi_{ij}} |\mathbf{J}(\mathbf{x})|^2 dx = \int_{S_1}^{S_2} \frac{(t_i - t_j)^2}{H^2(x)} H_{ij}(x) dx = g_{ij}^S (t_i - t_j)^2, \tag{5.2.6}$$

where

$$g_{ij}^S = \int_{S_1}^{S_2} \frac{dx}{H_{ij}(x)}, \tag{5.2.7}$$

and t_i, t_j are potentials on D_i, D_j, respectively, and S_1, S_2 are as defined on Figure 5.8.

Denote $I := \{\mathbf{x}_i, i = 1, \dots, K\} \setminus (S^+ \bigcup S^-)$, the set of internal nodes of the network. Here K is the total number of disks (including pseudo-disks) and $S^{\pm} := \{\mathbf{x}_i \mid \mathbf{x}_i \in \partial Q^{\pm}\}$. If we use (5.2.6) with (5.2.7) for all neighbors, then observation (5.2.1) would imply

$$A = \frac{1}{2} \min_{\phi \in V_p} \int_{Q_m} |\nabla(\mathbf{x})|^2 dx \approx \frac{1}{2} \int_{Q_m} |\mathbf{J}(\mathbf{x})|^2 dx \tag{5.2.8}$$

$$\sim E_d^S := \frac{1}{2} \min_{t \in T} \sum_{\Pi_{ij} \in \mathbb{P}} g_{ij}^S (t_i - t_j)^2,$$

where

$$T = \{t_i \mid t_i = 1 \text{ for } i \in S^+, t_i = -1 \text{ for } i \in S^-, \text{ and } t_i \in \mathbb{R} \text{ for } i \in I\}.$$

This is the modified network approximation, similar to the network approximation used in Section 3.4. The difference in this network approximation is that the energy of the discrete network E_d^S defined in (5.2.8) is determined by the choice of the specific fluxes g_{ij}^S, whereas in Section 3.4, only the asymptotic value of the specific fluxes is used.

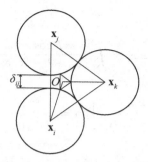

Figure 5.9 The worst-case scenario for the lower bound on $\beta_{ij}^{\min}$.

5.2.2 Asymptotic equivalence of fluxes given by formulas (2.5.8) and (5.2.7)

Formulas (2.5.8) and (5.2.7) introduce the specific (i.e., per unit difference, $t_i - t_j = 1$) interparticle fluxes between two neighboring disks. Now we establish a relation between the relative interparticle flux g_{ij}^{S} (2.5.8) and the specific flux g_{ij} (5.2.7) and demonstrate that the formulas (2.5.8) and (5.2.7) give the same leading term in the power expansion in δ_{ij} about 0. In order to demonstrate the asymptotic equivalence of formulas (2.5.8) and (5.2.7), we will need the following lemma.

Lemma 5.5 *The following lower bound for $\beta_{ij}^{\min}$ holds*

$$\beta_{ij}^{\min} \geq \max\left(0, \, 1 - \frac{1}{2}\left(1 + \frac{\delta_{ij}}{2R}\right)^2\right).$$ (5.2.9)

Proof For any distribution of centers $\mathbf{x}_i$, $\mathbf{x}_j$ and $\mathbf{x}_k$, the quantity $\beta_{ij}^{\min}$ is larger than the one shown in Figure 5.9. In the latter case

$$\beta_{ij}^{\min} = \sin \angle O\mathbf{x}_i\mathbf{x}_j = \sin\left(\frac{\pi}{2} - 2\angle O\mathbf{x}_k\mathbf{x}_i\right)$$

$$= 1 - 2\sin^2 \angle O\mathbf{x}_k\mathbf{x}_i = 1 - \frac{1}{2}\left(1 + \frac{\delta_{ij}}{2R}\right)^2.$$

$\square$

First, consider the case of a disk and a pseudo-disk. In this case $R_i = R$, $R_j = \infty$ and $H_{ij}(x) = \delta_{ij} + R - \sqrt{R^2 - x^2}$. With the change of variables

$$x = R\cos\alpha;$$

$$S_k^* = \arcsin\left(\frac{S_k}{R}\right), \, k = 1, 2;$$

$$-\frac{\pi}{2} \leq S_k^* \leq \frac{\pi}{2}.$$ (5.2.10)

we have

$$g_{ij}^S = \int_{S_1}^{S_2} \frac{dx}{\delta_{ij} + R - \sqrt{R^2 - x^2}} = \int_{S_1^*}^{S_2^*} \frac{R \cos \alpha \, d\alpha}{\delta_{ij} + R - R \cos \alpha}$$

$$= \int_{S_1^*}^{S_2^*} \left[-1 + \frac{\delta_{ij} + R}{\delta_{ij} + R - R \cos \alpha} \right] d\alpha = \int_{S_1^*}^{S_2^*} \left[-1 + \frac{\delta_{ij} + R}{\delta_{ij} + 2R \sin^2 \left(\frac{\alpha}{2} \right)} \right] d\alpha$$

$$= \int_{S_1^*}^{S_2^*} \left[-1 + 2 \frac{\dfrac{\delta_{ij} + R}{\delta_{ij}}}{\cos^2 \dfrac{\alpha}{2} + \left(1 + \dfrac{2R}{\delta_{ij}} \right) \sin^2 \left(\dfrac{\alpha}{2} \right)} \right] d\alpha$$

$$= \left[-\alpha + 2 \frac{\delta_{ij} + R}{\delta_{ij}} \left(1 + \frac{2R}{\delta_{ij}} \right)^{-1/2} \arctan \left(\left(1 + \frac{2R}{\delta_{ij}} \right)^{1/2} \tan \left(\frac{\alpha}{2} \right) \right) \right] \Bigg|_{S_1^*}^{S_2^*}$$

$$= \left[-\alpha + 2 \frac{\delta_{ij} + R}{\sqrt{\delta_{ij}^2 + 2R\delta_{ij}}} \arctan \left(\frac{\sqrt{\delta_{ij}^2 + 2R\delta_{ij}} \tan \left(\frac{\alpha}{2} \right)}{\delta_{ij}} \right) \right] \Bigg|_{S_1^*}^{S_2^*}.$$

For each $k = 1, 2$

$$0 < \frac{\pi}{2} - \arctan \left(\sqrt{\frac{2R + \delta_{ij}}{\delta_{ij}}} \tan \left(\frac{|S_k^*|}{2} \right) \right)$$

$$= \arctan \left(\sqrt{\frac{\delta_{ij}}{2R + \delta_{ij}}} \cot \left(\frac{|S_k^*|}{2} \right) \right) < \min \left(\sqrt{\frac{\delta_{ij}}{2R}} \cot \left(\frac{|S_k^*|}{2} \right), \frac{\pi}{2} \right)$$

$$\leq \min \left(\sqrt{\frac{\delta_{ij}}{2R}} \cdot \frac{1 + \sqrt{1 - (\beta_{ij}^{\min})^2}}{\beta_{ij}^{\min}}, \frac{\pi}{2} \right).$$

Using Lemma 5.2.9, we have

$$\frac{\pi}{2} - \arctan \left[\sqrt{\frac{2R + \delta_{ij}}{\delta_{ij}}} \tan \left(\frac{|S_k^*|}{2} \right) \right] \tag{5.2.11}$$

$$\leq \min \left(\sqrt{\frac{\delta_{ij}}{2R}} \left(5.5 + \frac{3\delta_{ij}}{R} \right), \frac{\pi}{2} \right).$$

Also

$$0 < \frac{\delta_{ij} + R}{\sqrt{\delta_{ij}^2 + 2R\delta_{ij}}} - \sqrt{\frac{R}{2\delta_{ij}}} = \frac{1}{\sqrt{\delta_{ij}}} \frac{\delta_{ij} + R - \sqrt{\frac{R\delta_{ij}}{2} + R^2}}{\sqrt{\delta + 2R}} \qquad (5.2.12)$$

$$= \sqrt{\delta_{ij}} \frac{\frac{3}{2}R + \delta_{ij}}{\delta_{ij} + R + \sqrt{\frac{R\delta_{ij}}{2} + R^2}} \sqrt{\delta_{ij} + 2R} < \frac{3}{4\sqrt{2}} \sqrt{\frac{\delta_{ij}}{R}},$$

and we obtain for any δ_{ij}

$$g_{ij} - g_{ij}^S \qquad\qquad (5.2.13)$$

$$= \pi\sqrt{\frac{2R}{\delta_{ij}}} - \left[-\alpha + \frac{2(\delta_{ij} + R)}{\sqrt{\delta_{ij}^2 + 2R\delta_{ij}}} \arctan\left(\frac{\sqrt{\delta_{ij}^2 + 2R\delta_{ij}} \tan\left(\frac{\alpha}{2}\right)}{\delta_{ij}} \right) \right]\Bigg|_{S_1^*}^{S_2^*}$$

$$= \left[\alpha + \left(\frac{2(\delta_{ij} + R)}{\sqrt{\delta_{ij}^2 + 2R\delta_{ij}}} - \sqrt{\frac{2R}{\delta_{ij}}} \right) \arctan\left(\frac{\sqrt{\delta_{ij}^2 + 2R\delta_{ij}} \tan\left(\frac{\alpha}{2}\right)}{\delta_{ij}} \right) \right]\Bigg|_{S_1^*}^{S_2^*}$$

$$+ \sqrt{\frac{2R}{\delta_{ij}}} \left[\frac{\pi}{2} - \arctan\left(\frac{\sqrt{\delta_{ij}^2 + 2R\delta_{ij}} \tan\left(\frac{\alpha}{2}\right)}{\delta_{ij}} \right) \right]\Bigg|_0^{-S_1^*}$$

$$+ \sqrt{\frac{2R}{\delta_{ij}}} \left[\frac{\pi}{2} - \arctan\left(\frac{\sqrt{\delta_{ij}^2 + 2R\delta_{ij}} \tan\left(\frac{\alpha}{2}\right)}{\delta_{ij}} \right) \right]\Bigg|_0^{S_2^*}.$$

Combining (5.2.11), (5.2.12) and (5.2.14)

$$0 < g_{ij} - g_{ij}^S \leq \pi + \pi \frac{3}{2\sqrt{2}} \sqrt{\frac{\delta_{ij}}{R}} + \min\left(\left(5.5 + \frac{3\delta_{ij}}{R}\right), \frac{\pi}{2}\sqrt{\frac{2R}{\delta_{ij}}} \right). \qquad (5.2.14)$$

The case of two disks with equal radii $R_i = R_j = R$ follows from (5.2.14), because this is equivalent to solving two problems for a disk and pseudo-disk with length $\delta_{ij}/2$. In this case, formula (5.2.11) becomes

$$g_{ij}^S = \left[-\frac{\alpha}{2} + \frac{\delta_{ij} + 2R}{\sqrt{\delta_{ij}^2 + 4R\delta_{ij}}} \arctan\left(\frac{\sqrt{\delta_{ij}^2 + 4R\delta_{ij}} \tan\left(\frac{\alpha}{2}\right)}{\delta_{ij}} \right) \right]\Bigg|_{S_1^*}^{S_2^*}. \qquad (5.2.15)$$

However, these new estimates are tighter than (5.2.14), therefore we can use (5.2.14) for an estimate independent of the radii of the disks. Observe that estimate

(5.2.14) is valid for any δ_{ij}. Estimate (5.2.14) implies that the term g_{ij}^S is the leading term in the asymptotic expansion of g_{ij}^S if δ_{ij} is small and the width $S_2^* - S_1^*$ is non-small. Observe that if δ_{ij} is small, then the width of the neck will not be small (see estimate (5.2.9) on $\beta_{ij}^{\min}$ below).

5.2.3 The discrete network

Since formulas (5.2.7) and (2.5.8) are asymptotically equivalent as $\delta_{ij} \to 0$, we can use (5.2.7) instead of (2.5.8). We will do this for all neighbors, not just those that are closely spaced. Such a modified choice of the specific fluxes allows us to derive tight variational bounds for the relative error of the modified network approximation. The relation (5.2.8) is a discrete network approximation, because we approximate the continuum minimization problem with a discrete minimization problem of a quadratic form, given by

$$I^S(\mathbf{t}) := \frac{1}{2} \sum_{\Pi_{ij} \in \mathbb{P}} g_{ij}^S (t_i - t_j)^2. \tag{5.2.16}$$

The unknowns in this minimization problem are the values of the discrete potentials $\{t_i, i \in I\}$ in the interior disks. The quadratic form is defined on a graph (network) where the vertices $\mathbf{x}_i$ are the centers of the disks D_i, and the edges e_{ij} are the necks Π_{ij}. This graph is the Delaunay triangulation for the centers of the disks with additions made at the boundaries. If the Voronoi cell V_i corresponding to the vertex $\mathbf{x}_i$ is adjacent to the boundary (that is, one of the sides of V_i lies on the boundary of Π) and the radius of the disk $R_i = R$ is smaller than the distance from $\mathbf{x}_i$ to this boundary, then we connect $\mathbf{x}_i$ with this boundary with a line that is perpendicular to the boundary. Denote the intersection of this perpendicular and the boundary by $\mathbf{x}_{i''}$ (note that $i'' > N$) and the line segment between $\mathbf{x}_i$ and $\mathbf{x}_{i''}$ by $e_{ii''}$. This modification adds vertices $\mathbf{x}_{i''}$, $i'' > N$, that lie on the boundary of Π (see Figure 5.7). These vertices correspond to the pseudo-disks of the triangle–neck partition.

The discrete potentials t_i at the "boundary" vertices $\mathbf{x}_i$, are prescribed on the horizontal boundaries by

$$t_i = 1 \text{ for } i \in S^+, \quad t_i = -1 \text{ for } i \in S^-. \tag{5.2.17}$$

The sets S^+ and S^- of indices in (5.2.17) correspond to the upper and lower boundaries of the domain Π. A vertex $\mathbf{x}_i$ is an upper (lower) boundary vertex if $\mathbf{x}_i$ lies on the upper (lower) boundary Π or $\mathbf{x}_i$ is the center of a disk D_i that intersects the upper (lower) boundary.

As a result, we obtain the following problem

$$\mathbb{P}\text{-model:} \quad \begin{cases} I^S(\mathbf{t}) = \dfrac{1}{2} \displaystyle\sum_{\Pi_{ij} \in \mathbb{P}} g_{ij}^S (t_i - t_j)^2 \to \min, \\[2mm] t_i = 1 \text{ for } i \in S^+, \quad t_i = -1 \text{ for } i \in S^-. \end{cases} \tag{5.2.18}$$

The discrete potentials t_i at the "interior" vertices $\mathbf{x}_i$, $i \in I$, where

$$I = \{\mathbf{x}_i, i = 1, \dots, N\} \setminus (S^+ \bigcup S^-)$$

are the only values which must be determined by the minimization problem (5.2.16), because the values t_i, $i \in S^+ \bigcup S^-$, are prescribed.

Definition 5.6 For a given distribution of disks D_i with centers $\{\mathbf{x}_i, i = 1, \dots, N\}$ the discrete network $\mathcal{G}$ is a set of vertices $\{\mathbf{x}_i, i = 1, \dots, M\}$, $M \geq N$, and edges e_{ij} between neighbors $\mathbf{x}_i$ and $\mathbf{x}_j$ of the modified Delaunay graph.

5.3 Perturbed network models

In conjunction with the network problem (5.2.18), we consider another network problem described by the energy form with the coefficients g_{ij} introduced by (2.5.8):

$$\mathbb{DP}\text{-model:} \quad \begin{cases} I(\mathbf{t}) := \dfrac{1}{2} \displaystyle\sum_{\Pi_{ij} \in \mathbb{P}} g_{ij}(t_i - t_j)^2 \to \min, \\ t_i = 1 \text{ for } i \in S^+, \ t_i = -1 \text{ for } i \in S^-. \end{cases} \quad (5.3.1)$$

This problem coincides with (5.2.18) in topology and differs only in the coefficients. The problem (5.3.1) can be considered as a perturbation of problem (5.2.18).

5.4 δ-**N** connectedness and δ-subgraphs

A necessary condition for the validity of (5.2.8) is the existence of a conducting spanning cluster – that is, there is at least one path on the graph $\mathcal{G}$ such that this path connects the top and the bottom boundaries of the domain Π and the distance δ_{ij} between every two consecutive vertices of this path $\delta_{ij} \leq \delta$, where δ is sufficiently small, see Kolpakov and Kolpakov (2010) and the final part of the proof of Lemma 3.12. A sufficient condition for the existence of a conducting spanning cluster is the δ-**N** connectedness property of the discrete network $\mathcal{G}$ which can be formulated in terms of δ-subgraphs of $\mathcal{G}$.

Definition 5.7 For any $\delta > 0$ the δ-subgraph $\mathcal{G}_\delta$ of the discrete network (graph) $\mathcal{G}$ is the subset of edges e_{ij} and their incident vertices $\mathbf{x}_i$ and $\mathbf{x}_j$ of $\mathcal{G}$, such that their length $\delta_{ij} \leq \delta$. For any subgraph a vertex is incident if it is an end-vertex of one of its edges.

The δ-**N** connectedness property of $\mathcal{G}$ is used extensively in this section, therefore, for completeness, in the rest of this section we give the precise graph-theoretical definitions related to this notion. Most of them are taken from Bollobás (1998).

Definition 5.8 A path of a (sub)graph $\mathcal{G}_\delta$ from $\mathbf{x}_0$ to $\mathbf{x}_n$ is an alternating sequence

$$\mathbf{x}_0, e_{01}, \mathbf{x}_1, e_{12}, \mathbf{x}_2, e_{23}, \ldots, \mathbf{x}_{n-1}, e_{n-1n}, \mathbf{x}_n,$$

of distinct vertices $\mathbf{x}_i$ and edges e_{ij}. Such a path has size n, and the vertices $\mathbf{x}_0$ and $\mathbf{x}_n$ are said to be the end-vertices. The vertices $\mathbf{x}_1, \ldots, \mathbf{x}_{n-1}$ are said to be the interior vertices.

Definition 5.9 An internal cycle C of $\mathcal{G}_\delta$ is an alternating sequence

$$\mathbf{x}_0, e_{01}, \mathbf{x}_1, e_{12}, \mathbf{x}_2, e_{23}, \ldots, \mathbf{x}_n, e_{n0}, \mathbf{x}_0,$$

of vertices and edges, such that

$$\mathbf{x}_0, e_{01}, \mathbf{x}_1, e_{12}, \mathbf{x}_2, e_{23}, \ldots, \mathbf{x}_n$$

is a path, and e_{n0} connects the vertices $\mathbf{x}_0$ and $\mathbf{x}_n$. Such a cycle has size $n + 1$.

Definition 5.10 A boundary cycle C of $\mathcal{G}_\delta$ is a path

$$\mathbf{x}_0, e_{01}, \mathbf{x}_1, e_{12}, \mathbf{x}_2, e_{23}, \ldots, \mathbf{x}_n,$$

such that the end-vertices $\mathbf{x}_0$ and $\mathbf{x}_n$ lie on the boundary of the domain Π. Such a boundary cycle has size n.

Note that the end-vertices $\mathbf{x}_0$ and $\mathbf{x}_n$ of a boundary cycle may lie on different parts of the boundary of the domain Π, for example left and upper.

Definition 5.11 A minimal cycle $C_{\min}$ is a (internal or boundary) cycle, such that for any two vertices $\mathbf{x}_i$ and $\mathbf{x}_j$ of this cycle the shortest path from $\mathbf{x}_i$ to $\mathbf{x}_j$ is a subset of the cycle $C_{\min}$. If the cycle $C_{\min}$ is a boundary cycle, we also require that for any interior point $\mathbf{x}_i$ of this cycle the shortest path from to $\mathbf{x}_i$ to any point $\mathbf{x}_k$ on the boundary is a subset of the cycle $C_{\min}$.

Note that we require for a minimal boundary cycle an additional condition. This condition guarantees that a boundary cycle cannot be shortened by connecting an interior point of this cycle with the boundary. Definition 5.11 is a formalization of the intuitive notion of a hole in a composite material. Each hole in a composite material is surrounded by a loop of conducting disks, see Figure 5.1. On the modified Delaunay graph this loop corresponds to an **N**-gon, which is a minimal cycle of this graph.

Definition 5.12 **N** is the upper-sided bound on the number of vertices in the minimal cycles of the (sub)graph $\mathcal{G}_\delta$ – that is, $\mathbf{N} = \max_{C_{\min} \subset \mathcal{G}_\delta} \operatorname{size}(C_{\min})$, where $C_{\min}$ is a minimal cycle and $\operatorname{size}(C_{\min})$ is the number of vertices in the cycle $C_{\min}$.

The degree of connectedness of the whole graph $\mathcal{G}$ can now be quantified in terms of the two parameters: δ and **N**, and an a priori relative error estimate for the discrete network approximation $\mathcal{G}$ is determined in terms of these parameters only.

Definition 5.13 The interior of a cycle C Int_C is the polygon whose boundary is the cycle C – that is, $\partial \text{Int}_C = C$.

Definition 5.14 A discrete network (graph) $\mathcal{G}$ is δ-**N** connected if (i) for every vertex $\mathbf{x}_i$ of the graph $\mathcal{G}$ there exists a minimal cycle C of its δ-subgraph $\mathcal{G}_\delta$ such that $\mathbf{x}_i \in \text{Int}_C$. (ii) the size of the largest minimal cycle of $\mathcal{G}_\delta$ is **N**.

Condition (i) in Definition 5.14 implies that if R, **N** and δ are not too large, then there exists at least one path in the graph $\mathcal{G}_\delta$ that connects the top and bottom boundaries of the domain Π. This can be shown by contradiction using the duality argument. Suppose there are no paths on the graph $\mathcal{G}_\delta$ that connect the top and the bottom boundaries. Then there exists a path in the whole domain Π such that it connects the left and the right boundaries of Π and does not intersect $\mathcal{G}_\delta$. The length of this path must be larger than the distance between the left and the right boundaries. On the other hand, it cannot be larger than the diameter of the largest minimal cycle. Here, by diameter of a cycle C, we mean the largest distance between any two vertices in C. Recall that the distance between any two vertices in $\mathcal{G}$ does not exceed $2R + \delta$, hence the inequality

$$2L \leq \frac{(2R + \delta)\mathbf{N}}{2} \tag{5.4.1}$$

must hold. But inequality (5.4.1) cannot hold if R, **N** and δ are not too large, therefore we have a contradiction. The existence of a path on $\mathcal{G}_\delta$ which connects the top and the bottom boundaries means the existence of the conducting spanning cluster, since the distance δ_{ij} between every two consecutive vertices of this path satisfies $\delta_{ij} \leq \delta$ by definition of $\mathcal{G}_\delta$.

5.5 Properties of the discrete network

Here we collect some results on the properties of the discrete network. The first lemma gives an upper bound on the number of necks and triangles that lie in the interior of any minimal cycle of the δ-subgraph in terms of the size of the largest minimal cycle **N**. Consider a δ-**N** connected discrete network $\mathcal{G}$ (see Definition 5.14). Consider a minimal cycle $C_{\min}$ of the δ-subgraph $\mathcal{G}_\delta$. Denote by $\#\Delta_{C_{\min}}$ the number of triangles Δ_{ijk} that lie in the interior of this minimal cycle $\Delta_{ijk} \subset \text{Int}_{C_{\min}}$. Similarly $\#\Pi_{C_{\min}}$ is the number of necks Π_{ij} that lie in the interior of this minimal cycle $\Pi_{ij} \subset \text{Int}_{C_{\min}}$, and $\#\mathbf{x}_{C_{\min}}$ is the number of vertices (centers of disks) $\mathbf{x}_i$ such that $\mathbf{x}_i \subset \text{Int}_{C_{\min}}$.

Lemma 5.15 *Suppose the discrete network $\mathcal{G}$ is δ-**N** connected. Then for any minimal cycle $C_{\min}$ of the δ-subgraph $\mathcal{G}_\delta$ the number of triangles $\#\Delta_{C_{\min}}$, and the* *number of necks* $\#\Pi_{C_{\min}}$ *that lie in the interior of* $C_{\min}$ *satisfy the bounds*

$$\#\Delta_{C_{\min}} \leq 2\left(N + \frac{2}{\pi\sqrt{3}}N^2\right), \tag{5.5.1}$$

$$\#\Pi_{C_{\min}} \leq 3\left(N + \frac{2}{\pi\sqrt{3}}N^2\right).$$

Proof Consider a minimal cycle $C_{\min}$. By Euler's formula (see van Lint and Wilson (2001)),

$$\#\Delta_{C_{\min}} - \#\Pi_{C_{\min}} + \#x_{C_{\min}} = 2.$$

Since each triangle has three edges and each edge belongs to at most two triangles, $3\#\Delta_{C_{\min}} \leq 2\#\Pi_{C_{\min}}$.

Therefore

$$\#\Delta_{C_{\min}} \leq 2(\#\Pi_{C_{\min}} - \#\Delta_{C_{\min}}) \leq 2\#x_{C_{\min}}, \tag{5.5.2}$$

$$\#\Pi_{C_{\min}} \leq x_{C_{\min}} + \#\Delta_{C_{\min}} \leq 3\#x_{C_{\min}}.$$

If $C_{\min}$ is a minimal cycle of $\mathcal{G}_\delta$, then by the δ-N connectedness of the graph, the length of this cycle is less than $N(2R + \delta)$. By the isoperimetric inequality (an area A surrounded by a curve of length L satisfies $4\pi A \leq L^2$), the area of the interior of an internal cycle is less than that of a disk of radius

$$R_0 = \frac{N(2R + \delta)}{2\pi}.$$

The area of the interior of a boundary cycle is similarly less than that of a half-circle at the boundary, but with a radius greater than R_0. Namely, in this case we have

$$R_0 \leq \frac{N(2R + \delta)}{\pi + 2} \leq \frac{N(2R + \delta)}{\pi}.$$

Therefore, an upper bound on the area of this half-circle is given by

$$\frac{\pi}{2}\left(\frac{N(2R + \delta)}{\pi}\right)^2 = 2\frac{N^2\left(R + \frac{\delta}{2}\right)^2}{\pi}.$$

This also serves as an upper bound on the area of the interior of any cycle. Let us then compute the maximal number of disks of radius R that can be placed inside a half-disk with radius $N(2R + \delta)/\pi$. Suppose this number is K. Since $C_{\min}$ is a minimal cycle each disk in the interior must be separated from any other disk by δ. Therefore the total area covered by K disks with the (densest possible) hexagonal packing is $2\sqrt{3}\left(R + \frac{\delta}{2}\right)^2 K$. By our assumption that all centers of the disks lie inside the domain Π, at least half of the disk lies in the interior of $C_{\min}$. Hence, the total area covered by the portions of the K disks inside Π is $\sqrt{3}\left(R+\dfrac{\delta}{2}\right)^2 K$. Therefore we have

$$\sqrt{3}\left(R+\frac{\delta}{2}\right)^2 K \le 2\frac{\mathbf{N}^2\left(R+\dfrac{\delta}{2}\right)^2}{\pi},$$

so

$$K \le \frac{2}{\pi\sqrt{3}}\mathbf{N}^2.$$

The total number of vertices $\mathbf{x}_i \in \{\mathbf{x}\}_{C_{\min}}$ thus satisfies

$$\#\mathbf{x}_{C_{\min}} \le K + \mathbf{N} \le \mathbf{N} + \frac{2}{\pi\sqrt{3}}\mathbf{N}^2.$$

By (5.5.2), we conclude

$$\#\Delta_{C_{\min}} \le 2\left(\mathbf{N} + \frac{2}{\pi\sqrt{3}}\mathbf{N}^2\right),$$

and

$$\#\Pi_{C_{\min}} \le 3\left(\mathbf{N} + \frac{2}{\pi\sqrt{3}}\mathbf{N}^2\right). \qquad \square$$

The discrete network is a connected graph in the sense that there is a path between each vertex $\mathbf{x}_i$ and a boundary vertex $\mathbf{x}_j \in S^+ \bigcup S^-$. This implies that the discrete minimization problem has a unique solution, because if a graph is connected, then any local minimizer of the quadratic form (5.2.16) is the (unique) global minimizer. Therefore, we have

Lemma 5.16 *There is a unique solution $\{t_i^S, i \in I\}$ of the discrete minimization problem (5.2.18). This solution satisfies a discrete analog of the Euler–Lagrange equations, given by*

$$\sum_{j\in N_i} g_{ik}^S(t_i^S - t_j^S) = 0, \, i \in I, \tag{5.5.3}$$

where N_i refers to the set of indices of neighbors of the vertex $\mathbf{x}_i$.

Since the quadratic form (5.2.16) is positive-definite the proof follows immediately from linear algebra.

Similar to the fluxes through the horizontal boundaries on the right-hand side of (5.1.7), denote by P^+ and P^- the discrete fluxes through the disks (including pseudo-disks) S^+ and S^-, respectively. That is,

$$P^+ := \sum_{i \in S^+} \sum_{j \in N_i} g_{ij}^S (1 - t_j), \tag{5.5.4}$$

$$P^- := \sum_{i \in S^-} \sum_{j \in N_i} g_{ij}^S (-1 - t_j).$$

Here we have used that $t_i = 1$ if $i \in S^+$ and $t_i = -1$ if $i \in S^-$.
   Then,

$$E_d^S = \frac{1}{2}(P^+ - P^-) = \frac{1}{2} \sum_{\Pi_{ij} \in \mathbb{P}} g_{ij}^S (t_i^S - t_j^S)^2. \tag{5.5.5}$$

For the proof, see Lemma 3.13.

Lemma 5.17 (Discrete maximum principle for a cycle). *Suppose* $\mathbf{t} = \{t_i, i \in I\}$
*is the solution of the $\mathbb{D}$-model (5.2.18). For any (internal or boundary) cycle C of
$\mathcal{G}$ define*

$$t_{max} = \max\{t_i^S, \mathbf{x}_i \in C\},$$

$$t_{min} = \min\{t_i^S, \mathbf{x}_i \in C\}.$$

*Then for any vertex $\mathbf{x}_k$ with potential t_k such that $\mathbf{x}_k \in Int_C$, that is $\mathbf{x}_k$ belongs to
the interior of the cycle C (as in Definition 5.13), we have*

$$t_{min} \le t_k^S \le t_{max}.$$

Proof The proof of the discrete maximum principle is by contradiction. From
(5.5.3), the following "mean value theorem" is obtained:

$$t_i^S \sum_{k \in N_i} g_{ik} = \sum_{k \in N_i} g_{ik} t_k^S, \, i \in I. \tag{5.5.6}$$

Suppose not all the values t_i for points $\mathbf{x}_i$ in the interior of the cycle are the
same and the maximum of t_i is achieved in the proper interior of the cycle at
$\mathbf{x}_k \in Int_C$, $\mathbf{x}_k \notin C$. Then we have $t_i^S \le t_k^S$ for any $\mathbf{x}_i \in Int_C$.
   Therefore, equation (5.5.6) is satisfied only if $t_i \equiv t_k$ whenever vertices $\mathbf{x}_i$ and
$\mathbf{x}_k$ are connected by a neck Π_{ik}. Hence the maximum is also achieved at all the
points $\mathbf{x}_i$ which are connected with $\mathbf{x}_k$ by a neck Π_{ik}. Since the graph is connected,
induction over all the points in the interior of the cycle implies the contradiction:
the values t_i for all the points in the interior of the cycle are the same.   □

   As a corollary of the discrete maximum principle, we have the following
lemma.

Lemma 5.18 *If the discrete network $\mathcal{G}$ is δ-N connected, then for any minimal cycle $C_{\min}$ of the δ-subgraph $\mathcal{G}_\delta$ and a vertices $\mathbf{x}_k \in Int_{C_{\min}}$ and $\mathbf{x}_l \in Int_{C_{\min}}$ we have*

$$(t_k^S - t_l^S)^2 \le \mathbf{N} \sum_{\Pi_{ij} \in C_{\min}} (t_i^S - t_j^S)^2 \qquad (5.5.7)$$

(the number $\mathbf{N}$ is introduced by Definition 5.12).

Proof By the discrete maximum principle,

$$(t_k^S - t_l)^2 \le (t_{\max}^S - t_{\min}^S)^2, \qquad (5.5.8)$$

where

$$t_{\max} = \max\{t_i^S \mid \mathbf{x}_i \in C_{\min}\},$$

$$t_{\min} = \min\{t_i^S \mid \mathbf{x}_i \in C_{\min}\}.$$

Suppose the maximum $t_{\max}$ and the minimum $t_{\min}$ are achieved at the vertices $x' \in C_{\min}$ and $x'' \in C_{\min}$, respectively. Since both vertices belong to the minimal cycle $C_{\min}$, there is a path contained in this minimal cycle with size $s \le \mathbf{N}$ (where s here means the number of edges in this path). Therefore, by the triangle inequality, t_i, $\mathbf{x}_i \in C$,

$$(t_{\max} - t_{\min})^2 \le \mathbf{N} \sum_{\Pi_{ij} \in C_{\min}} (t_i^S - t_j^S)^2,$$

which, when inserted in (5.5.8), yields (5.5.7). $\qquad\square$

5.6 Variational error estimates

Our approach to variational error estimates is to construct trial fields that give rise to tight upper- and lower-sided bounds, when particles are close to touching. For both lower- and upper-sided bounds, the trial functions $\phi(\mathbf{x}) \in V$ and $\mathbf{v}(\mathbf{x}) \in W$ are constructed separately in the necks and triangles. The piecewise-differentiable trial function $\phi(\mathbf{x}) \in V$ is chosen to be linear in the neck and a linear interpolation in the triangles.

The trial field $\mathbf{v} \in W$ is chosen so that it is equal to zero everywhere except in the necks. In the necks it is equal to the local flux defined by (5.2.2), where t_i are determined as solutions of the discrete minimization problem (5.2.18). In addition, the trial field $\mathbf{v}$ is constructed in such a way that it is divergence-free in whole domain Q_m. Now, we construct the refined lower- and upper-sided bounds.

A successful choice of the trial functions leads to tight bounds.

In this section, we construct the upper- and the lower-sided bounds for a composite when its discrete network $\mathcal{G}$ is δ-N connected (see Definition 5.14) and we estimate the difference between the upper and the lower bounds in terms of δ, $\mathbf{N}$ and the energies $E_d^S = I^S(\mathbf{t}^S)$ and $E_d = I(\mathbf{t})$ of the discrete minimization problems (equations (5.2.18) and (5.3.1)). This allows one to obtain an estimate on the relative error between the effective conductivity (3.2.10) and the energy of the discrete network (5.3.1).

Certainly our trial functions give upper and lower variational bounds on the effective conductivity for any distribution of disks; however, our bounds are tight only under the assumption that "almost all" the distances δ_{ij} between neighbors are sufficiently small (δ-**N** connectedness).

Definition 5.19 A distribution of disks D_i, $i = 1, \ldots, N$, satisfies the δ-**N** close packing condition if its discrete network $\mathcal{G}$ is δ-**N** connected.

Here is the idea of the construction of the variational bounds. Our goal is to construct two trial functions, which mimic the behavior of the solution to the minimization problem (5.2.18) for the $\mathbb{DP}$-model. The domain Π consists of disks D_i, necks Π_{ij} and triangles Δ_{ijk} where the necks and triangles are determined by the triangle–neck partition. The function $\phi \in V_p$ and the function $\mathbf{v} \in W_p^0$ are defined first on the disks, then in the necks and the triangles. Suppose t_i^S are the values of the solution of the minimization problem (5.3.1) for the $\mathbb{DP}$-model. The function $\phi \in V_p$ is chosen so that it takes the values t_i^S on the disks D_i, it is equal to ± 1 on the boundaries $S^\pm$ and it is linear in the necks. Observe that the formulation of the problem is rotationally invariant, therefore in the construction of the trial functions we routinely locally rotate the plane so that in the local coordinates a particular neck Π_{ij} is always aligned with the vertical direction. The function ϕ should be piecewise-differentiable by definition of the space V_p (3.1.1). Therefore on triangles it is defined by linear interpolation. The function $\mathbf{v} \in W_p^0$ intuitively is the flux of ϕ, and it is chosen on disks and necks taking into account the discrete fluxes of the discrete minimization problem. The function $\mathbf{v}$ does not have to be piecewise-differentiable, hence we simply define it to be zero on the triangles.

The idea of the error estimate is to show that both the upper and lower bound are very close to the energy of the $\mathbb{DP}$-model. In order to show that the energy of the $\mathbb{P}$-model (5.2.16) is a good approximation as well we compute the difference between the solution of the $\mathbb{DP}$-model and the solution of the $\mathbb{P}$-model.

5.7 The refined lower-sided bound

Following the method used in Chapter 3, the trial function $\mathbf{v}(\mathbf{x})$ for the lower-sided bound is chosen to be zero everywhere except the necks Π_{ij} between adjacent disks; however, in our case, we use (5.2.7) for the specific fluxes instead of (2.5.8).

Lemma 5.20 *The lower-sided bound on A in terms of $\{g_{ij}^S\}$ and the parameters of the solution of the discrete minimization problem is*

$$E_d^S = \frac{1}{2} \sum_{\Pi_{ij} \in \mathbb{P}} g_{ij}^S (t_i^S - t_j^S)^2 \leq A, \qquad (5.7.1)$$

where $\mathbf{t}^S = \{t_i^S, i \in I\}$ *are the values of the discrete potentials of the solution of the discrete minimization problem (5.2.16).*

The left-hand side in (5.7.1) is always positive, which reflects the physics of the problem. The analogous lower-sided bound in Chapter 3 is positive and it is sufficiently tight for the δ-3 close packing condition only. Our bound allows us to handle a general distribution of disks that satisfies the δ-**N** close packing condition for any **N**.

Consider two neighbors centered at $\mathbf{x}_i$ and $\mathbf{x}_j$. Suppose the potentials t_i^S and t_j^S on the disks D_i and D_j, respectively, are the solutions to the minimization problem (5.2.16). Rotate the domain so that in the local coordinates the neck between them is aligned with the direction of the Oy-axis (as in Figure 5.8).

Define

$$\mathbf{v}(\mathbf{x}) = \begin{cases} \left(0, \dfrac{t_i^S - t_j^S}{H_{ij}(x)}\right) & \text{in the neck } \Pi_{ij}, \\ (0, 0) & \text{otherwise,} \end{cases} \qquad (5.7.2)$$

where $H_{ij}(x)$ is the distance between the disks. Since, for a piecewise constant function the divergence-free condition amounts to checking that the normal components of $\mathbf{v}(\mathbf{x})$ match along the discontinuity, we see that our trial function $\mathbf{v}(\mathbf{x})$ is divergence-free. This problem was discussed in detail in Section 3.6.2.

The matching condition

$$\int_{\partial D_i} \mathbf{v}(\mathbf{x})\mathbf{n}dx = 0$$

is satisfied (as in Chapter 3) due to (5.2.7) and (5.5.3). The insulating condition $\mathbf{v}(\mathbf{x})\mathbf{n} = 0$ on ∂Q_{lat} is satisfied by (5.7.2). Hence $\mathbf{v} \in W_p^0$. Observe that for the trial function (5.7.2) the fluxes through the upper and the lower boundary of Π are exactly the discrete fluxes P^+ and P^-:

$$P^+ = \int_{\partial Q^+} \mathbf{v}(\mathbf{x})\mathbf{n}dx, \quad P^- = \int_{\partial Q^-} \mathbf{v}(\mathbf{x})\mathbf{n}dx.$$

For the function (5.7.2) we have

$$\int_{Q_m} \mathbf{v}(\mathbf{x})^2 dx = \sum_{\Pi_{ij} \in \mathbb{P}} g_{ij}^S (t_i^S - t_j^S)^2,$$

because $\mathbf{v}(\mathbf{x}) \equiv 0$ on every triangle and the Dirichlet integral over a neck is

$$\int_{\Pi_{ij}} \mathbf{v}(\mathbf{x})^2 dx = (t_i - t_j)^2 \int_{S_1}^{S_2} \frac{dx}{H_{ij}(x)} = g_{ij}^S (t_i^S - t_j^S)^2.$$

Using (5.5.5) we have

$$-\frac{1}{2}\int_{Q_m} \mathbf{v}(\mathbf{x})^2 dx + \int_{\partial Q^+ \cup \partial Q^-} \phi^0(y)\mathbf{v}(\mathbf{x})\mathbf{n}dx = E_d^S.$$

By the first inequality in (5.1.12) we have (5.7.1).

5.8 The refined upper-sided bound

Lemma 5.21 *The upper-sided bound on A in terms of g_{ij}^S and the parameters of the solution of the discrete minimization problem is*

$$A \leq \frac{1}{2} \sum_{\Pi_{ij} \in \mathbb{P}} \left[g_{ij}^S + C_{ij} \right] (t_i^S - t_j^S)^2 = E_d^S + \frac{1}{2} \sum_{\Pi_{ij} \in \mathbb{P}} C_{ij} (t_i^S - t_j^S)^2, \qquad (5.8.1)$$

where the constants

$$C_{ij} \leq \frac{|\ln(1 - \beta_{ij}^{\max})| + \pi + \ln 2}{6} + \frac{4}{\sqrt{1 - (\beta_{ij}^{\max})^2}} \qquad (5.8.2)$$

depend on the relative neck thicknesses $\beta_{ij}^{\max}$ only.
Moreover if $\beta_{ij}^{\max} \leq \beta < 1$, then $C_{ij} \leq C$ with some $C = C(\beta) < \infty$.

To obtain this refined upper-sided estimate, we will construct a special trial function $\overline{\phi}(\mathbf{x})$ for the general upper-sided estimate (see (5.1.12))

$$A \leq \frac{1}{2} \int_{Q_m} |\nabla \overline{\phi}(\mathbf{x})|^2 d\mathbf{x}. \qquad (5.8.3)$$

5.9 Construction of trial function for the upper-sided bound

In this section, we construct a trial function and estimate the Dirichlet integrals in (5.8.3) by using a two-step procedure. In the first step, we construct the trial function in the necks. In the second step, we construct the trial function in the triangles.

5.9.1 Construction of the trial function and estimation of the Dirichlet integral in the necks

Consider a piecewise continuous trial function $\overline{\phi}(\mathbf{x})$. Similar to Chapter 3, the function $\overline{\phi}(\mathbf{x})$ is linear in y in the neck Π_{ij} with the values t_i and t_j on the boundary of the disks ∂D_i and ∂D_j (see Figure 5.8). Then on the neck Π_{ij}

$$\overline{\phi}(\mathbf{x}) = t_i^S + \frac{(t_j^S - t_i^S)\left(y + \dfrac{H(x)}{2}\right)}{H(x)} = t_i^S + (t_j^S - t_i^S)\left[\frac{y}{H(x)} + \frac{1}{2}\right],$$

$$y \in \left[-\frac{H(x)}{2}, \frac{H(x)}{2} \right]. \qquad (5.9.1)$$

The gradient of $\overline{\phi}(\mathbf{x})$ (5.9.1) on the neck Π_{ij} is given by

$$\frac{\partial \overline{\phi}}{\partial x}(\mathbf{x}) = -(t_j^S - t_i^S)\frac{yH'(x)}{H^2(x)}, \qquad (5.9.2)$$

$$\frac{\partial \overline{\phi}}{\partial y}(\mathbf{x}) = \frac{t_j^S - t_i^S}{H(x)}.$$

By (5.9.2)

$$\int_{\Pi_{ij}} \left(\frac{\partial\overline{\phi}}{\partial y}\right)^2 dxdy = \int_{S_1}^{S_2} \frac{(t_j^S - t_i^S)^2}{H^2(x)} H(x)dx = (t_j^S - t_i^S)^2 g_{ij}^S, \qquad (5.9.3)$$

where g_{ij}^S is defined in (5.2.7).

Also

$$\int_{\Pi_{ij}} \left(\frac{\partial\overline{\phi}}{\partial y}\right)^2 dxdy = \int_{\Pi_{ij}} (t_j^S - t_i^S)^2 \left[\frac{yH_{ij}'(x)}{H_{ij}^2(x)}\right]^2 dxdy \qquad (5.9.4)$$

$$= \int_{S_1}^{S_2} (t_j^S - t_i^S)^2 \left[\frac{H_{ij}'(x)}{H_{ij}^2(x)}\right]^2 \frac{y^3}{3}\bigg|_{-H_{ij}(x)/2}^{H_{ij}(x)/2} dx$$

$$= \frac{1}{12}(t_j^S - t_i^S)^2 \int_{S_1}^{S_2} \frac{H_{ij}'^2(x)}{H(x)} dx.$$

Since $H_{ij}(x) = \delta_{ij} + 2R - 2\sqrt{R^2 - x^2}$, therefore $H_{ij}'(x) = \dfrac{2x}{\sqrt{R^2 - x^2}}$ and

$$\frac{H_{ij}'^2(x)}{H_{ij}(x)} = \frac{4x^2}{R^2 - x^2} \cdot \frac{1}{\delta_{ij} + 2R - 2\sqrt{R^2 - x^2}} \qquad (5.9.5)$$

$$\le \frac{2}{R^2 - x^2} \cdot \frac{x^2}{R - \sqrt{R^2 - x^2}} = \frac{2}{R^2 - x^2}(R + \sqrt{R^2 - x^2})$$

$$= \frac{2R}{R^2 - x^2} + \frac{2}{\sqrt{R^2 - x^2}},$$

$$\int_{S_1}^{S_2} \frac{2R}{R^2 - x^2} dx + \int_{S_1}^{S_2} \frac{2}{\sqrt{R^2 - x^2}} dx = \left(\ln\frac{R + x}{R - x} + 2\arcsin\frac{x}{R}\right)\bigg|_{S_1}^{S_2} \qquad (5.9.6)$$

then from (5.9.4), (5.9.5) and (5.9.6) we have

$$\int_{\Pi_{ij}} \left(\frac{\partial\overline{\phi}}{\partial y}\right)^2 dxdy = \frac{1}{12}(t_j^S - t_i^S)^2 \int_{-S_1}^{S_2} \frac{H_{ij}'^2(x)}{H_{ij}(x)} dx$$

$$\le \frac{1}{12}(t_j^S - t_i^S)^2 \left(2\ln\frac{2}{1 - \beta_{ij}^{\max}} + 2\pi\right)$$

$$\le \frac{|\ln(1 - \beta_{ij}^{\max})| + \pi + \ln 2}{6}(t_j^S - t_i^S)^2,$$

where $\beta_{ij}^{\max}$ is defined by (5.2.1).

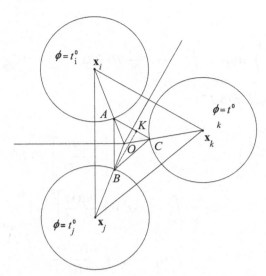

Figure 5.10 Typical $\triangle ABC \equiv \triangle_{ijk}$; half-necks are hatched.

Combining (5.9.3) and (5.9.7), the value of the integral $\displaystyle\int_{\Pi_{ij}} |\nabla\overline{\phi}(\mathbf{x})|^2 d\mathbf{x}$ is

$$\int_{\Pi_{ij}} |\nabla\overline{\phi}(\mathbf{x})|^2 d\mathbf{x} \le (t_j^S - t_i^S)^2 \left[g_{ij} + \frac{|\ln(1 - \beta_{ij}^{\max})| + \pi + \ln 2}{6} \right]. \quad (5.9.7)$$

5.9.2 Construction of the trial function and estimation of the Dirichlet integral in the triangles

Each triangle is bounded by three necks. In Figure 5.10 we show a typical case with three disks centered at $\mathbf{x}_i$, $\mathbf{x}_j$ and $\mathbf{x}_k$, three half-necks and the $\triangle ABC = \triangle_{ijk}$ bounded by these half-necks. Suppose in $\triangle ABC$ the side $|AC|$ is the longest, and the side $|BC|$ is the shortest.

Lemma 5.22 *The following inequality holds for the function (5.9.1):*

$$\int_{\triangle_{ijk}} |\nabla\overline{\phi}(\mathbf{x})|^2 d\mathbf{x} \le \frac{2}{\sqrt{1 - \beta^2}} \left((t_i^S - t_j^S)^2 + (t_k^S - t_i^S)^2 \right), \quad (5.9.8)$$

where $\beta = \max(\beta_{ij}^{\max}, \beta_{kj}^{\max}, \beta_{ki}^{\max})$.

Proof By construction the Voronoi tessellation induces the triangle neck partition of the domain Π. Each triangle is (typically) bounded by three (half)necks. In Figure 5.10 we show a typical case with three disks centered at $\mathbf{x}_i$, $\mathbf{x}_j$ and $\mathbf{x}_k$, three half-necks and the $\triangle ABC \equiv \triangle_{ijk}$ bounded by these half-necks.

The potential $\phi(\mathbf{x})$ is defined at the vertices of $\triangle ABC$

$$\phi(A) = t_i^S, \quad \phi(B) = t_j^S, \quad \phi(C) = t_k^S. \quad (5.9.9)$$

Since $\overline{\phi}(\mathbf{x})$ is piecewise continuously differentiable, define $\overline{\phi}(\mathbf{x})$ on $\triangle ABC$ by linear interpolation of these three values. Suppose in $\triangle ABC$, the side $|AC|$ is the longest, and the side $|BC|$ is the shortest (see Figure 5.10). Then,

$$0 < \angle \mathbf{x}_j \mathbf{x}_i \mathbf{x}_k = \angle BAC \le \frac{\pi}{3}, \tag{5.9.10}$$

$$1 \le \frac{|AC|}{|AB|} \le 2,$$

$$\beta = \beta_{jk}^{\max},$$

and the end-vertex of the perpendicular BK on the base AC lies between the points A and C. Suppose that $\overline{\phi}(K) = t_0$. Then we have

$$\frac{t_0 - t_i^S}{t_k^S - t_i^S} = \frac{|AB| \cos \angle BAC}{|AC|},$$

therefore

$$t_0 = t_i^S + (t_k^S - t_i^S)\frac{|AB| \cos \angle BAC}{|AC|}.$$

Denoting by $|ABC|$ the area of $\triangle ABC$ we have the following computation chain:

$$\int_{ABC} |\nabla \overline{\phi}(\mathbf{x})|^2 d\mathbf{x}$$

$$= \left(\left[\frac{t_0 - t_j^S}{|BK|} \right]^2 + \left[\frac{t_i^S - t_k^S}{|AC|} \right]^2 \right) |ABC|$$

$$= \frac{|AC| \cdot |AB| \sin \angle BAC}{2}$$

$$\times \left(\left[\frac{(t_i^S - t_j^S) + (t_k^S - t_i^S)\frac{|AB| \cos \angle BAC}{|AC|}}{|AB| \sin \angle BAC} \right]^2 + \left[\frac{t_i^S - t_k^S}{|AC|} \right]^2 \right)$$

$$= \frac{|AC||AB| \sin \angle BAC}{2}$$

$$\times \left(\left[\frac{(t_i^S - t_j^S)}{|AB| \sin \angle BAC} + \frac{(t_k^S - t_i^S)|AB| \cos \angle BAC}{|AC||AB| \sin \angle BAC} \right]^2 + \left[\frac{t_i^S - t_k^S}{|AC|} \right]^2 \right)$$

which, using Cauchy's inequality, is not greater than

$$|AC| \cdot |AB| \sin \angle BAC \left(\frac{(t_i^S - t_j^S)^2}{|AB|^2 \sin^2 \angle BAC} + \frac{(t_k^S - t_i^S)^2(1 + \cos^2 \angle BAC)}{|AC|^2 \sin \angle BAC} \right)$$

$$= \frac{1}{\sin \angle BAC} \left((t_i^S - t_j^S)^2\frac{|AC|}{|AB|} + (t_k^S - t_i^S)^2\frac{|AB|}{|AC|} \right).$$

This is, noting (5.9.10), not greater than $\dfrac{2}{\sin \angle BAC}\left((t_i^S - t_j^S)^2 + (t_k^S - t_i^S)^2 \right)$.

Hence,

$$\int_{ABC} |\nabla\overline{\phi}(\mathbf{x})|^2 d\mathbf{x} \le \frac{2}{\sin \angle BAC}\left((t_i^S - t_j^S)^2 + (t_k^S - t_i^S)^2\right).$$

In terms of the parameter $\beta = \beta_{jk}^{\max}$ we obtain the bound

$$\frac{1}{\sin \angle BAC} = \frac{1}{\sin\left(\frac{\pi}{2} - \angle\mathbf{x}_k\mathbf{x}_jO\right)} \le \frac{1}{\sqrt{1 - (\beta_{jk}^{\max})^2}}.$$

Therefore,

$$\int_{ABC} |\nabla\overline{\phi}(\mathbf{x})|^2 d\mathbf{x} \le \frac{2}{\sqrt{1 - \beta^2}}\left((t_i^S - t_j^S)^2 + (t_k^S - t_i^S)^2\right)$$

if $\beta < 1$.   □

Combining equations (5.9.8) and (5.9.7) and summing over all necks and triangles, we have (5.8.1).

5.9.3 Upper bounds on the relative half-neck widths

In order to make use of equation (5.8.1), we need to obtain an upper bound on $\beta_{ij}^{\max}$.

Lemma 5.23 *The following upper bounds for $\beta_{ij}^{\max}$ hold: if $\mathbf{N} = 3$*

$$\beta_{ij}^{\max} \le 1 - \frac{1}{2}\left(\frac{R}{R + \frac{\delta}{2}}\right)^2; \tag{5.9.11}$$

if $\mathbf{N} = 4$

$$\beta_{ij}^{\max} \le \sqrt{1 - \frac{1}{2}\left(\frac{R}{R + \frac{\delta}{2}}\right)^2}; \tag{5.9.12}$$

if $\mathbf{N}$ is arbitrary

$$\beta_{ij}^{\max} \le 1 - 2\left(\frac{R}{\left(R + \frac{\delta}{2}\right)\mathbf{N}}\right)^2. \tag{5.9.13}$$

Proof The upper bound on $\beta_{ij}^{\max}$ depends on the distribution of the other disks. By definition, all vertices O_k of the Voronoi tessellation (see Figure 5.5) are centers of circumcirles of the triangles $\Delta\mathbf{x}_i\mathbf{x}_j\mathbf{x}_k$ (in the degenerate cases, instead of a triangle $\Delta\mathbf{x}_i\mathbf{x}_j\mathbf{x}_k$, we may have n-gons), which contain no other vertices $\mathbf{x}_i$ of the Delaunay

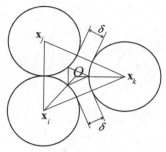

Figure 5.11 The worst-case scenario for the upper-sided bound on $\beta_{ij}^{\max}$ when $\mathbf{N} = 3$.

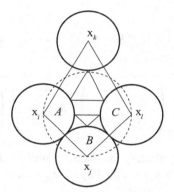

Figure 5.12 The worst-case scenario for the upper-sided bound on $\beta_{ij}^{\max}$ when $\mathbf{N} = 4$.

graph $\mathcal{G}$. The upper bound on $\beta_{ij}^{\max}$ is determined by the estimate on the maximal possible diameter of these circumcircles.

In the case $\mathbf{N} = 3$ the largest $\beta_{ij}^{\max}$ is shown on Figure 5.11. Here the half-neck between the disks D_i and D_j is the largest possible. Since

$$\frac{\pi}{2} - \angle O\mathbf{x}_i\mathbf{x}_j = \angle \mathbf{x}_i\mathbf{x}_k\mathbf{x}_j$$

therefore

$$\beta_{ij} = \sin \angle O\mathbf{x}_i\mathbf{x}_j = 1 - 2\sin^2 \angle \mathbf{x}_i\mathbf{x}_k O$$

$$= 1 - 2\left(\frac{R}{2R + \delta_{ij}}\right)^2 = 1 - \frac{1}{2}\left(\frac{R}{R + \dfrac{\delta_{ij}}{2}}\right)^2.$$

In the case $\mathbf{N} = 4$ the largest half-neck β_{ij} between disks D_i and D_j is shown in Figure 5.12. In this case $|\mathbf{x}_k\mathbf{x}_i| = |\mathbf{x}_l\mathbf{x}_k| = 2R + \delta_{ij}$. Therefore the maximal possible diameter of the circumcircle is $\sqrt{2}(2R + \delta_{ij})$. Since $|\mathbf{x}_i\mathbf{x}_j| \leq 2R$ and

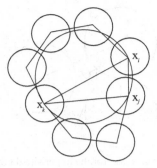

Figure 5.13 The worst-case scenario for the upper-bound on $\beta_{ij}^{\max}$ when **N** is arbitrary.

noting that (see Figure 5.12) $\beta_{ij}^{\max}R = R \sin \angle x_l x_i x_j$ and $\cos x_l x_i x_j = \dfrac{\beta_{ij}^{\max}R}{2R}$, we have

$$\beta_{ij}^{\max} = \sqrt{1 - \left(\frac{R}{\sqrt{2}\dfrac{2R+\delta_{ij}}{2}}\right)^2} = \sqrt{1 - \frac{1}{2}\left(\frac{R}{R+\dfrac{\delta_{ij}}{2}}\right)^2}.$$

In the case when **N** is arbitrary the diameter ρ of the maximal circumcircle is bounded by a half of the length of the largest minimal cycle. Thus,

$$\rho \le \left(R + \frac{\delta_{ij}}{2}\right)\mathbf{N}.$$

The largest half-neck in Figure 5.13 is between x_i and x_j. The neck width is maximized when the disks D_i and D_j are touching. Recall that this half-neck is determined by the center of the circumcircle O (this point is the vertex of the Voronoi tessellation used to construct this neck). Therefore, to calculate the half-neck width, one only needs to know the position of the points x_i, x_j, and x_k. The half-neck width will be the same for any position of x_k on the circumcircle. Therefore, without loss of generality, we can assume that x_k lies on the line going through O and the midpoint between x_i and x_j (see Figure 5.14). Then,

$$\beta_{ij}^{\max} = \sin \angle x_i x_j x_k$$

$$= \cos \angle x_i x_k x_j = 1 - 2\sin^2 \frac{x_i x_k x_j}{2}$$

$$\le 1 - 2\left(\frac{R}{\left(R+\dfrac{\delta}{2}\right)\mathbf{N}}\right)^2. \qquad \Box$$

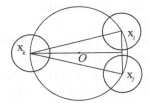

Figure 5.14 Position of the points $\mathbf{x}_k$, $\mathbf{x}_i$ and $\mathbf{x}_j$.

By using Lemma 5.23, it is possible to improve the estimates (5.9.8),

Corollary 5.24 *Assume* $\beta = \max(\beta_{ij}^{\max}, \beta_{kj}^{\max}, \beta_{ki}^{\max})$.
If $\mathbf{N} = 3$

$$\int_{\Delta_{ijk}} |\nabla\overline{\phi}(\mathbf{x})|^2 d\mathbf{x} \le \frac{4}{\sqrt{3}}\left(1 + \frac{\delta}{2R}\right)\left((t_i^S - t_j^S)^2 + (t_k^S - t_i^S)^2\right).$$

If $\mathbf{N} = 4$

$$\int_{\Delta_{ijk}} |\nabla\overline{\phi}(\mathbf{x})|^2 d\mathbf{x} \le 2\sqrt{2}\left(1 + \frac{\delta}{2R}\right)\left((t_i^S - t_j^S)^2 + (t_k^S - t_i^S)^2\right).$$

If $\mathbf{N}$ *is arbitrary*

$$\int_{\Delta_{ijk}} |\nabla\overline{\phi}(\mathbf{x})|^2 d\mathbf{x} \le \frac{\mathbf{N}}{\sqrt{1 - \dfrac{1}{\mathbf{N}^2}}}\left(1 + \frac{\delta}{2R}\right)\left((t_i^S - t_j^S)^2 + (t_k^S - t_i^S)^2\right).$$

5.10 The network approximation theorem with an error estimate independent of the total number of particles

Theorem 5.2 *If for a distribution of disks* $\{D_i, i = 1, \ldots, N\}$ *its discrete network is* δ-$\mathbf{N}$ *connected then the relative error*

$$\frac{|A - E_d^S|}{E_d^S} \le C(\mathbf{N})\sqrt{\frac{\delta}{R}}, \tag{5.10.1}$$

where $C(\mathbf{N}) < \infty$ *depends on* $\mathbf{N}$, *only.*

Proof As $\delta_{ij} \to 0$, we know that $g_{ij}^S = O(\delta^{-1/2})$ and therefore if there is a conducting spanning cluster then (see Lemma 3.12) the energy $E_d^S = O(\delta^{-1/2})$. Hence (5.10.1) will follow if we can "absorb" all the $O(1)$ terms in (5.8.1) as the "smaller order corrections" into $O\left(\sqrt{\dfrac{R}{\delta}}\right)$ terms and derive the estimate

$$E_d^S \le A \le E_d^S + O(1) \le E_d^S\left(1 + C\sqrt{\frac{\delta}{R}}\right), \tag{5.10.2}$$

where the first inequality in (5.10.2) follows from (5.7.1). The second inequality is also immediate if all necks are short: $\delta_{ij} \leq \delta$.

If not all the necks are short then for the not short necks $g_{ij}^S = O(1)$ and therefore C_{ij} in (5.8.1) are compatible in magnitude with g_{ij}^S.

The key observation here is that the centers of any two neighbors D_k and D_l must lie inside a minimal cycle $C_{\min}$. Therefore by Lemma 5.18 the values of the potentials t_k^S and t_l^S at these disks satisfy

$$(t_k^S - t_l^S)^2 \leq \mathbf{N} \sum_{\Pi_{ij} \in C_{\min}} (t_i^S - t_j^S)^2. \tag{5.10.3}$$

By Lemma 5.15 we have a bound that depends on $\mathbf{N}$ only, on the total number of vertices, necks and triangles, that lie inside any minimal cycle $C_{\min}$. Therefore for the trial function $\phi(\mathbf{x})$ for the upper-sided bound from Section 5.9, the Dirichlet integral over all the triangles and necks that lie inside any $C_{\min}$ is a smaller order correction to the energy of the minimal cycle $C_{\min}$:

$$\sum_{\Pi_{ij}, e_{ij} \in C_{\min}} g_{ij}^S (t_i - t_j)^2.$$

Therefore we have (see formula (5.8.1))

$$A \leq \frac{1}{2} \sum_{\Pi_{ij} \in \mathbb{P}} \left[g_{ij}^S + C \right] (t_i^S - t_j^S)^2.$$

Here, $C = C(\mathbf{N})$ (i.e., C depends on $\mathbf{N}$ only) due to equation (5.8.2) and Lemma 5.9.11. Then

$$A \leq \frac{1}{2} \sum_{\Pi_{ij}, e_{ij} \notin \mathbb{P}_\delta} \left[g_{ij}^S + C(\mathbf{N}) \right] (t_i^S - t_j^S)^2 + \frac{1}{2} \sum_{\Pi_{ij}, e_{ij} \in \mathbb{P}_\delta} \left[g_{ij}^S + C(\mathbf{N}) \right] (t_i^S - t_j^S)^2,$$

where $\mathbb{P}_\delta$ is the set of necks and triangles in the δ-subgraph of $\mathbb{D}$.

Due to equation (5.10.3), we can absorb $\sum_{\Pi_{ij}, e_{ij} \notin \mathbb{P}_\delta} C(t_i^S - t_j^S)^2$ into the sum over $\Pi_{ij}, e_{ij} \in \mathbb{P}_\delta$. Thus, we obtain

$$A \leq \frac{1}{2} \sum_{\Pi_{ij}, e_{ij} \notin \mathbb{P}_\delta} g_{ij}^S (t_i^S - t_j^S)^2 + \frac{1}{2} \sum_{\Pi_{ij}, e_{ij} \in \mathbb{P}_\delta} \left[g_{ij}^S + C(\mathbf{N}) \right] (t_i^S - t_j^S)^2$$

$$\leq \frac{1}{2} \sum_{\Pi_{ij}, e_{ij} \notin \mathbb{P}_\delta} g_{ij}^S (t_i^S - t_j^S)^2 + \frac{1}{2} \sum_{\Pi_{ij}, e_{ij} \in \mathbb{P}_\delta} g_{ij}^S \left(1 + C(\mathbf{N}) \sqrt{\frac{\delta}{R}} \right) (t_i^S - t_j^S)^2$$

$$\leq \left(1 + C(\mathbf{N}) \sqrt{\frac{\delta}{R}} \right) \frac{1}{2} \sum_{\Pi_{ij} \in \mathbb{P}} g_{ij}^S (t_i^S - t_j^S)^2.$$

Hence we obtain (5.10.2). $\qquad\square$

5.11 Estimation of the constant in the network approximation theorem

In Section 5.10, we presented the network approximation theorem with an error estimate depending on **N** only (thus, independent of the total number of inclusions). The main goal of this section is to derive better estimates for the constant $C(\mathbf{N})$ in (5.10.1).

The lemmas presented below in this section can be understood as improved approximation theorems.

Lemma 5.3 *Under the* δ*-**N** close packing condition (see Definition 5.19) for* $\delta \leq \dfrac{R}{32}$ *the relative error is*

$$\frac{|A - E_d^S|}{E_d^S} \leq 2.56 \mathbf{N}^4 \sqrt{\frac{\delta}{R}},$$

where the effective conductivity is

$$A = \frac{1}{2} \int_{Q_m} |\nabla \phi(\mathbf{x})|^2 dx = \frac{1}{2} \min_{\tilde{\phi} \in V_p} \int_{Q_m} |\nabla \tilde{\phi}(\mathbf{x})|^2 dx$$

and the energy of the $\mathbb{DP}$*-model is*

$$E_d = \frac{1}{2} \sum_{\Pi_{ij} \in \mathbb{P}} g_{ij}(t_i - t_j)^2 = \frac{1}{2} \min_{\tilde{\mathbf{t}} \in T} \sum_{\Pi_{ij} \in \mathbb{P}} g_{ij}(\tilde{t}_i - \tilde{t}_j)^2,$$

where $\mathbf{t} = \{t_i, \, i \in I\}$ *is solution of the problem (5.3.1).*

The factor $\dfrac{1}{32}$ in Lemma 5.3 is chosen for simplicity of presentation and it is not essential. If our estimates on $|g_{ij}^S - g_{ij}|$ in Section 5.2.2 and the proof of Lemma 5.3 are modified, the statement of Lemma 5.3 is true for any $\delta \leq R$.

Proof By Lemma 5.15

$$\#\Delta_{C_m} \leq 2 \left(\mathbf{N} + \frac{2}{\pi\sqrt{3}} \mathbf{N}^2 \right), \tag{5.11.1}$$

$$\#\Pi_{C_m} \leq 3 \left(\mathbf{N} + \frac{2}{\pi\sqrt{3}} \mathbf{N}^2 \right).$$

By the analog of Lemma 5.18 for the $\mathbb{DP}$-model for every $x_k \in \text{Int}_{C_m}$ and $x_l \in \text{Int}_{C_m}$ we have

$$(t_k^S - t_l^S)^2 \leq \mathbf{N} \sum_{\Pi_{ij} \in C_m} (t_i^S - t_j^S)^2. \tag{5.11.2}$$

Using the estimates (5.11.1), (5.11.2) and (5.9.13) in Lemma 5.21 we have

$$A \leq \frac{1}{2} \sum_{\Pi_{ij} \in \mathbb{P}} \left[g_{ij}^S + \frac{|\ln(1-\beta)| + \pi + \ln 2}{6} \right] (t_i^S - t_j^S)^2$$

$$+ \frac{1}{2} \frac{4}{\sqrt{1-\beta^2}} \sum_{\Pi_{ij} \in \mathbb{P}} (t_i^S - t_j^S)^2$$

which, using Lemma 5.18 and equation (5.5.1), gives

$$A \leq \frac{1}{2} \sum_{\Pi_{ij}, e_{ij} \notin \mathbb{P}_\delta} g_{ij}^S (t_i^S - t_j^S)^2 + \frac{1}{2} \sum_{\Pi_{ij}, e_{ij} \in \mathbb{P}_\delta} (t_i^S - t_j^S)^2$$

$$\times \left[g_{ij}^S + \mathbf{N} \left(\mathbf{N} + \frac{2}{\pi\sqrt{3}} \mathbf{N}^2 \right) \left(2 \frac{4}{\sqrt{1-\beta^2}} + 3 \frac{|\ln(1-\beta)| + \pi + \ln 2}{6} \right) \right]$$

$$\leq \frac{1}{2} \sum_{\Pi_{ij}, e_{ij} \notin \mathbb{P}_\delta} g_{ij}^S (t_i^S - t_j^S)^2 + \frac{1}{2} \sum_{\Pi_{ij}, e_{ij} \in \mathbb{P}_\delta} g_{ij}^S \left(1 + 2.56 \mathbf{N}^4 \sqrt{\frac{\delta}{R}} \right) (t_i^S - t_j^S)^2$$

$$\leq \left(1 + 2.56 \mathbf{N}^4 \sqrt{\frac{\delta}{R}} \right) \frac{1}{2} \sum_{\Pi_{ij} \in \mathbb{P}} g_{ij}^S (t_i^S - t_j^S)^2 = \left(1 + 2.56 \mathbf{N}^4 \sqrt{\frac{\delta}{R}} \right) E_d^S.$$

Here, $2.56 \mathbf{N}^4$ is a uniform upper bound on

$$\sqrt{\frac{R}{\delta}} \frac{1}{g_{ij}^S} \left(\mathbf{N} + \frac{2}{\pi\sqrt{3}} \mathbf{N}^2 \right) \left(2 \frac{4}{\sqrt{1-\beta^2}} + 3 \frac{|\ln(1-\beta)| + \pi + \ln 2}{6} \right).$$

Therefore,

$$E_d^S \leq A \leq \left(1 + 2.56 \mathbf{N}^4 \sqrt{\frac{\delta}{R}} \right) E_d^S \qquad (5.11.3)$$

or

$$0 \leq A - E_d^S \leq 2.56 \mathbf{N}^4 \sqrt{\frac{\delta}{R}} E_d^S.$$

Hence,

$$\frac{|A - E_d^S|}{E_d^S} \leq 2.56 \mathbf{N}^4 \sqrt{\frac{\delta}{R}}. \qquad \square$$

Lemma 5.4 *Under the δ-$\mathbf{N}$ close packing condition for $\delta \leq \dfrac{R}{32}$ the relative error is*

$$\frac{|A - E_d|}{E_d} \leq 9.82 \mathbf{N}^4 \sqrt{\frac{\delta}{R}},$$

where the effective conductivity is

$$A = \frac{1}{2} \int_{Q_m} |\nabla \phi(\mathbf{x})|^2 dx = \frac{1}{2} \min_{\tilde{\phi} \in V_p} \int_{Q_m} |\nabla \tilde{\phi}(\mathbf{x})|^2 dx$$

and the energy of the $\mathbb{P}$-model is

$$E_d = \frac{1}{2} \min_{\tilde{\mathbf{t}}} \sum_{\Pi_{ij} \in \mathbb{P}} g_{ij} (\tilde{t}_i - \tilde{t}_j)^2.$$

Proof Since $g_{ij}^S \le g_{ij}$, see (5.2.14), we have

$$\frac{1}{2} \sum_{\Pi_{ij} \in \mathbb{P}} g_{ij}^S (t_i^S - t_j^S)^2 = E_d^S \le E_d.$$

Also, we have

$$E_d^S \le \frac{1}{2} \sum_{\Pi_{ij} \in \mathbb{P}} g_{ij}^S (t_i^S - t_j^S)^2 + \frac{1}{2} \sum_{\Pi_{ij}} |g_{ij}^S - g_{ij}| (t_i^S - t_j^S)^2. \tag{5.11.4}$$

The length of the largest neck δ_{ij} can be bounded by means of the maximal size of the minimal cycles of the δ-subgraph $\mathcal{G}_\delta$. We have

$$\delta_{ij} \le \mathbf{N} \left(R + \frac{\delta}{2} \right) \le \frac{5}{4} \mathbf{N} R. \tag{5.11.5}$$

Similar to the proof of (5.10.1), by (5.5.7), (5.5.1), (5.11.5) and (5.2.14) inequality (5.11.4) gives

$$E_d^S \le \frac{1}{2} \sum_{\Pi_{ij}, e_{ij} \notin \mathbb{P}_\delta} g_{ij}^S (t_i^S - t_j^S)^2 + \frac{1}{2} \sum_{\Pi_{ij}, e_{ij} \in \mathbb{P}_\delta} (t_i^S - t_j^S)^2$$

$$\times \left[g_{ij}^S + 3\mathbf{N} \left(\mathbf{N} + \frac{2}{\pi \sqrt{3}} \mathbf{N}^2 \right) \left(\pi + \pi \frac{3}{2\sqrt{2}} \sqrt{\frac{5}{4}} \mathbf{N} + 6 \right) \right]$$

$$\le \frac{1}{2} \sum_{\Pi_{ij}, e_{ij} \notin \mathbb{P}_\delta} g_{ij}^S (t_i^S - t_j^S)^2 + \frac{1}{2} \sum_{\Pi_{ij}, e_{ij} \in \mathbb{P}_\delta} g_{ij}^S \left(1 + 9.82 \mathbf{N}^4 \sqrt{\frac{\delta}{R}} \right) (t_i^S - t_j^S)^2$$

$$\le \left(1 + 9.82 \mathbf{N}^4 \sqrt{\frac{\delta}{R}} \right) \frac{1}{2} \sum_{\Pi_{ij} \in \mathbb{P}} g_{ij}^S (t_i^S - t_j^S)^2 = \left(1 + 9.82 \mathbf{N}^4 \sqrt{\frac{\delta}{R}} \right) E_d^S.$$

Therefore

$$E_d \le \left(1 + 9.82 \mathbf{N}^4 \sqrt{\frac{\delta}{R}} \right) E_d^S. \tag{5.11.6}$$

By (5.11.4), (5.11.6) and (5.11.3)

$$|A - E_d| \leq \max(9.82\mathbf{N}^4, 2.56\mathbf{N}^4)\sqrt{\frac{\delta}{R}}E_d \leq 9.82\mathbf{N}^4\sqrt{\frac{\delta}{R}}E_d.$$

Hence

$$\frac{|A - E_d|}{E_d} \leq 9.82\mathbf{N}^4\sqrt{\frac{\delta}{R}}. \qquad \qquad \square$$

Asymptotically as $\delta \to 0$ both models (the $\mathbb{P}$-model and $\mathbb{DP}$-model) give the same values of the effective conductivity, in the sense that

$$E_d^S = O\left(\sqrt{\frac{R}{\delta}}\right), \; |E_d^S - A| = O(1),$$

$$E_d = O\left(\sqrt{\frac{R}{\delta}}\right), \; |E_d - A| = O(1),$$

$$|E_d^S - E_d| = O(1) \text{ as } \delta \to 0.$$

Since in the $\mathbb{DP}$-model we take into account more geometry of the original continuous problem, it may seem that the $\mathbb{DP}$-model gives a better approximation to the effective conductivity than the $\mathbb{P}$-model. However, due to the errors that arise when we integrate perpendicular to the necks (see (5.9.4)) the errors of both models are comparable. More specifically, the $\mathbb{P}$-model approximates the Dirichlet integral in a neck

$$\int_{\Pi_{ij}} |\nabla \phi(\mathbf{x})|^2 d\mathbf{x} = \int_{\Pi_{ij}} \left(\frac{\partial \phi}{\partial x}\right)^2 d\mathbf{x} + \int_{\Pi_{ij}} \left(\frac{\partial \phi}{\partial y}\right)^2 d\mathbf{x} \qquad (5.11.7)$$

by $g_{ij}(t_i - t_j)^2$, where g_{ij} is given by the Keller asymptotic formula (2.5.8). This asymptotic formula neglects the value of the first term on the right-hand side of (5.11.7) and approximates the second term of (5.11.7). On the other hand, in the $\mathbb{DP}$–model we compute the second term of the Dirichlet integral (5.11.7) exactly, but we neglect the first term of it. When $\delta \to 0$ the value of this second term in (5.11.7) is comparable to the error of the Keller approximation (2.5.8) of the first term, hence the $\mathbb{DP}$–model does not provide a better approximation of the effective conductivity than the $\mathbb{P}$–model. This means that for the purpose of estimating the effective conductivity of a composite material where particles are almost touching each other both models are equivalently good; however, the $\mathbb{DP}$–model also provides a rigorously justified lower-sided variational bound on the effective conductivity even if δ is not small, because $E_d \leq A$.

5.12 A posteriori numerical error

The main goal of this section is to give a rigorous quantitative justification of the discrete network model (5.2.8) by means of a priori error estimates. However, the use of our trial functions for the upper- and lower-sided bounds gives a numerical a posteriori error. We must solve the discrete network problem (5.2.16), construct the trial functions $\phi \in V_p$ and $\mathbf{v} \in W_p^0$ and evaluate explicitly the left-hand side and the right-hand side of the upper- and lower-sided bound (5.1.12).

The evaluation of these dual bounds is not computationally expensive, because we use simple trial functions – they are given by explicit analytic formulas on the necks by the formula (5.9.1), along with

$$
\begin{aligned}
g_{ij}^S &= \int_{S_1}^{S_2} \frac{dx}{H(x)} \\
&= \left[-\alpha + 2\frac{\delta_{ij} + R}{\sqrt{\delta_{ij}^2 + 2R\delta_{ij}}} \arctan\left(\frac{\sqrt{\delta_{ij}^2 + 2R\delta_{ij}} \tan\left(\frac{\alpha}{2}\right)}{\delta_{ij}} \right) \right]_{\arcsin(S_1/R)}^{\arcsin(S_2/R)}.
\end{aligned}
$$

and they are linear interpolations on the triangles.

Also the use of the a posteriori error widens the range of the characteristic distance δ, where the discrete network gives a good approximation. This section implements this idea for numerical simulations of a randomized hexagonal lattice. Our numerical experiments consist of three parts: numerical simulations of a randomized hexagonal distribution of disks; numerical evaluations of the dual variational bounds; and statistics on the data.

The distribution of disks is implemented by randomization of a periodic hexagonal lattice of disks of equal radii $R_i = R = 0.02$ on a square domain $\Pi = [-1, 1] \times [-1, 1]$ with a volume fraction V, and then removal of some fraction V_r of these disks from this distribution. For fixed V and V_r this algorithm creates a distribution of disks with the volume fraction $V_0 = V - V_r$.

For a given random distribution ω of disks we compute E_d^S (formula (5.2.16)), the energy of the discrete network. After the energy $E_d^S(\omega)$ is computed, we also construct the trial function $\phi(\mathbf{x}, \omega)$ for the upper-sided bound as in the previous section (we compute $\phi(\mathbf{x}, \omega)$ in the necks in accordance with (5.9.1) and then continue it into triangles), and then we compute

$$
I(\omega) = \frac{1}{2} \int_{Q_m} |\nabla \phi(\mathbf{x})|^2 d\mathbf{x}
$$

for this trial function. The argument ω indicates that the function (the quantity) depends on the random distribution of disks. The set of all random distributions is denoted by Ω. Therefore we have in accordance with (5.7.1) and (5.8.3) $E_d^S(\omega) \leq A \leq I(\omega)$. Hence $E_d^S(\omega)$ and $I(\omega)$ are a posteriori lower- and upper-sided bounds,

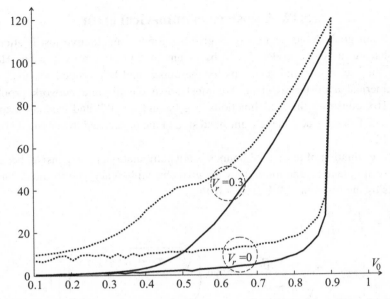

Figure 5.15 $\mathbb{E}E_d^S$ (solid lines) and $\mathbb{E}I$ (dotted lines) as functions of the volume fraction $V_0 = 0.105, \ldots, 0.905$. The lower two lines are for the case with no holes ($V_r = 0$); the upper two lines are for the case with holes ($V_r = 0.3$).

respectively, for the effective conductivity of a composite material with a given distribution ω of disks.

The simulations are done for various values of V_r incremented by 0.05. For fixed V and V_r there were $|\Omega| = 80$ simulations (random generations of the disk distribution). For the expected value we use the notation:

$$\mathbb{E}I = \frac{1}{|\Omega|} \sum_{k=1}^{n} I(\omega),$$

where $I(\omega)$ is the result of the simulation $\omega \in \Omega$ with fixed V and V_r and the number of simulations is $|\Omega| = 80$. The expected value of $E_d^S(\omega)$ is

$$\mathbb{E}E_d^S = \frac{1}{|\Omega|} \sum_{k=1}^{n} E_d^S(\omega),$$

Here we present the results of the numerical simulations that show the dependence of the effective conductivity on the presence of holes in the matrix.

In Figure 5.15 we plot $\mathbb{E}E_d^S$ (solid lines) and $\mathbb{E}I$ (dotted lines) as functions of the volume fraction $V_0 = 0.105, \ldots, 0.905$.

Note that $V_0 = 0.905$ is very close to the densest possible (hexagonal) packing. Indeed, the volume fraction of the periodic hexagonal packing when disks touch is $\pi \dfrac{\sqrt{3}}{6} \approx 0.9068 \dots$.

The volume fraction of removed disks V_r takes two values: $V_r = 0$ and $V_r = 0.3$. If the total volume fraction of the particles is fixed, then the increase of the volume fraction of holes in the material implies that the interparticle distance δ decreases. Hence percolation effects play a more significant role. Therefore we expect that for highly packed composite materials the relative error of our discrete model is smaller in the case when there are holes compared with the case when there are no holes. Indeed, when $V_0 > 0.6$ the relative a posteriori error for composite materials with holes is up to eight times better than the relative error for composite materials with no holes for the same volume fraction.

When $V_0 \leq 0.35$ the effective conductivities of a material with holes and of a material without holes are numerically very close, at least the computed a posteriori error does not allow one to distinguish between these composite materials. For such volume fractions there are no percolation effects in both cases.

Observe that in the presence of holes the a posteriori error of the network approximation $E_{\mathrm{apost}} = \mathbb{E}I - \mathbb{E}E_d^S$ is significantly larger than in the case when there are no holes. For example, when $V_0 = 0.5$ in the presence of holes the a posteriori error is 3.5 times larger than the error in the case when there are no holes; however, for the same volume fraction of particles the relative a posteriori error estimate is better in the presence of holes. Additionally, when $V_0 = 0.6$, the relative a posteriori error estimate in the presence of holes in the conducting cluster is up to eight times better than for uniformly highly packed composite materials. Hence, we observe numerically that our network approximation works better for irregular geometric patterns, which are not quasi-hexagonal, that is when the typical number of nearest neighbors can vary significantly.

We emphasize that the network approximation presented in this section is no longer asymptotic in nature. We decompose the Dirichlet integral (5.1.8) for the effective conductivity into two parts: the network approximation, which is a quadratic form, and the error term. The network approximation accounts for all fluxes between the neighboring particles, where neighbors are defined via Voronoi tessellation. If the fluxes are small (neighbors are not closely spaced), then the corresponding coefficients in the modified network approximation g_{ij}^S are not significant but, unlike in Section 3.1, we do not need to introduce any cut-off distance in the numerical implementation.

The approach developed in this section allowed us to consider generic geometrical arrays of particles which satisfy the δ-**N** close packing condition. This condition allows for strongly non-uniform geometrical arrays when a significant fraction of the particles does not participate in the conducting cluster. We have shown numerically how such non-uniformity affects the error estimate and therefore the quality of the approximation. We observe that irregularity in the geometrical distribution of the particles in all simulations consistently lead to a significant (up to 10 times) increase in the effective conductivity at the same total volume fraction of the particles.

Thus we conclude that the approximation proposed in this section provides a very efficient computational tool for evaluation of the effective properties of high-contrast composite materials, which is capable of capturing the effects of irregular geometrical arrays with good control of the approximation error.

6

Network method for nonlinear composites

In this chapter we explain how to construct a discrete network for nonlinear high-contrast densely-packed composites. We use this presentation to demonstrate the so-called perforated medium technique in the analysis of high-contrast composites. We also use this investigation to demonstrate how the discrete network approximation arises from the interplay between geometry and asymptotic analysis. More specifically, the key mathematical feature of partial differential equations that describe high-contrast densely-packed composites is that their solutions exhibit asymptotically singular behavior, when particles are close to touching (high concentration). The singularities of the solutions occur exactly in the necks between almost touching particles. The location of these singularities can be characterized naturally by the geometric patterns of the distribution of the particles in the materials. Thus a geometric construction of a network is completely natural. As it is illustrated in this book, a rigorous mathematical justification is based on geometric and asymptotic arguments. These two arguments are coupled together. As a result, most of the constructions of asymptotic discrete network approximations for high-contrast composites are complicated, thus they are not attractive for practitioners. It is possible, however, to separate the geometric and the asymptotic arguments. It makes the construction of the network more transparent, and allows us to strengthen some of the previous results. In particular, it turns out that the validity of discrete network approximations could be verified for a class of composites, which is larger than the one that satisfies the δ-N close-packing condition.

The methods presented in this chapter are still based on the variational approach and the constructions of the dual asymptotically tight bounds. In the construction of these upper and lower bounds we make three observations, which allow this construction to be less technical. The first observation is the idea of a *perforated medium*. A perforated medium is *not* a discrete network, but there is a connection: we can interpret the discrete network approximation as an equivalence relation between the effective properties of two composites – the original one and the one that has inclusions and necks only. We may call such a medium perforated, because we replace parts of the material in the matrix by voids. The second observation is directly related to the first: if the voids are so big that each neck touches only two inclusions, then we do not need to account for any of the "non-local" conditions (e.g., condition (3.6.3)) when we construct the trial function for the lower bound. These conditions will be satisfied automatically as a consequence of an iterative minimization lemma. This lemma provides a simple construction of this trial function for the lower bound, and it is the key reason to introduce the perforated medium. The last observation is related to an extension of the trial field for the upper bound. This construction was shown in this book twice: in Sections 3.6.6–3.6.10 and in Section 5.8. In both cases the trial field is constructed in the necks, and then it is extended to the rest of the matrix, e.g., into triangles. It is unnecessary to make this extension as explicit as was done in Sections 3.6.6–3.6.10 and in Section 5.8. One can use results about optimal Lipschitz extensions that guarantee a uniform upper bound on the gradient of the trial function outside the necks in terms of an estimate of this gradient at the lateral boundaries of the necks.

As in the rest of this book, we will discuss a two-dimensional composite. The results in three dimensions, however, are completely analogous to the two-dimensional case, and they can be found in Novikov (2009).

6.1 Formulation of the mathematical model

As above, we will treat the problem under consideration as an electric conductivity problem. But now, we consider a composite with a nonlinear matrix. We geometric characteristics of our composite are the same as in Section 3.1.1. As before, the domain $\Pi = [-L, L] \times [-1, 1]$ (see Figure 3.1) contains inclusions $\{D_i, i = 1, ..., N\}$ and the matrix $Q_m = \Pi \setminus \bigcup_{i=1}^{N} D_i$.

The matrix is a nonlinear homogeneous medium, and we will assume that the current–electric field relation is the power law:

$$\mathbf{J} = |\mathbf{E}|^{p-2}\mathbf{E}, \quad p = 3, 4, 5, \ldots. \tag{6.1.1}$$

The electric field is determined by a potential $\mathbf{E} = \nabla\phi(\mathbf{x})$, and the current field is divergence-free, $\mathrm{div}\,\mathbf{J} = 0$. So in the matrix the electric potential satisfies

$$\mathrm{div}(|\nabla\phi|^{p-2}\nabla\phi) = 0. \tag{6.1.2}$$

The inclusions are assumed to be ideally conducting (3.2.4). There are no sinks or sources inside the disks, which amounts to

$$\int_{\partial D_i} \mathbf{J}(\mathbf{x})\mathbf{n}dx = \int_{\partial D_i} |\mathbf{E}(\mathbf{x})|^{p-2}\mathbf{E}(\mathbf{x})\mathbf{n}dx = \int_{\partial D_i} |\nabla\phi(\mathbf{x})|^{p-2}\frac{\partial\phi}{\partial\mathbf{n}}(\mathbf{x})dx = 0.$$
$$\tag{6.1.3}$$

As it is natural, when bulk effective properties are discussed, we impose the constant applied field boundary conditions: Dirichlet boundary conditions on the top and the bottom boundaries (3.2.6), and Neumann boundary conditions on the lateral boundaries (3.2.7). As earlier, we will analyze the total flux (3.1.15). Computations identical to those in Section 3.1.2 show that for the power-law nonlinearity the identity (3.1.13) holds with $a(\mathbf{x}) = |\nabla \phi(\mathbf{x})|^{p-2}$. Therefore the total flux A through the composite in the Oy-direction is

$$A = \int_{\partial Q^+} \mathbf{J}(\mathbf{x}) \mathbf{n} d\mathbf{x} = \frac{1}{2} \int_{Q_m} \mathbf{J}(\mathbf{x}) \nabla \phi(\mathbf{x}) d\mathbf{x} = \frac{1}{2} \int_{Q_m} |\nabla \phi(\mathbf{x})|^p d\mathbf{x}. \qquad (6.1.4)$$

Similar to the linear case, it is well-known Jikov *et al.* (1994) that the problem (6.1.2)–(6.1.4) with (3.2.4), (3.2.6), (3.2.7), admits a variational formulation. Let the potential $\phi(\mathbf{x})$ be a function from the functional space:

$$V = \{\phi \in W^{1,p}(Q_m) \ : \ \phi(\mathbf{x}) = t_i^{\phi} \in \mathbb{R} \text{ on } D_i, \ i = 1, ..., N; \qquad (6.1.5)$$
$$\phi(\mathbf{x}) = 1 \text{ on } \partial Q^+, \phi(\mathbf{x}) = -1 \text{ on } \partial Q^-\}$$

and define an energy functional:

$$I(\phi) = \frac{1}{2} \int_{Q_m} |\nabla \phi(\mathbf{x})|^p d\mathbf{x}, \ \phi \in V. \qquad (6.1.6)$$

Then the flux (6.1.4) is the minimum of this energy functional:

$$A = \min_{\widetilde{\phi} \in V} I(\widetilde{\phi}); \qquad (6.1.7)$$

moreover, the minimizer ϕ of (6.1.7) satisfies (3.2.4), (3.2.6) because $\phi \in V$, and (6.1.2), (6.1.3), (3.2.7) because these are the Euler–Lagrange equations of the variational formulation (6.1.7).

6.2 A two-step construction of the network

The basic steps in the construction of the network are the same as in Chapter 3. More specifically, the discrete network is the Delaunay triangulation, associated with the centers of the inclusions: the vertices are the centers of the inclusions and the edges are the edges of the Delaunay triangulation. In addition to N vertices which correspond to inclusions we assign the vertex $\mathbf{x}_{N+1}$ to the whole upper boundary and $\mathbf{x}_{N+2}$ to the whole lower boundary, respectively. This representation by a network clearly shows geometric patterns in a particulate medium, for example clusters of particles. In order to define the discrete energy we assign a potential t_i to each vertex $\mathbf{x}_i$ and assign the local energy $g_{ij} = g_{ij}(\Pi_{ij}, t_i, t_j)$ to each edge Π_{ij}. Then the discrete energy is

$$A_d = \frac{1}{2} \min_{t_i, i=1,...,N} \sum_{\Pi_{ij}} g_{ij}|t_i - t_j|^p, \ t_{N+1} = 1, \ t_{N+2} = -1, \qquad (6.2.1)$$

where the summation is over all edges Π_{ij}.

The justification that the total flux A (6.1.4) is well-approximated by A_d will be done in two steps. In the first step we show that we can approximate the effective conductivity of our composite by the effective conductivity of a composite that has only inclusions and necks between them. Therefore we may call this the domain partitioning step. Our second step is to approximate the effective conductivity in the necks. We may call this step asymptotic, because we only use here the smallness of δ, the distance between inclusions.

6.2.1 The domain partitioning step

In the domain partitioning step we show that a good approximation to the effective conductivity arises if take into account only nearest-neighbor pairwise interactions. Mathematically this means that it suffices to take into account only the interactions of inclusions that are connected by an edge in the Delaunay triangulation, as is done in (6.2.1). Our first theorem shows the error of such an approximation.

Theorem 6.3 *The weights g_{ij} in (6.2.1) could be chosen so that the flux A (6.1.4) could be approximated by the discrete energy A_d (6.2.1) in the sense that*

$$\frac{1}{2} \min_{t_i, i=1,...,N} \sum_{\Pi_{ij}} g_{ij}|t_i - t_j|^p \leq A \leq \frac{1}{2} \min_{t_i, i=1,...,N} \sum_{\Pi_{ij}} (g_{ij} + C)|t_i - t_j|^p, \quad (6.2.2)$$

where $t_{N+1} = 1$, $t_{N+2} = -1$, and the constant $C = C(R, L)$ depends on the radius of the inclusions R, and the size of the composite L.

We will prove Theorem 6.3 using variational upper and lower bounds. The choice of g_{ij} in (6.2.2) is not unique, because a sufficiently small constant could be subtracted from g_{ij}, and (6.2.2) will still hold. Our choice of g_{ij} has a natural physical interpretation in terms of perforated media. This logic behind our choice of g_{ij} is as follows. As it follows from the asymptotic analysis in this chapter, for power-law media with $p \geq 3$ the dominant contribution to the flux A comes from areas that could be characterized geometrically. Similarly to the linear case discussed in Chapter 3, these areas are necks between closely spaced disks (the shaded regions in Figure 5.6). This means in our set-up:

$$A = \frac{1}{2} \int_{Q_m} |\nabla \phi(\mathbf{x})|^p d\mathbf{x} \approx \frac{1}{2} \sum_{\Pi_{ij}} \int_{\Pi_{ij}} |\nabla \phi(\mathbf{x})|^p d\mathbf{x}, \quad (6.2.3)$$

where Π_{ij} is the neck between the i-th and j-th disk (see Figure 5.6). Consider a *perforated medium*, a composite whose matrix has holes, with corresponding perforated domain

$$\Pi_o = Q_m \setminus \bigcup_{i,j,k} \Delta_{ikj} = \bigcup_{i,j} \Pi_{ij}, \quad (6.2.4)$$

where Δ_{ikj} are triangles in the triangle–neck partition (see Figure 5.6). Then (6.2.3) could be rewritten as

$$A = \frac{1}{2} \int_{Q_m} |\nabla\phi(\mathbf{x})|^p d\mathbf{x} \approx \frac{1}{2} \int_{\Pi_o} |\nabla\phi(\mathbf{x})|^p d\mathbf{x}. \tag{6.2.5}$$

In a perforated medium we require that the necks Π_{ij} do not overlap: if $j \neq k$ then $\Pi_{ij} \cap \Pi_{ik} = \emptyset$. We can construct Π_{ij} using, e.g., the triangle–neck partition, presented in Section 5.2.1. The stringent conditions of the triangle–neck partition can, however, be relaxed. We will only need to have the necks to be wide enough. This is due to the fact that we simplify the construction of trial functions outside of the necks using results about optimal Lipschitz extensions. These results do not require any explicit construction of trial functions and we discuss them in more detail in Section 6.3.3.

Suppose ϕ_o is the minimizer of the functional

$$\min_{\phi \in V_o} \int_{\Pi_o} |\nabla\phi(\mathbf{x})|^p d\mathbf{x}, \tag{6.2.6}$$

with the space

$$V_o = \left\{ \phi \in W^{1,p}(\Pi_o) \mid \phi(\mathbf{x}) = t_i^\phi \text{ on } \partial D_i, \ \phi(x, \pm 1) = \pm 1 \right\}. \tag{6.2.7}$$

We always have

$$A = \frac{1}{2} \int_{\Pi_o} |\nabla\phi(\mathbf{x})|^p d\mathbf{x} \geq A_d = \frac{1}{2} \int_{\Pi_o} |\nabla\phi_o(\mathbf{x})|^p d\mathbf{x}. \tag{6.2.8}$$

The quantity on the right-hand side of (6.2.8) has a natural physical interpretation: it is the total electric flux for the perforated medium. Thus on the one hand inequality (6.2.8) is physically intuitive: the total flux for a composite is always larger than the same composite, perforated by voids. On the other hand, (6.2.8) provides an easy and useful lower bound for A. Indeed, since the necks do not overlap, the global minimization problem over Π_o can be split into two consecutive problems: one of them is on a single neck Π_{ij}, and the other one is a minimization problem of discrete variables t_i. This observation is the key step in proving Theorem 6.3, and we state it in more detail in the next lemma.

Lemma 6.4 (Iterative minimization lemma). *Consider a high-contrast power-law composite. Let Π_o be its perforated domain. Then*

$$A_d = \frac{1}{2} \min_{\phi \in V_o} \int_{\Pi_o} |\nabla\phi(\mathbf{x})|^p d\mathbf{x} = \frac{1}{2} \min_{t_i, i=1,\ldots,N} \sum_{\Pi_{ij}} g_{ij} |t_i - t_j|^p, \tag{6.2.9}$$

with $t_{N+1} = 1, t_{N+2} = -1$, where

$$g_{ij} = \min_{\phi \in V_{\Pi_{ij}}} \int_{\Pi_{ij}} |\nabla\phi(\mathbf{x})|^p d\mathbf{x}, \ \phi \in V_{\Pi_{ij}}, \tag{6.2.10}$$

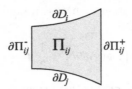

Figure 6.1 Neck Π_{ij} and its boundary.

with the space

$$V_{\Pi_{ij}} = \left\{ \phi \in W^{1,p}(\Pi_{ij}) \mid \phi(\mathbf{x}) = 1/2 \text{ } on \text{ } \partial D_i, \phi(\mathbf{x}) = -1/2 \text{ } on \text{ } \partial D_i \right\}. \quad (6.2.11)$$

Moreover the minimizer ϕ_{ij} of (6.2.10) satisfies

$$\text{div}(|\nabla \phi_{ij}|^{p-2} \nabla \phi_{ij}) = 0 \text{ } in \text{ } \Pi_{ij}, \quad (6.2.12)$$

$$\frac{\partial \phi_{ij}}{\partial \mathbf{n}}(\mathbf{x}) = 0 \text{ } on \text{ } \partial \Pi_{ij}^{\pm},$$

where $\partial \Pi_{ij}^{\pm}$ are the "lateral" boundaries of a neck as depicted on Figure 6.1.

For clarity of presentation we put all the technical details of the proof of Theorem 6.3 in Section 6.3. In particular, we prove Lemma 6.4 in Section 6.3.1. In this lemma we define g_{ij} by (6.2.10). These are exactly the same g_{ij} as in (6.2.2) in Theorem 6.3. Hence Theorem 6.3 states that the total flux in the original domain is approximately the same as the total flux in the perforated domain. Further, the first inequality in (6.2.2) is simply a reformulation of (6.2.8) using (6.2.10). Thus only the second inequality in (6.2.2) requires work. The idea of the proof of the second inequality in (6.2.2) is to extend ϕ_o, the minimizer of (6.2.6) outside the necks and use it as a trial function for the upper variational bound. This will be done in Section 6.3.3 using the optimal Lipschitz extensions. A preliminary step for these extensions is to obtain a $W^{1,\infty}$-bound on ϕ_0 on $\partial \Pi_{ij}^{\pm}$, the lateral boundaries of the necks. Since $\phi_o(\mathbf{x}) = (t_i + t_j) + (t_j - t_i)\phi_{ij}(\mathbf{x})$ on Π_{ij}, we need to obtain $W^{1,\infty}$ estimates for ϕ_{ij}, and this is done in Section 6.3.2.

We emphasize the key distinction between our construction here and those presented in Chapters 3 and 5: in our domain partitioning step we *do not use the asymptotic parameter δ* when we construct our trial functions. As a consequence, we *will not* use the dual variational principle in the proof of Theorem 6.3.

6.2.2 The asymptotic step

Theorem 6.3 says that if we take into account only the effect of pairwise inter-actions between inclusions, then we make an error that is independent of the characteristic interparticle distance. This theorem is therefore ineffective, if the total energy (flux) A is itself of the order of this error. The effective conductiv-ity, however, becomes large when the characteristic interparticle distance is small, because the weights g_{ij} in (6.2.2) tend to ∞ as δ_{ij}, the distance between the i-th and the j-th inclusions, tends to zero. The next lemma provides an explicit asymptotic behavior of g_{ij}.

Lemma 6.5 (**asymptotics of** g_{ij}) *Suppose the weights g_{ij} are chosen as in* (6.2.10). *Consider any two closely spaced inclusions D_i and D_j, as depicted in Figure 5.8. Suppose the interparticle distance between them is δ_{ij}. Then*

$$\frac{g_{ij} - g_{ij}^o}{g_{ij}^o} \to 0, \ \text{as } \delta_{ij} \to 0,$$

with the constants

$$g_{ij}^o = \mathbf{G}(\delta_{ij}) = \frac{\pi a_p}{R^{p-2}} \left(\frac{R}{\delta_{ij}} \right)^{p-3/2}, \tag{6.2.13}$$

and a_p are the coefficients of the Taylor series

$$\frac{1}{\sqrt{1-x}} = 1 + \frac{x}{2} + \frac{3x^2}{8} + \cdots + a_p x^{p-2} + \cdots \tag{6.2.14}$$

For a neck Π_{ij} between an inclusion B_i and the flat boundary $j = N$ or $j = N + 1$, g_{ij}^o is given by

$$g_{ij}^o = 2^{p-1} \mathbf{G}(2\delta_{ij}). \tag{6.2.15}$$

Using (6.2.14), we find that, e.g.,

$$g_{ij} = \begin{cases} \dfrac{\pi}{2R} \left(\dfrac{R}{\delta_{ij}} \right)^{3/2}, & p = 3, \\[3mm] \dfrac{3\pi}{8R^2} \left(\dfrac{R}{\delta_{ij}} \right)^{5/2}, & p = 4. \end{cases}$$

The proof of Lemma 6.5 uses dual variational principles, because we will need here to construct explicitly trial functions which provide tight bounds as $\delta \to 0$. The *asymptotic argument* relies on Lemma 6.5, and it allows us to conclude that if, as in Chapter 3, all interparticle distances are sufficiently small, then A_d approximates A, as described in the following theorem. It turns out that it is useful to modify the direct and dual variational principles in order to avoid boundary integrals. This technical novelty is presented below in (6.4.11).

Theorem 6.6 *Consider a high-contrast power-law composite. Suppose the distribution of inclusions satisfies the close-packing condition: for all Π_{ij} we have that $\delta_{ij} \leq \delta$. Let the weights g_{ij} be chosen as in* (6.2.10). *Then*

$$\frac{|A - A_d|}{A_d} \leq C\delta^{p-3/2}, \tag{6.2.16}$$

where the constant C may depend on R and L only.

A few remarks are in order about Theorem 6.6. Firstly, the result estimates the relative error of two asymptotically large quantities: the effective conductivity and the energy of the discrete network. The latter is a simple polynomial. Its evaluation is simpler, especially in three dimensions, than the evaluation of (6.1.4). This fact is the main attractive feature of the discrete network approximation. Secondly, for composites with a nonlinear constitutive relation, the constants g_{ij} may depend on the values of t_i and t_j. In our case g_{ij} can be chosen *independent* of t_i. Finally, the main advantage of the discrete network approximation can be seen when the concentration of inclusions is high; for moderate concentrations the discrete network approximation should be used together with other methods of characterization of the effective properties of composites.

If we use g_{ij}^o from (6.2.13) instead of g_{ij} in the functional (6.2.1), then with this choice we will obtain worse error estimates, which are summarized in the next corollary.

Corollary 6.7 *Consider a high-contrast power-law composite. Suppose for all* Π_{ij} *we have that* $\delta_{ij} \leq \delta$. *Let*

$$A_d^o = \frac{1}{2} \min_{t_i, i=1,\dots,N} \sum_{\Pi_{ij}} g_{ij}^o |t_i - t_j|^p,$$

where the weights g_{ij}^o *are chosen as in* (6.2.13). *Then*

$$\frac{|A - A_d^o|}{A_d^o} \leq C\delta, \tag{6.2.17}$$

where the constant C may depend on R and L only.

Note that g_{ij}^o in (6.2.13) are only the leading asymptotic terms of g_{ij}, as $\delta_{ij} \to 0$. All the singular terms of g_{ij} are not computed here. This is not a drawback of the method; on the contrary it is the main advantage of our two-step construction of the discrete network. Indeed, if we need to obtain better accuracy of the discrete network approximation, we only need to improve the results in Lemma 6.5 by calculating other singular terms in the asymptotic expansion of g_{ij}. Once other singular terms of g_{ij} are found, we immediately obtain an improved Theorem 6.6 and Corollary 6.7.

Further corollaries from Theorem 6.6 could be obtained, when we modify the close-packing condition, and use the δ–N close-packing condition, discussed in Chapter 5 (see, e.g., Figure 5.1). Our proof of Theorem 6.6 allows us to avoid working with the δ–N close-packing condition and obtain a result which is more general; it is applicable in three dimensions, and, at the same time, it has a natural physical interpretation. The result states that Theorem 6.6 and Corollary 6.7 hold for discrete networks that contain δ-percolation clusters. We state it in more detail in the next theorem, which we precede with the following definition.

Definition 6.8 We say that a discrete network with vertices $\mathbf{x}_i$, $i = 1, \ldots, N$, contains a δ-percolation cluster, if there is a sequence of vertices connecting the top and the bottom boundary of the domain, such that their consecutive interparticle distances are less than δ. More specifically, recall that the bottom boundary is indexed by $N + 1$, and the top boundary is indexed by $N + 2$. Then there exist vertices $\mathbf{x}_{i_1}, \mathbf{x}_{i_2}, \ldots, \mathbf{x}_{i_k}$, such that $\delta_{N+1,i_1} \leq \delta$, $\delta_{k_1,N+2} \leq \delta$, and $\delta_{i_j,i_j+1} \leq \delta$, $j = 1, 2, \ldots, k - 1$.

Theorem 6.9 *Consider a high-contrast power-law composite. Suppose the distribution of inclusions satisfies the close-packing condition: for all Π_{ij} we have that $\delta_{ij} \leq \delta$. Let the weights g_{ij}, g_{ij}^o be chosen as in (6.2.10) and (6.2.13), respectively. Then*

$$\frac{|A - A_d|}{A_d} \leq C\delta^{p-3/2}, \tag{6.2.18}$$

and

$$\frac{|A - A_d^o|}{A_d^o} \leq C\delta, \tag{6.2.19}$$

where the constant C may depend on R and L only.

We will prove Theorem 6.9 only, because Theorem 6.6 and Corollary 6.7 are its easy particular cases. The proof contains two conceptually different arguments: verification of the asymptotics stated in Lemma 6.5, and verification that the existence of a δ-percolation cluster is sufficient for validity of the network approximation. We discuss these two arguments in Sections 6.4.1 and 6.4.2, respectively.

6.3 Proofs for the domain partitioning step

In this section we give proofs related to the first step of our two-step approximation method.

6.3.1 Iterative minimization lemma

Here we prove Lemma 6.4. Once the proof is complete it follows from (6.2.8) that the first inequality in (6.2.2) of Theorem 6.3 is true.

Proof Let

$$I_o(\phi) = \frac{1}{2} \int_{\Pi_o} |\nabla \phi(\mathbf{x})|^p d\mathbf{x}, \quad \phi \in V_o, \tag{6.3.1}$$

where

$$V_o = \left\{ \phi \in W^{1,p}(\Pi_o) \,|\, \phi(\mathbf{x}) = t_i \text{ on } \partial D_i, \, \phi(x, \pm 1) = \pm 1 \right\}. \tag{6.3.2}$$

The function $f(x) = |x|^p$ is convex, and $f(x) \to \infty$ as $|x| \to \infty$. Hence (Ladyzhenskaya and Ural'tseva, 1968) there is a unique ϕ_o, the minimizer of (6.3.1). We start by showing that ϕ_o, the minimizer of (6.3.1), is the solution of the corresponding Euler–Lagrange equations

$$\text{div}(|\nabla\phi_o|^{p-2}\nabla\phi_o) = 0 \text{ in } \Pi_o \qquad (6.3.3)$$

$$\phi_o(\mathbf{x}) = t_i^o, \text{ on } D_i, i = 1, \ldots, N, \qquad (6.3.4)$$

$$\int_{\partial D_i} |\nabla\phi_o(\mathbf{x})|^{p-2}\frac{\partial\phi_o}{\partial\mathbf{n}}(\mathbf{x})d\mathbf{x} = 0, \qquad (6.3.5)$$

$$\phi_o(x, \pm 1) = \pm 1, \qquad (6.3.6)$$

$$\frac{\partial\phi_o}{\partial\mathbf{n}}(\pm L, y) = 0, \qquad (6.3.7)$$

$$\frac{\partial\phi_o}{\partial\mathbf{n}}(\mathbf{x}) = 0 \text{ on } \Pi_{ij}^{\pm} \ (i, j = 1, 2, \ldots, N). \qquad (6.3.8)$$

As above, we use the notation $\mathbf{x} = (x, y)$.

The convexity of the function $f(x) = |x|^p$ also implies that for any $\psi \in W^{1,p}(\Pi_o)$, $\psi(\mathbf{x}) \not\equiv 0$ with $\psi(\mathbf{x}) = \psi_i$, on D_i $i = 1, \ldots, N$, and $\psi(x, \pm 1) = 0$ we have

$$I_o(\phi_o + \delta\psi) > I_o(\phi_o) + \delta\langle F'(\phi_o), \psi\rangle,$$

where $F'(\phi_o)$ is the Gâteaux differential as in Section 1.4,

$$\langle F'(\phi_o), \psi\rangle = \frac{p}{2}\int_{\Pi_o} |\nabla\phi_o(\mathbf{x})|^{p-2}\nabla\phi_o(\mathbf{x})\nabla\psi(\mathbf{x})d\mathbf{x}.$$

If ϕ_o is a minimizer of $I_o(\phi)$, we must have $\langle F'(\phi_o), \psi\rangle = 0$ for all $\psi \in W^{1,p}(\Pi_o)$. We verify that if ϕ_o solves (6.3.3)–(6.3.8), then $\langle F'(\phi_o), \psi\rangle = 0$ for all $\psi \in W^{1,p}(\Pi_o)$. Indeed

$$\int_{\Pi_o} |\nabla\phi_o(\mathbf{x})|^{p-2}\nabla\phi_o(\mathbf{x})\nabla\psi(\mathbf{x})dx$$

$$= \int_{y=\pm 1} \psi(\mathbf{x})\nabla\phi_o(\mathbf{x})|^{p-2}\frac{\partial\phi_o}{\partial\mathbf{n}}(\mathbf{x})dx$$

$$+ \int_{x=\pm L} \psi(\mathbf{x})|\nabla\phi_o(\mathbf{x})|^{p-2}\frac{\partial\phi_o}{\partial\mathbf{n}}(\mathbf{x})dx$$

$$+ \sum_i \left(\psi_i(\mathbf{x})\int_{\partial D_i} |\nabla\phi_o(\mathbf{x})|^{p-2}\frac{\partial\phi_o}{\partial\mathbf{n}}(\mathbf{x})d\mathbf{x}\right)$$

$$+ \sum_{i,j} \int_{\partial\Pi_{ij}^{\pm}} \psi(\mathbf{x})|\nabla\phi_o(\mathbf{x})|^{p-2}\frac{\partial\phi_o}{\partial\mathbf{n}}(\mathbf{x})dx$$

$$- \sum_{i,j} \int_{\Pi_{ij}} \psi(\mathbf{x})\text{div}(|\nabla\phi_o(\mathbf{x})|^{p-2}\nabla\phi_o(\mathbf{x}))dx.$$

The first term on the right-hand side vanishes, because $\psi(x, \pm 1) = 0$. The others vanish due to conditions (6.3.7), (6.3.5), (6.3.8), (6.3.3), respectively.

Denote

$$\phi_{ij}(\mathbf{x}) = \frac{\phi_o(\mathbf{x}) - (t_i^o + t_j^o)}{t_j^o - t_i^o}.$$

Then ϕ_{ij} solves

$$\operatorname{div}(|\nabla \phi_{ij}|^{p-2} \nabla \phi_{ij}) = 0 \text{ in } \Pi_{ij}, \tag{6.3.9}$$

$$\phi_{ij}(\mathbf{x}) = 1/2 \text{ on } \partial D_i, \quad \phi_{ij}(\mathbf{x}) = -1/2 \text{ on } \partial D_j \tag{6.3.10}$$

$$\frac{\partial \phi_{ij}}{\partial \mathbf{n}}(\mathbf{x}) = 0 \text{ on } \Pi_{ij}^{\pm}. \tag{6.3.11}$$

As in (6.3.9) we verify that ϕ_{ij} is the minimizer of the functional

$$I_{ij}(\phi) = \int_{\Pi_{ij}} |\nabla \phi(\mathbf{x})|^p d\mathbf{x}, \quad \phi \in V_{ij}, \tag{6.3.12}$$

where

$$V_{ij} = \left\{ \phi \in W^{1,p}(\Pi_{ij}) \mid \phi(\mathbf{x})|_{\partial D_i} = 1/2, \ \phi(\mathbf{x})|_{\partial D_j} = -1/2, \right\}. \tag{6.3.13}$$

Therefore

$$\int_{\Pi_{ij}} |\nabla \phi_o(\mathbf{x})|^p d\mathbf{x} = g_{ij} |t_i^o - t_j^o|^p,$$

where

$$g_{ij} = \int_{\Pi_{ij}} |\nabla \phi_{ij}(\mathbf{x})|^p d\mathbf{x} = \min_{\phi \in V_{ij}} I_{ij}(\phi).$$

Thus

$$I_o(\phi_o) = \frac{1}{2} \sum_{\Pi_{ij}} \int_{\Pi_{ij}} |\nabla \phi_{ij}(\mathbf{x})|^p d\mathbf{x} = \frac{1}{2} \sum_{\Pi_{ij}} g_{ij} |t_i^o - t_j^o|^p.$$

We finally observe that the form

$$F(t_1, \ldots, t_N) = \frac{1}{2} \sum_{\Pi_{ij}} g_{ij} |t_i - t_j|^p, \quad t_{N+1} = 1, \ t_{N+2} = -1$$

is convex in each variable t_i, $i = 1, \ldots, N$. Further, $F \to \infty$, as $\sum_i |t_i| \to \infty$.

Thus F has a unique critical point, a global minimum. We claim that $t_1^o, \ldots, t_N^o$ is the minimizer of F, and

$$I_o(\phi_o) = F(t_1^o, \ldots, t_N^o) = \frac{1}{2} \min_{t_i, i=1, \ldots, N} \sum_{\Pi_{ij}} g_{ij} |t_i - t_j|^p,$$

with $t_{N+1} = 1$, $t_{N+2} = -1$.

Indeed, similar to (3.4.11) the necessary and sufficient conditions for the unique critical point,

$$\frac{\partial F}{\partial t_i}(t_1, \ldots, t_N) = 0, i = 1, \ldots, N$$

imply the Kirchhoff equations,

$$\sum_{j \in n_i} g_{ij}(t_i - t_j)|t_i - t_j|^{p-2} = 0, \quad i = 1, \ldots, N, \tag{6.3.14}$$

where n_i are the indices of the neighbors of the vertex $\mathbf{x}_i$. We, therefore, will complete the proof of the lemma once we verify (6.3.14) for $t_1^o, t_2^o, \ldots, t_N^o$. The derivation of (6.3.14) is as follows. From (6.3.5) we derive

$$0 = \int_{\partial D_i} |\nabla \phi_o(\mathbf{x})|^{p-2} \frac{\partial \phi_o}{\partial \mathbf{n}}(\mathbf{x}) d\mathbf{x} = \sum_{j \in n_i} b_{ij}(t_i^o - t_j^o)|t_i^o - t_j^o|^{p-2},$$

where

$$b_{ij} = \int_{\partial D_i \cap \Pi_{ij}} |\nabla \phi_{ij}(\mathbf{x})|^{p-2} \frac{\partial \phi_{ij}}{\partial \mathbf{n}}(\mathbf{x}) d\mathbf{x}.$$

We compute

$$b_{ij} = \int_{\partial D_i \cap \Pi_{ij}} \left(\phi_{ij}(\mathbf{x}) + \frac{1}{2} \right) |\nabla \phi_{ij}(\mathbf{x})|^{p-2} \frac{\partial \phi_{ij}}{\partial \mathbf{n}}(\mathbf{x}) d\mathbf{x}$$

$$= - \int_{\Pi_{ij}} |\nabla \phi_{ij}(\mathbf{x})|^p d\mathbf{x} = -g_{ij}.$$

Thus (6.3.14) holds for $t_1^o, t_2^o, \ldots, t_N^o$.   □

6.3.2 Boundary gradient estimates for the perforated medium

We use ϕ_o, the solution of (6.3.9), as a trial function for the lower bound. In the next subsection we extend ϕ_o into triangles, and use it as a trial function for the upper bound. In order to obtain a useful extension, we need boundary gradient estimates for ϕ_o in the perforated medium. The objective of this subsection is to analyze the properties of the perforated medium in order to obtain this crucial estimate. Namely, we show that

$$|\nabla \phi_{ij}(\mathbf{x})| \leq C \tag{6.3.15}$$

for $\mathbf{x} \in \partial \Pi_{ij}$, where ϕ_{ij} is the solution of (6.3.9), and the constant $C = C(R)$ depends on the radius of inclusions R only.

We start with a geometric observation. Typically a neck Π_{ij} is not symmetric with respect to the line connecting the centers of the disks D_i and D_j. An example of a neck is given on Figure 5.8, where we used the local coordinate system when the centers of both disks lie on the y-axis. In this coordinate system the width of the left half-neck is $|S_1|$, $S_1 < 0$, and the width of the right half-neck is $|S_2|$,

$S_2 > 0$. Note that the inequalities $S_1 < 0$, and $S_2 > 0$ are not true in general. We define the (possibly negative) minimal relative half-neck width as

$$\beta_{ij} = \min\left(-\frac{S_1}{R}, \frac{S_2}{R}\right), \quad -1 < \beta_{ij} < 1. \tag{6.3.16}$$

Lemma 6.10 *For any triangle–neck partition, a neck Π_{ij} has either non-small δ_{ij}: $\delta_{ij} > R/100$, or non-small $\beta_{ij} > 1/3$.*

We remark that the numbers 100 and $1/3$ in the above lemma arise as some explicit universal constants. They are not optimal, and they illustrate the dichotomy: either a neck is not too short, or it is not too thin.

Proof Consider the construction of the Delaunay triangulation, and the triangle–neck partition, as depicted in Figure 5.6. For a pair of vertices $\mathbf{x}_i$ and $\mathbf{x}_j$ let α_{ij} be the smallest angle opposite to $\mathbf{x}_i\mathbf{x}_j$: for each $\mathbf{x}_i\mathbf{x}_j$ there are two angles opposite to $\mathbf{x}_i\mathbf{x}_j$; in Figure 5.6 they are $\angle\mathbf{x}_i\mathbf{x}_l\mathbf{x}_j$ and $\angle\mathbf{x}_i\mathbf{x}_k\mathbf{x}_j$. For definiteness, assume $\alpha_{ij} = \angle\mathbf{x}_i\mathbf{x}_k\mathbf{x}_j$. Note that $\beta_{ij} = \cos(\alpha_{ij}) = \sin(\pi/2 - \alpha_{ij})$. If $\delta_{ij} \leq R/100$, then we claim that we have an upper bound $\cos(\alpha_{ij}) > 1/3$. Indeed, if $\delta_{ij} \leq R/100$ then we have lower bounds $|\mathbf{x}_k\mathbf{x}_j| \geq 200|\mathbf{x}_i\mathbf{x}_j|/201$, and $|\mathbf{x}_k\mathbf{x}_i| \geq 200|\mathbf{x}_i\mathbf{x}_j|/201$. Thus α_{ij} could be at most $\pi/3 + \varepsilon$, for some $\varepsilon \ll 1$. The estimate $\beta_{ij} = \cos(\alpha_{ij}) > 1/3$ now follows from the law of cosines. $\qquad\square$

Observe that if a neck Π_{ij} has a non-small δ_{ij}, $\delta_{ij} > R/100$, then the crucial estimate (6.3.15) is immediate. Hence we are left to show the following result.

Lemma 6.11 *Consider a neck Π_{ij} such that its relative half-neck width is $\beta_{ij} > 1/3$, as in Figure 6.1. Suppose ϕ satisfies*

$$\text{div}(|\nabla\phi|^{p-2}\nabla\phi) = 0 \quad in\ \Pi_{ij}, \tag{6.3.17}$$

$$\frac{\partial\phi(\mathbf{x})}{\partial\mathbf{n}} = 0 \quad on\ \partial\Pi_{ij}^{\pm},$$

$$\phi(\mathbf{x}) = 1/2\ on\ \partial D_i,\ \phi(\mathbf{x}) = -1/2 \quad on\ \partial D_j.$$

Then

$$|\nabla\phi(\mathbf{x})| \leq C \tag{6.3.18}$$

for $\mathbf{x} \in \partial\Pi_{ij}^{\pm}$, where the constant $C = C(R)$ depends on R only.

In this lemma we used ϕ instead of ϕ_{ij} in order to simplify the notation.

Proof Since ϕ satisfies homogeneous Neumann conditions on the lateral boundaries, we can extend ϕ periodically by even reflection on a domain Ω, obtained from Π_{ij} by mirror reflections along lateral boundaries, see the top domain in

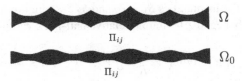

Figure 6.2 (Top) Ω, periodization of the neck Π_{ij}, and (bottom) Ω_0, smoothing of the boundary.

Figure 6.2. Further, it suffices to show the estimate (6.3.18) for ϕ that solves

$$\text{div}(|\nabla\phi|^{p-2})\nabla\phi) = 0 \text{ in } \Omega_0, \qquad (6.3.19)$$

$$\phi(\mathbf{x}) = 1/2 \text{ on the top part of } \partial\Omega_0,$$

$$\phi(\mathbf{x}) = -1/2 \text{ on the bottom part of } \partial\Omega_0$$

on a family of smooth periodic domains Ω_0 that approximate the periodization of Π_{ij} arbitrarily well. An example of Ω_0 is depicted in the bottom of Figure 6.2.

We will use Bernstein-type arguments to prove (6.3.18) for ϕ. This argument requires ϕ to be twice differentiable, but we know (Lieberman, 1988) that ϕ is only $C^{1,\alpha}(\bar{\Omega}_0)$. Thus we prove (6.3.18) for ϕ using the standard technique: we first regularize (6.3.19) by considering

$$\text{div}((\varepsilon + |\nabla\phi^\varepsilon|^{p-2})\nabla\phi^\varepsilon) = 0 \text{ in } \Omega_0, \qquad (6.3.20)$$

$$\phi^\varepsilon(\mathbf{x}) = 1/2 \text{ on the top part of } \partial\Omega_0,$$

$$\phi^\varepsilon(\mathbf{x}) = -1/2 \text{ on the bottom part of } \partial\Omega_0,$$

for $0 < \varepsilon < 1$ and then pass to the limit $\varepsilon \to 0$. Once $\varepsilon > 0$ equation (6.3.20) is uniformly elliptic, and therefore Ladyzhenskaya and Ural'tseva (1968) we can use classical elliptic regularity theory as follows. On Ω_0 the solution ϕ^ε is $C^\infty(\bar{\Omega}_0)$. Therefore we can differentiate ϕ^ε twice. Since $\phi^\varepsilon \to \phi$, in $C^{1,\alpha'}(\bar{\Omega}_0)$ for some $0 < \alpha' < \alpha$ (Lieberman, 1988), it suffices to obtain (6.3.18) for ϕ^ε so that the constant in (6.3.18) is independent of $\varepsilon > 0$ from (6.3.20).

Since ϕ^ε is periodic, it could be thought of as a solution of $\text{div}(\varepsilon + |\nabla\phi^\varepsilon|^{p-2}\nabla\phi^\varepsilon) = 0$ on a compact domain. Thus we reduce the original problem to the following. We need to show that on a domain $\overline{\Pi}_{ij}$ (see Figure 6.3) there is $\phi \in C^\infty(\overline{\Pi}_{ij})$, that solves

$$\text{div}((\varepsilon + |\nabla\phi|^{p-2})\nabla\phi) = 0 \text{ in } \overline{\Pi}_{ij}, \qquad (6.3.21)$$

$$\frac{\partial\phi}{\partial\mathbf{n}}(\mathbf{x}) = 0 \text{ on } \partial\overline{\Pi}_{ij}^{\pm},$$

$$\phi(\mathbf{x}) = 1/2 \text{ on } \partial\overline{\Pi}_{ij}^{\text{top}},$$

$$\phi(\mathbf{x}) = -1/2 \text{ on } \partial\overline{\Pi}_{ij}^{\text{bottom}}$$

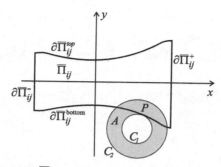

Figure 6.3 Neck $\overline{\Pi}_{ij}$ with smoothed boundary, and an annulus A.

and it suffices to show that (6.3.18) holds for such ϕ, where the constant C is independent of δ_{ij}, ε and smoothing of the boundary. For notational convenience we have omitted the dependence of ϕ in (6.3.21) on δ_{ij}, ε and smoothing of the boundary.

Following, e.g., Evans and Gangbo (1999) we will now apply a Bernstein-type argument. Let

$$U(\mathbf{x}) = f(x)|\nabla\phi(\mathbf{x})|^2 + \lambda\phi^2(\mathbf{x}),$$

where we will choose later the constant λ and the function $f(x)$ (we recall the notation $\mathbf{x} = (x, y)$). Note that the latter is a function of the horizontal variable x only. Denote

$$a_{ij} = (p-1)\varepsilon\delta_{ij} + \delta_{ij}|\nabla\phi|^{p-2} + (p-2)|\nabla\phi|^{p-4}\frac{\partial\phi}{\partial x_i}\frac{\partial\phi}{\partial x_j} \quad (i,j = 1,2).$$

Observe that

$$\sum_{i,j=1,2} \frac{\partial}{\partial x_i}\left(a_{ij}\frac{\partial\phi^2}{\partial x_j}\right) = 2(p-1)(\varepsilon|\nabla\phi|^2 + |\nabla\phi|^p), \qquad (6.3.22)$$

because

$$\sum_{i,j=1,2} \frac{\partial}{\partial x_i}\left(((p-1)\varepsilon + |\nabla\phi|^{p-2})\frac{\partial\phi^2}{\partial x_i} + (p-2)|\nabla\phi|^{p-4}\frac{\partial\phi}{\partial x_i}\frac{\partial\phi}{\partial x_j}\frac{\partial\phi^2}{\partial x_j}\right)$$

$$= 2\sum_{i,j=1,2} \frac{\partial}{\partial x_i}\left(\phi((p-1)\varepsilon + |\nabla\phi|^{p-2})\frac{\partial\phi}{\partial x_i} + (p-2)\phi|\nabla\phi|^{p-4}\frac{\partial\phi}{\partial x_i}\frac{\partial\phi}{\partial x_j}\frac{\partial\phi}{\partial x_j}\right)$$

$$= 2(p-1)\sum_{i=1,2} \frac{\partial}{\partial x_i}\left(\phi(\varepsilon + |\nabla\phi|^{p-2})\frac{\partial\phi}{\partial x_i}\right) = 2(p-1)(\varepsilon|\nabla\phi|^2 + |\nabla\phi|^p).$$

Let us show that

$$
\sum_{i,j=1,2} \frac{\partial}{\partial x_i} \left(a_{ij} \frac{\partial}{\partial x_j} |\nabla\phi|^2 \right) = 2(\varepsilon + |\nabla\phi|^{p-2})|\nabla^2\phi|^2
$$

$$
+ \frac{p-2}{2}|\nabla\phi|^{p-4}\left|\nabla|\nabla\phi|^2\right|^2 \geq |\nabla\phi|^{p-2}|\nabla^2\phi|^2,
$$

(6.3.23)

where

$$
|\nabla^2\phi|^2 = \sum_{i,j=1,2} \left(\frac{\partial^2\phi}{\partial x_i x_j} \right)^2.
$$

Differentiate div $\left((\varepsilon + |\nabla\phi|^{p-2})\nabla\phi\right) = 0$ with respect to x_k to obtain

$$
\sum_{i,j=1,2} \frac{\partial}{\partial x_i} \left((\varepsilon + |\nabla\phi|^{p-2})\frac{\partial^2\phi}{\partial x_i x_k} + (p-2)|\nabla\phi|^{p-4}\frac{\partial\phi}{\partial x_i}\frac{\partial\phi}{\partial x_j}\frac{\partial^2\phi}{\partial x_j x_k} \right) = 0.
$$

Multiply the above by $\dfrac{\partial\phi}{\partial x_k}$ and obtain

$$
\sum_{i,j=1,2} \frac{\partial\phi}{\partial x_k}\frac{\partial}{\partial x_i} \left((\varepsilon + |\nabla\phi|^{p-2})\frac{\partial^2\phi}{\partial x_i x_k} + (p-2)|\nabla\phi|^{p-4}\frac{\partial\phi}{\partial x_i}\frac{\partial\phi}{\partial x_j}\frac{\partial^2\phi}{\partial x_j x_k} \right) = 0.
$$

This implies that

$$
\sum_{i,j=1,2} \frac{\partial}{\partial x_i} \left((\varepsilon + |\nabla\phi|^{p-2})\frac{\partial}{\partial x_i}\left|\frac{\partial\phi}{\partial x_k}\right|^2 + (p-2)|\nabla\phi|^{p-4}\frac{\partial\phi}{\partial x_i}\frac{\partial\phi}{\partial x_j}\frac{\partial}{\partial x_j}\left|\frac{\partial\phi}{\partial x_k}\right|^2 \right)
$$

$$
= 2\sum_{i,j=1,2} \frac{\partial^2\phi}{\partial x_i x_k} \left((\varepsilon + |\nabla\phi|^{p-2})\frac{\partial^2\phi}{\partial x_i x_k} + (p-2)|\nabla\phi|^{p-4}\frac{\partial\phi}{\partial x_i}\frac{\partial\phi}{\partial x_j}\frac{\partial^2\phi}{\partial x_j x_k} \right).
$$

Sum over k to obtain

$$
\sum_{i,j=1,2} \frac{\partial}{\partial x_i} \left((\varepsilon + |\nabla\phi|^{p-2})\frac{\partial}{\partial x_i}|\nabla\phi|^2 + (p-2)|\nabla\phi|^{p-4}\frac{\partial\phi}{\partial x_i}\frac{\partial\phi}{\partial x_j}\frac{\partial}{\partial x_j}|\nabla\phi|^2 \right)
$$

$$
= 2(\varepsilon + |\nabla\phi|^{p-2})|\nabla^2\phi|^2 + \frac{p-2}{2}|\nabla\phi|^{p-4}\left|\nabla|\nabla\phi|^2\right|^2,
$$

which is (6.3.23).

Combining (6.3.22) and (6.3.23) for $U(\mathbf{x}) = f(x)|\nabla\phi(\mathbf{x})|^2 + \lambda\phi^2(\mathbf{x})$ we obtain

$$\sum_{i,j=1,2} \frac{\partial}{\partial x_i}\left(a_{ij}\frac{\partial U}{\partial x_j}\right) = |\nabla\phi|^2 \sum_{i,j=1,2} \frac{\partial}{\partial x_i}\left(a_{ij}\frac{\partial f}{\partial x_j}\right) + f \sum_{i,j=1,2} \frac{\partial}{\partial x_i}\left(a_{ij}\frac{\partial}{\partial x_j}|\nabla\phi|^2\right)$$

$$+ 2\sum_{j=1,2} a_{1j}\frac{\partial}{\partial x_j}|\nabla\phi|^2 f'(x) + 2(p-1)\lambda|\nabla\phi|^p$$

$$\geq 2f(x)|\nabla\phi|^{p-2}|\nabla^2\phi|^2 + 2(p-1)\lambda|\nabla\phi|^p$$

$$- C|f'(x)|\,|\nabla\phi|^{p-1}\,|\nabla^2\phi| - C|f''(x)|\,|\nabla\phi|^p$$

$$\geq \left(\sqrt{f(x)(p-1)\lambda} - C|f'(x)|\right)|\nabla\phi|^{p-1}\,|\nabla^2\phi|$$

$$+ ((p-1)\lambda - C|f''(x)|)|\nabla\phi|^p,$$

where we have used the Cauchy–Schwarz inequality in the last step. The constants above are universal, in particular they do not depend on ε.

We are now ready to choose $f(x)$ and λ. Recall that in our notation the distance between ∂B_i and ∂B_j is the smallest at $x = 0$ and it equals δ_{ij} there, see Figure 5.8. Noting that $R/6 \leq \beta_{ij}/2$, we set $f(x) \equiv 0$, for $|x| \leq R/12$, $f(x) = (|x| - R/12)^4$ for $R/12 \leq |x| \leq R/6$. We can further define $f(x)$ to be smooth, $R^4/12^4 \leq f(x) \leq 1$ for $|x| \geq R/6$, and $f(x) \equiv 1$ near $\mathbf{x} \in \partial\Pi^{\pm}_{ij}$. Then we can choose a large universal constant λ so that

$$\left(\sqrt{f(x)(p-1)\lambda} - C|f'(x)|\right)|\nabla\phi|^{p-1}\,|\nabla^2\phi| + ((p-1)\lambda - C|f''(x)|)|\nabla\phi|^p > 0.$$

With this choice of λ we obtain

$$\sum_{i,j=1,2} \frac{\partial}{\partial x_i}\left(a_{ij}\frac{\partial U}{\partial x_j}\right) > 0,$$

with

$$\sum_{i,j=1,2} a_{ij}\zeta_i\zeta_j > \varepsilon\left(|\zeta_1|^2 + |\zeta_2|^2\right),$$

for any (ζ_1, ζ_2). Therefore $U((\mathbf{x})) = f(x)|\nabla\phi(\mathbf{x})|^2 + \lambda\phi^2(\mathbf{x})$ satisfies the maximum principle: the maximum of U is achieved at the boundary of $\overline{\Pi}_{ij}$. Since U satisfies the maximum principle on a periodic domain Ω_0 (see Figure 6.2) as well, its maximum is achieved on the top or the bottom boundaries of $\overline{\Pi}_{ij}$. Using $-1/2 \leq \phi(\mathbf{x}) \leq 1/2$, we obtain

$$\max_{\mathbf{x}\in\partial\overline{\Pi}^{\pm}_{ij}} |\nabla\phi(\mathbf{x})|^2 \leq \max_{\mathbf{x}\in\overline{\Pi}_{ij}} U(\mathbf{x}) \leq C + \max_{\mathbf{x}\in\partial\overline{\Pi}^{\text{top}}_{ij},\partial\overline{\Pi}^{\text{bottom}}_{ij}} f(x)|\nabla\phi(\mathbf{x})|^2. \quad (6.3.24)$$

Hence we only need to obtain estimates for $f(x)|\nabla\phi(\mathbf{x})|^2$ on the top and the bottom boundaries. We will obtain them by using barriers for $\nabla\phi(\mathbf{x})$. Since $f(x) \equiv 0$, for $|x| \leq R/12$, we need to estimate $\nabla\phi(\mathbf{x})$, for $R/12 \leq |x| \leq R$ only. Fix any point $P \in \partial\overline{\Pi}^{\text{bottom}}_{ij}$, such that $R/12 \leq |x| \leq R$. Consider two concentric circles C_1 and C_2 of radii $R/600$ and $R/1200$, respectively, such that its interior circle is tangent to $\partial \overline{\Pi}_{ij}^{\text{bottom}}$ at the point P and the center of the circle does not lie in $\overline{\Pi}_{ij}$ (see Figure 6.3). Note that the annulus A between these two circles does not intersect the top boundary $\partial \overline{\Pi}_{ij}^{\text{top}}$ for any choice of $P \in \partial \overline{\Pi}_{ij}^{\text{bottom}}$ with $R/12 \leq |x| \leq R$. Therefore the function ψ, the solution of

$$\text{div}((\varepsilon + |\nabla \psi|^{p-2})\nabla \psi) = 0 \text{ in } A,$$
$$\psi(\mathbf{x}) = -1/2 \text{ on } C_1,$$
$$\psi(\mathbf{x}) = 1/2 \text{ on } C_2,$$

will be a super-solution for ϕ in $A \cap \overline{\Pi}_{ij}$. Lastly $\psi(\mathbf{x}) = \phi(\mathbf{x}) = -1/2$ at P, and thus ψ is a barrier for ϕ at the point P. The solution of (6.3.25) can be found explicitly; it is a function that depends on r, the distance from the center of the annulus. It is, however, sufficient to note that $|\nabla \psi(\mathbf{x})| \leq C$, $C = C(R)$, and C is independent of ε. Therefore we obtain

$$\max_{\mathbf{x} \in \partial \overline{\Pi}_{ij}^{\text{bottom}}} f(x)|\nabla \phi(\mathbf{x})|^2 \leq C, \ C = C(R).$$

Using the last inequality and a similar inequality at the top boundary in (6.3.24), we obtain (6.3.18). $\qquad \square$

6.3.3 Optimal Lipschitz extensions

In order to obtain the second inequality (6.2.2) in Theorem 6.3 we need to construct a good trial function for the upper bound. We will use $\phi_o(\mathbf{x})$, the solution of (6.3.9), as this trial function in the perforated domain. Thus it suffices to show that we can appropriately extend $\phi_o(\mathbf{x})$ into triangles Δ_{ijk}. Instead of explicit constructions as in Sections 3.6.6–3.6.10 and in Section 5.8, we will use the Kirszbraun theorem (see Section 1.6) to obtain an appropriate extension.

We now complete the proof of Theorem 6.3.

Proof of the second inequality (6.9) Consider an arbitrary set of potentials on the inclusions $t_1, \ldots, t_N$. On a neck Π_{ij} we define a trial function

$$\phi_*(\mathbf{x}) = \phi_{ij}(\mathbf{x})(t_i - t_j) + (t_i + t_j)/2, \tag{6.3.25}$$

where $\phi_{ij}(\mathbf{x})$ is the solution of (6.3.3)–(6.3.8). We will use ϕ_* as a trial function for the upper bound (6.2.2) in the perforated domain. It remains to extend it into triangles. First suppose that all triangles in the triangle–neck partition do not have too obtuse angles. For example, we may assume that for any triangle Δ_{ijk} all its angles are less than $2\pi/3$. Then the distance from any vertex of any triangle to the opposite side of this triangle is bounded from below by a constant that depends on R only. Thus using the crucial estimate (6.3.15) we conclude

$$|\phi_*(\mathbf{x}_1) - \phi_*(\mathbf{x}_2)| \leq \mathcal{L}|\mathbf{x}_1 - \mathbf{x}_2| \, |t_i - t_j| \tag{6.3.26}$$

for any points $\mathbf{x}_1, \mathbf{x}_2 \in \partial \Pi_{ij}$, where $\mathcal{L}$ depends on the constant in (6.3.15), the radius R and the largest size of the angles in the triangle–neck partition. By the Kirszbraun theorem there is a Lipschitz continuous extension $\phi_{ijk}^{\text{triangle}}$ of ϕ_* into each triangle Δ_{ijk} such that

$$|\nabla \phi_{ijk}^{\text{triangle}}(\mathbf{x})| \leq C \max\left(|t_i - t_j|, |t_i - t_k|, |t_j - t_k|\right), \quad \mathbf{x} \in \Delta_{ijk},$$

where $C = \mathcal{L}$ from (6.3.26).

Consider a trial field $\tilde{\phi}(\mathbf{x})$ such that

$$\tilde{\phi}(\mathbf{x}) = \begin{cases} \phi_*(\mathbf{x}) & \text{if } \mathbf{x} \in \Pi_o, \\ \phi_{ijk}^{\text{triangle}}(\mathbf{x}) & \text{if } \mathbf{x} \in \Delta_{ijk}. \end{cases} \tag{6.3.27}$$

By construction $\tilde{\phi} \in V$ (defined by (6.1.5)), thus using the direct variational principle (6.1.6), (6.1.7) on the whole domain Q we obtain

$$A \leq \frac{1}{2} \sum_{\Pi_{ij}} g_{ij} |t_i - t_j|^p + C \sum_{\Delta_{ijk}} \left(|t_i - t_j| + |t_i - t_k| + |t_j - t_k|\right)^p |\Delta_{ijk}|$$

$$\leq \frac{1}{2} \sum_{\Pi_{ij}} \left(g_{ij} + C\right) |t_i - t_j|^p,$$

where the constants depend on R and L.

If we do not have control over the size of the largest angle in a triangle, then the estimate (6.3.26) does not follow from (6.3.15), and the argument should be modified as follows. If an edge $\mathbf{x}_i \mathbf{x}_j$ is not opposite to an obtuse angle larger than $2\pi/3$, then we proceed as in the previous step and define $\tilde{\phi}(\mathbf{x}) = \phi_*(\mathbf{x})$ on Π_{ij} as defined in (6.3.25). If a triangle Δ_{ijk} does not have obtuse angles larger than $2\pi/3$, then we also proceed as in the previous step and define $\tilde{\phi}(\mathbf{x})$ on Δ_{ijk} using the Kirszbraun theorem as in (6.3.27). If, however, in the triangle Δ_{ijk} the edge $\mathbf{x}_i \mathbf{x}_j$ is opposite to the obtuse angle $\angle \mathbf{x}_i \mathbf{x}_k \mathbf{x}_j > 2\pi/3$, then we construct $\tilde{\phi}(\mathbf{x})$ differently. Consider the curvilinear hexagon D_{ijkm} that consists of triangles Δ_{ijk}, Δ_{ijm}, and the neck Π_{ij} between them. By construction of the triangle–neck partition, the vertex $\mathbf{x}_m$ lies outside the circumscribed circle of the triangle Δ_{ijk}. Using the fact that $\angle \mathbf{x}_i \mathbf{x}_k \mathbf{x}_j > 2\pi/3$, we estimate the radius of this circumscribed circle to be at least $2R$. Thus D_{ijkm} has the following property. Suppose the vertices of D_{ijkm} are denoted as A, B, C, D, E, and F. Pick any vertex, say, A. Suppose AB and AF are adjacent curvilinear edges. The other four edges are BC, CD, DE, and EF. Then the distance from A to BC, CD, DE, and EF is bounded from below by a constant that depends on R only. This property and (6.3.15) imply that

$$|\phi_*(_1) - \phi_*(\mathbf{x}_2)| \leq \mathcal{L}|\mathbf{x}_1 - \mathbf{x}_2| \max\left(|t_i - t_j|, |t_i - t_k|, |t_j - t_k, |t_i - t_m|, |t_j - t_m|\right)$$

$$\leq C|\mathbf{x}_1 - \mathbf{x}_2| \max\left(|t_i - t_k|, |t_j - t_k, |t_i - t_m|, |t_j - t_m|\right),$$

for any points $\mathbf{x}_1, \mathbf{x}_2 \in \partial D_{ijkm}$, where $\mathcal{L}$ depends on the constant in (6.3.15), and R. By the Kirszbraun theorem applied to the entire domain D_{ijkm}, we may define $\tilde{\phi}(\mathbf{x})$ in the entire rhombus D_{ijkm}, so that

$$|\nabla \phi(\mathbf{x})| \leq C \max \left(|t_i - t_k|, |t_j - t_k, |t_i - t_m|, |t_j - t_m| \right), \ \mathbf{x} \in D_{ijkm}.$$

Note that in this modified construction we replaced ϕ_* on the neck Π_{ij} with the trial function that comes from the Kirszbraun theorem. Since $\angle \mathbf{x}_i \mathbf{x}_k \mathbf{x}_j > 2\pi/3$, this implies that

$$g_{ij} \leq C,$$

where $C = C(R)$. Therefore as before we have

$$A \leq \frac{1}{2} \sum_{\Pi_{ij}} g_{ij} |t_i - t_j|^p + C \sum_{\Delta_{ijk}} \left(|t_i - t_j| + |t_i - t_k| + |t_j - t_k| \right)^p |\Delta_{ijk}|$$

$$+ C \sum_{D_{ijkm}} \left(|t_i - t_k| + |t_j - t_k| + |t_i - t_m| + |t_j - t_m| \right)^p |D_{ijkm}|$$

$$\leq \frac{1}{2} \sum_{\Pi_{ij}} \left(g_{ij} + C \right) |t_i - t_j|^p. \qquad \square$$

6.4 Proofs for the asymptotic step

In this section we give proofs related to the second step of our two-step approximation method.

6.4.1 Asymptotics in the necks

We recall that for a pair of vertices $\mathbf{x}_i$ and $\mathbf{x}_j$, connected by the edge Π_{ij}, the *interparticle distance* is the distance between the corresponding inclusions:

$$\delta_{ij} = |\mathbf{x}_i - \mathbf{x}_j| - 2R. \qquad (6.4.1)$$

Similarly, the interparticle distance between an inclusion and one of the special (boundary) vertices $\mathbf{x}_{N+1}$ and $\mathbf{x}_{N+2}$ is set as the distance between the inclusion and the boundary. When the concentration of inclusions is high, we expect that a *typical* $\delta_{ij} \ll 1$, and, therefore, we are interested in asymptotic values of g_{ij} as $\delta_{ij} \to 0$.

Using the Legendre transform, it is possible to represent g_{ij} as the maximum value of the dual functional. A lower bound on g_{ij} in (6.2.10) can then be obtained using variational duality. In detail, following the steps in Section 3.2.4 we first use the Legendre transform in $\mathbb{R}^2$ (see Section 1.3.1) which establishes the duality

$$\frac{1}{p} |\mathbf{x}|^p = \max_{\mathbf{v} \in \mathbb{R}^2} \left(\mathbf{v}\mathbf{x} - \frac{1}{q} |\mathbf{v}|^q \right), \ \mathbf{x} \in \mathbb{R}^2, \ q = \frac{p}{p-1}. \qquad (6.4.2)$$

Then we apply (6.4.2) to the integrand in (6.2.10), and obtain

$$g_{ij} = \min_{\phi \in V_{\Pi_{ij}}} \int_{\Pi_{ij}} |\nabla \phi(\mathbf{x})|^p d\mathbf{x} = p \min_{\phi \in V_{\Pi_{ij}}} \int_{\Pi_{ij}} \max_{\mathbf{v} \in R^2} \left(\mathbf{v}(\mathbf{x}) \nabla \phi(\mathbf{x}) - \frac{1}{q} |\mathbf{v}(\mathbf{x})|^q \right) d\mathbf{x}$$

$$= \max_{\mathbf{v} \in W_{\Pi_{ij}}} \min_{\phi \in V_{\Pi_{ij}}} \int_{\Pi_{ij}} (p\mathbf{v}(\mathbf{x}) \nabla \phi(\mathbf{x}) - (p-1)|\mathbf{v}(\mathbf{x})|^q) d\mathbf{x}$$

$$= \max_{\mathbf{v} \in W_{\Pi_{ij}}} \left(p \min_{\phi \in V_{\Pi_{ij}}} \left[\int_{\Pi_{ij}} \mathbf{v}(\mathbf{x}) \nabla \phi(\mathbf{x}) d\mathbf{x} \right] - (p-1) \int_{\Pi_{ij}} |\mathbf{v}(\mathbf{x})|^q d\mathbf{x} \right),$$

with the space

$$W_{\Pi_{ij}} = \left\{ \mathbf{v} \in L^q(\Pi_{ij}, \mathbb{R}^2) \,|\, \mathrm{div}\, \mathbf{v} = 0 \text{ in } \Pi_{ij}, \mathbf{v}(\mathbf{x})\mathbf{n} = 0 \text{ on } \partial \Pi_{ij}^{\pm} \right\}. \quad (6.4.3)$$

Using the boundary conditions $\phi(\mathbf{x}) = \pm 1/2$ on $\partial P_{ij}^{\pm}$, we compute

$$\min_{\phi \in V_{\Pi_{ij}}} \left[\int_{\Pi_{ij}} \mathbf{v}(\mathbf{x}) \nabla \phi(\mathbf{x}) d\mathbf{x} \right] = \frac{1}{2} \int_{\partial D_i \cup \partial D_j} \mathbf{v}(\mathbf{x})\mathbf{n} d\mathbf{x}.$$

Introducing the new function,

$$\mathbf{u}(\mathbf{x}) = |\mathbf{v}(\mathbf{x})|^{\frac{2-p}{p-1}} \mathbf{v}(\mathbf{x}),$$

we obtain

$$g_{ij} = \max_{\mathbf{u} \in V_{\Pi_{ij}}^*} \left[\frac{p}{2} \int_{\partial D_i \cup \partial D_j} |\mathbf{u}(\mathbf{x})|^{p-2} \mathbf{u}(\mathbf{x})\mathbf{n}\, d\mathbf{x} - (p-1) \int_{\Pi_{ij}} |\mathbf{u}(\mathbf{x})|^p d\mathbf{x} \right],$$
$$(6.4.4)$$

with the space

$$V_{\Pi_{ij}}^* = \left\{ \mathbf{u} \in L^p(\Pi_{ij}, \mathbb{R}^2) \,|\, \mathrm{div}(|\mathbf{u}|^{p-2}\mathbf{u}) = 0, \mathbf{u}(\mathbf{x})\mathbf{n} = 0 \text{ on } \partial \Pi_{ij}^{\pm} \right\}. \quad (6.4.5)$$

Moreover, $\mathbf{u}_{ij}$, the maximizer of the right-hand side of (6.4.4), satisfies $\mathbf{u}_{ij}(\mathbf{x}) = \nabla \phi_{ij}(\mathbf{x})$, where ϕ_{ij} solves (6.3.3).

The above $L^p(\Pi_{ij}, \mathbb{R}^2) = L^p(\Pi_{ij})^2$ denotes the space of vector–valued functions $\mathbf{u} = (v, w)$, such that $v(\mathbf{x}) \in L^p(\Pi_{ij})$ and $w(\mathbf{x}) \in L^p(\Pi_{ij})$.

We will prove a slightly stronger version of Lemma 6.5. More specifically, we have the following result.

Lemma 6.12 *Consider a neck Π_{ij} between two inclusions D_i and D_j. Suppose g_{ij} is given by (6.2.10). As $\delta_{ij} \to 0$*

$$g_{ij} = g_{ij}^o + O(f(\delta_{ij})), \quad (6.4.6)$$

where the constants $g_{ij}^o = \mathbf{G}(\delta_{ij})$ are given in (6.2.13), and

$$f(\delta) = \delta^{-p+(d+3)/2}. \quad (6.4.7)$$

Consider a neck Π_{ij} between an inclusion D_i and the flat boundary $j = N$ or $j = N + 1$. As $\delta_{ij} \to 0$,

$$g_{ij} = g_{ij}^o + O(f(\delta_{ij})),$$

where g_{ij}^o is given by

$$g_{ij}^o = 2^{p-1} \mathbf{G}(2\delta_{ij}). \tag{6.4.8}$$

Proof Consider a neck Π_{ij} between two inclusions D_i and D_j. Pick any $\phi \in V_{\Pi_{ij}}$ and $\mathbf{u} \in V_{\Pi_{ij}}^*$, with $\mathrm{div}(|\mathbf{u}|^{p-2}\mathbf{u}) = 0$. Integrating by parts $\phi \mathrm{div}(|\mathbf{u}|^{p-2}\mathbf{u}) = 0$, we obtain the following identity:

$$\frac{1}{2} \int_{\partial D_i \cup \partial D_j} |\mathbf{u}(\mathbf{x})|^{p-2}\mathbf{u}(\mathbf{x})\mathbf{n}d\mathbf{x} = \int_{\Pi_{ij}} |\mathbf{u}(\mathbf{x})|^{p-2}\nabla\phi(\mathbf{x})\mathbf{u}(\mathbf{x})d\mathbf{x}. \tag{6.4.9}$$

By the direct (6.2.10) and the dual (6.4.4) variational principles we have

$$\left[\frac{p}{2} \int_{\partial D_i \cup \partial D_j} |\mathbf{u}(\mathbf{x})|^{p-2}\mathbf{u}(\mathbf{x})\mathbf{n}d\mathbf{x} - (p-1) \int_{\Pi_{ij}} |\mathbf{u}(\mathbf{x})|^p d\mathbf{x} \right] \tag{6.4.10}$$

$$\leq g_{ij} \leq \int_{\Pi_{ij}} |\nabla\phi(\mathbf{x})|^p d\mathbf{x}.$$

Using identity (6.4.9) in the last inequality we obtain the following variational bounds: for any $\phi \in V_{\Pi_{ij}}$ and $\mathbf{u} \in V_{\Pi_{ij}}^*$

$$\int_{\Pi_{ij}} \left(p|\mathbf{u}(\mathbf{x})|^{p-2}\nabla\phi(\mathbf{x})\mathbf{u}(\mathbf{x}) - (p-1)|\mathbf{u}(\mathbf{x})|^p \right) d\mathbf{x} \leq g_{ij} \leq \int_{\Pi_{ij}} |\nabla\phi(\mathbf{x})|^p d\mathbf{x}. \tag{6.4.11}$$

Note that in principle ϕ on the left-hand side of (6.4.11) and ϕ on the right-hand side of (6.4.11) may be different. We, however, will choose them to be the same, and we will use (6.4.11) in order to obtain upper and lower bounds on g_{ij} simultaneously. The advantage of (6.4.11) compared to (6.4.10) is computational. Namely, all integrals in (6.4.11) are area integrals, whereas in (6.4.10) we have one boundary integral. It turns out that when we use standard mathematical software, evaluation of area integrals is easier to perform (see identities (6.4.14) and (6.4.15) below).

Evaluating (6.4.11) with

$$\phi(\mathbf{x}) = \frac{y}{H(x)}, \quad \mathbf{u}(\mathbf{x}) = \left(0, \frac{1}{H(x)}\right) \tag{6.4.12}$$

we obtain $g_{ij}^l \leq g_{ij} \leq g_{ij}^u$ for

$$g_{ij}^l = \int_{\Pi_{ij}} \frac{1}{[H(x)]^p}dx, \quad g_{ij}^u = \int_{\Pi_{ij}} \left[\frac{[yH'(x)]^2}{[H(x)]^4} + \frac{1}{[H(x)]^2} \right]^{p/2} dx. \tag{6.4.13}$$

It suffices to show that

$$|g_{ij}^u - g_{ij}^l| = \int_{\Pi_{ij}} \left(\left[\frac{[yH'(x)]^2}{[H(x)]^4} + \frac{1}{[H(x)]^2} \right]^{p/2} - \frac{1}{[H(x)]^p} \right) d\mathbf{x} \quad (6.4.14)$$

$$\leq \int_{\Pi_{ij}} \frac{[yH'(x)]^2}{[H(x)]^{p+2}} d\mathbf{x} = O(f(\delta_{ij})),$$

and

$$g_{ij}^l = \int_{\Pi_{ij}} \frac{1}{[H(x)]^p} d\mathbf{x} = g_{ij}^o + O(f(\delta_{ij})), \quad (6.4.15)$$

where g_{ij}^u and g_{ij}^l are defined by (6.4.13). Using Lemma 6.10 we know that as $\delta_{ij} \to 0$, the width of Π_{ij} stays bounded away from zero and the line segment $\mathbf{x}_i \mathbf{x}_j$ that connects the centers of inclusions always crosses Π_{ij}. We then verify (6.4.14) and (6.4.15) by lengthy, but straightforward computations using standard mathematical software.

The argument for the boundary necks is the same, whereas the last estimate (6.4.8) is obtained by observing that two identical boundary necks of height δ_{ij} glued along the flat horizontal boundary give one neck between two inclusions, but of height $2\delta_{ij}$. $\square$

6.4.2 Connectivity of networks

We use the notation

$$A_d = \min_{t_i, i=1,\dots,N} I(t_1, \dots, t_N), \quad I(t_1, \dots, t_N) = \frac{1}{2} \sum_{\Pi_{ij}} g_{ij} |t_i - t_j|^p, \quad (6.4.16)$$

$$A_d^o = \min_{t_i, i=1,\dots,N} I^o(t_1, \dots, t_N), \quad I^o(t_1, \dots, t_N) = \frac{1}{2} \sum_{\Pi_{ij}} g_{ij}^o |t_i - t_j|^p. \quad (6.4.17)$$

We start by restating the nonlinear analog of Lemma 3.2 in Chapter 3.

Lemma 6.13 (Discrete maximum principle) *If all $g_{ij} > 0$ (or all $g_{ij}^o > 0$), then $\{t_1, \dots t_N\}$, the unique solution of the minimization problem (6.4.16) or (6.4.17), satisfies $-1 \leq t_k \leq 1$ for any t_k.*

Proof Let us consider (6.4.16). We first assume that a minimizer does not satisfy $1 \leq t_k \leq 1$ for some k. This implies that there exists k_i, $i = 1, \dots, m$, such that $t_{k_i} > 1$, or $t_{k_i} < -1$. Consider the case $t_{k_i} > 1$. Let

$$M = \min_{i=1,\dots,m} t_{k_i} - 1,$$

and

$$s_i = \begin{cases} t_i, & \text{if } i \notin \{k_1, k_2, \dots, k_m\} \\ t_i - M, & \text{if } i \notin \{k_1, k_2, \dots, k_m\}. \end{cases}$$

We verify that

$$I(s_1, \ldots, s_N) < I(t_1, \ldots, t_N),$$

thus $\{t_1, \ldots t_N\}$ is not a minimizer, and we must have $1 \le t_k \le 1$. The case $t_{k_i} < -1$ is similar. $\qquad\square$

The existence of the percolating cluster implies that the energy of the discrete network is asymptotically large, and this suffices to prove Theorem 6.9 as follows.

Proof of Theorem 6.9 Consider the set $\{t_1, t_2, \ldots t_N\}$, the unique solution of the minimization problem (6.4.16). By Lemma 6.13 (the discrete maximum principle) it satisfies $-1 \le t_k \le 1$. By Theorem 6.3 with g_{ij} given by (6.2.10) we obtain

$$A_d \le A \le \frac{1}{2} \min_{t_i, i=1,\ldots,N} \sum_{\Pi_{ij}} (g_{ij} + C) |t_i - t_j|^p, \tag{6.4.18}$$

with $t_{N+1} = 1$, $t_{N+2} = -1$, and the constant C in (6.4.18) depends on R and L only. Using $-1 \le t_k \le 1$ inequalities (6.4.18) imply

$$A_d \le A \le A_d + C. \tag{6.4.19}$$

the existence of the percolating cluster and (6.2.13) implies that

$$A_d \ge C\delta^{-p+(d+1)/2}, \quad A_d^o \ge C\delta^{-p+(d+1)/2}. \tag{6.4.20}$$

Thus

$$\frac{|A - A_d|}{A_d} \le C\delta^{p-(d+1)/2}, \tag{6.4.21}$$

which means that the estimate (6.2.18) in Theorem 6.9 holds if there is a δ-percolating cluster. The proof of (6.2.19) requires more work, mainly because, in contrast to the linear case considered in Chapter 3, the quantity

$$|g_{ij} - g_{ij}^o| \to \infty, \quad \text{as } \delta \to 0, \quad \text{for } \Pi_{ij} \in \mathbb{D}_\delta,$$

where $\mathbb{D}_\delta$ the set of δ-small necks:[1] $\mathbb{D}_\delta = \{\Pi_{ij} | \delta_{ij} \le \delta\}$. We also denote

$$g^o(\delta) = \min_{\Pi_{ij} \in \mathbb{D}_\delta} g_{ij}^o.$$

From (6.4.14), (6.4.15), and (6.2.13) we estimate

$$|g_{ij} - g_{ij}^o| \le C\delta^{-p+(d+3)/2}$$

if $\Pi_{ij} \notin \mathbb{D}_\delta$. This means that

$$(1 - C\delta)\, g_{ij}^o \le g_{ij} \le (1 + C\delta)\, g_{ij}^o, \quad \text{if } \Pi_{ij} \in \mathbb{D}_\delta.$$

---

[1] A δ-percolating cluster is a subset of $\mathbb{D}_\delta$, but $\mathbb{D}_\delta$ may contain other necks.

Hence for any $\{t_1, \ldots t_N\}$ that satisfy $-1 \leq t_k \leq 1$, we obtain

$$(1 - C\delta)I^o(t_1, \ldots, t_N) - C\delta^{-p+(d+3)/2} \leq I(t_1, \ldots, t_N)$$
$$\leq (1 + C\delta)I^o(t_1, \ldots, t_N) + C\delta^{-p+(d+3)/2},$$

which implies

$$(1 - C\delta)A_d^o - C\delta^{-p+(d+3)/2} \leq A_d \leq (1 + C\delta)A_d^o + C\delta^{-p+(d+3)/2}.$$

Using the last inequality, estimates (6.4.21) and (6.4.20) we obtain (6.2.19). $\quad\square$

7

Network approximation for potentials of bodies

In the previous chapters it was demonstrated that the original continuum problem can be approximated as the interparticle distance $\delta \to 0$ by a finite-dimensional network problem in the sense of closeness of energy or total flux.

In this chapter we demonstrate that the potentials of nodes determined from the network problem approximate the potentials of disks determined from the continuum problem as $\delta \to 0$ (Kolpakov, 2006b). We note that this problem is non-trivial when one analyzes *non-periodic* arrays of disks. In fact, in a periodic array all disks are equivalent. From this equivalence, it follows that the difference of potential between two disks D_i and D_j is $\mathbf{E}(\mathbf{x}_i - \mathbf{x}_j)$, where $\mathbf{x}_i$ and $\mathbf{x}_j$ are the coordinates of the centers of the disks D_i and D_j, and $\mathbf{E}$ is the average gradient of the field applied to the periodic array.

7.1 Formulation of the problem of approximation of potentials of bodies

As in Chapter 3, we consider a system of non-overlapping and non-touching disks $\{D_i, i = 1, \ldots, N\} \subset \Pi$ occupied by perfectly conducting bodies. As above, we denote by $Q_m = \Pi \backslash \bigcup_{i=1}^{N} D_i$ the space outside the disks.

We analyze the boundary-value problem

$$\Delta \phi = 0 \text{ in } Q_m; \tag{7.1.1}$$

$$\phi(\mathbf{x}) = t_i \text{ on } D_i, \ i = 1, \ldots, N; \tag{7.1.2}$$

$$\int_{\partial D_i} \frac{\partial \phi}{\partial n}(\mathbf{x}) d\mathbf{x} = 0, \ i \in I; \tag{7.1.3}$$

$$\frac{\partial \phi}{\partial n}(\mathbf{x}) = 0 \text{ on } \partial Q_{\text{lat}}; \tag{7.1.4}$$

$$\phi(\mathbf{x}) = 1 \text{ on } \partial Q^+, \ \phi(\mathbf{x}) = -1 \text{ on } \partial Q^-. \tag{7.1.5}$$

As above, the unknowns in (7.1.1)–(7.1.5) are the function $\phi(\mathbf{x})$ and real numbers $\{t_i, i \in I\}$ (the potentials of the internal disks $\{D_i, i \in I\}$, see (3.4.9)). In the previous chapters, we analyzed the characteristics depending on the function $\phi(\mathbf{x})$ determined in the domain Q_m (energy, total flux through the composite). In this chapter, the main object of our analysis is the potentials $\{t_i, i \in I\}$ of the disks $\{D_i, i \in I\}$.

The boundary-value problem (7.1.1)–(7.1.5) is equivalent to the minimization problem

$$I(\phi) = \frac{1}{2} \int_{Q_m} |\nabla \phi(\mathbf{x})|^2 d\mathbf{x} \to \min, \qquad (7.1.6)$$

considered on the set of functions

$$V_p = \left\{ \phi \in H^1(Q_m) \,|\, \phi(\mathbf{x}) = t_i^\phi \in \mathbb{R} \text{ in } D_i, \ i = 1, \ldots, N; \qquad (7.1.7)\right.$$

$$\left. \phi(\mathbf{x}) = 1 \text{ on } \partial Q^+, \phi(\mathbf{x}) = -1 \text{ on } \partial Q^- \right\},$$

where $\{t_i, i \in I\}$ are not fixed constants (generally, they are different in different bodies).

The flux A corresponding to the problem (7.1.1)–(7.1.5) is introduced as

$$A = I(\phi^{\text{cont}}) = \min_{\phi \in V_p} I(\phi), \qquad (7.1.8)$$

where

$$I(\phi) = \frac{1}{2} \int_{Q_m} |\nabla \phi(\mathbf{x})|^2 d\mathbf{x}$$

and $\phi^{\text{cont}}(\mathbf{x})$ means a solution of the boundary-value problem (7.1.1)–(7.1.5) (or the minimization problem (7.1.6)). We denote by $\{t_i^{\text{cont}}, i \in I\}$ the corresponding values of the variables $\{t_i, i \in I\}$ determined from the solution of the boundary-value problem (7.1.1)–(7.1.5).

The heuristic network problem corresponding to the continuum problem (7.1.1)–(7.1.5) was constructed in Section 3.4. It is associated with the network

$$\mathcal{G} = \{\mathbf{x}_i, g_{ij}; \ i, j = 1, \ldots, K\};$$

with the nodes $\{\mathbf{x}_i, i \in K\}$ corresponding to the disks (including pseudo-disks, K is the total number of disks and pseudo-disks, see Section 3.2.5) $\{D_i, i \in K\}$ and specific transport characteristics $g_{ij} = \sqrt{\dfrac{R}{\delta_{ij}}}$.

We restrict our consideration to the case of a system of disks satisfying the uniform close-packing condition (see Definition 3.9) in order to concentrate on the principal idea of asymptotic analysis of the problem and not digress into issues related to the topology of sets of bodies (the general case was discussed in Kolpakov and Kolpakov (2010)).

The fluxes arising in the network must satisfy the Kirchhoff equations for the internal nodes of the network $\mathcal{G}$:

$$\sum_{j \in N_i} g_{ij}(t_i - t_j) = 0, \ i \in I; \tag{7.1.9}$$

(N_i means the set of nodes of the network $\mathcal{G}$ adjacent to the i-th node) supplemented with the boundary conditions

$$t_i = 1 \quad \text{for } i \in S^+, t_i = -1 \quad \text{for } i \in S^-. \tag{7.1.10}$$

We denote the solution of the finite-dimensional network problem (7.1.9), (7.1.10) by $\{t_i^{\text{net}}, i \in K\}$.

The quantities $\mathbf{t}^{\text{cont}} = \{t_i^{\text{cont}}, i \in K\}$, which are actual potentials determined from the solution of the (continuous) boundary-value problem (7.1.1)–(7.1.5), and $\mathbf{t}^{\text{net}} = \{t_i^{\text{net}}, i \in K\}$ which are potentials of the network nodes determined from the solution of the finite-dimensional (network) problem (7.1.9), (7.1.10), are not equal, in general. At the same time, numerical experiments (Kolpakov, 2007; Kolpakov and Kolpakov, 2010) provide us with strong arguments in favor of the coincidence of the potentials of the particles determined from the solution of the continuum problem and the potentials of the nodes of the network.

In this chapter we present a proof of the asymptotic coincidence of the potentials $\mathbf{t}^{\text{cont}}$ and $\mathbf{t}^{\text{net}}$ under the condition of uniformly dense packing of the particles. Namely, we prove that $|\mathbf{t}^{\text{cont}} - \mathbf{t}^{\text{net}}| \to 0$ as $\delta \to 0$.

In Chapter 3 it was demonstrated (see Theorem 3.14) that under the close-packing condition the following aproximation holds

$$A - A_d = O\left(\sqrt{\frac{\delta}{R}}\right) A_d, \tag{7.1.11}$$

where the flux $A = I(\phi^{\text{cont}})$ is defined by (7.1.8) and

$$A_d = \frac{1}{4} \sum_{i \in K} \sum_{j \in N_i} g_{ij}(t_i^{\text{net}} - t_j^{\text{net}})^2$$

is the energy corresponding to the network problem (7.1.9), (7.1.10). It was also demonstrated (see Theorem 3.14) that under the uniform close-packing condition the fluxes A and A_d have the orders of $\sqrt{\dfrac{R}{\delta}}$ (under the uniform close-packing condition all the interparticle distances δ_{ij} have the same order δ).

7.2 Network approximation theorem for potentials

This section contains a three-step proof of the approximation theorem for the potentials of bodies determined from the solution of the continuum and the network problems.

7.2.1 Step A. An auxiliary boundary-value problem

We consider the following boundary-value problem:

$$\Delta\phi = 0 \text{ in } Q_m; \tag{7.2.1}$$

$$\phi(\mathbf{x}) = t_i^{\text{net}} \text{ in } D_i, \ i \in I; \tag{7.2.2}$$

$$\frac{\partial\phi}{\partial\mathbf{n}}(\mathbf{x}) = 0 \text{ on } \partial D_{\text{lat}}; \tag{7.2.3}$$

$$\phi(\mathbf{x}) = 1 \text{ on } \partial Q^+, \ \phi(\mathbf{x}) = -1 \text{ on } \partial Q^-, \tag{7.2.4}$$

where $\mathbf{t}^{\text{net}} = \{t_i^{\text{net}}, i \in I\}$ is a solution of the network problem (7.1.9), (7.1.10). Since $\{t_i^{\text{net}}, i \in I\}$ in (7.2.2) are known, this is a classical electrostatic problem for a system of bodies, see Brown (1956); Tamm (1979). We denote the solution of the problem (7.2.1)–(7.2.4) by $\phi^{\text{net}}(\mathbf{x})$.

The problem (7.2.1)–(7.2.4) is external Dirichlet in which we give the disks $\{D_i, i \in I\}$ values of the potentials determined from the solution of the network problem (7.1.9), (7.1.10). Note that the original problem (7.1.1)–(7.1.5) is not a Dirichlet problem.

Lemma 7.1 *The following inequality holds*

$$\left|I(\phi^{\text{net}}) - A_d\right| = O\left(\sqrt{\frac{\delta}{R}}\right) A_d \text{ as } \delta \to 0. \tag{7.2.5}$$

In the inequality (7.2.5)

$$I(\phi^{\text{net}}) = \frac{1}{2}\int_{Q_m} |\nabla\phi^{\text{net}}(\mathbf{x})|^2 d\mathbf{x}$$

is the value of the Dirichlet integral corresponding to the problem (7.2.1)–(7.2.4).

Proof The function $\phi^{\text{net}}(\mathbf{x})$ is a solution of the minimization problem

$$I(\phi) = \frac{1}{2}\int_Q |\nabla\phi(\mathbf{x})|^2 d\mathbf{x} \to \min \tag{7.2.6}$$

on the set of functions

$$V^{\text{net}} = \left\{\phi(\mathbf{x}) \in H^1(Q) \mid \phi(\mathbf{x}) = t_i^{\text{net}} \text{ on } D_i, \ i \in I; \tag{7.2.7}\right.$$

$$\left.\phi(\mathbf{x}) = 1 \text{ on } \partial Q^+, \ \phi(\mathbf{x}) = -1 \text{ on } \partial Q^-\right\}.$$

The dual problem for (7.2.6) is

$$J(\mathbf{v}) \to \max, \ \mathbf{v} \in W_p, \ \text{div}\,\mathbf{v} = 0, \tag{7.2.8}$$

where

$$J(\mathbf{v}) = -\frac{1}{2}\int_{Q_m} \mathbf{v}^2(\mathbf{x})dx + \sum_{i \in I} t_i^{\text{net}} \int_{\partial D_i} \mathbf{v}(\mathbf{x})\mathbf{n}dx + \int_{\partial Q^+} \mathbf{v}(\mathbf{x})\mathbf{n}dx - \int_{\partial Q^-} \mathbf{v}(\mathbf{x})\mathbf{n}dx,$$

and

$$W_p = \left\{ \mathbf{v}(\mathbf{x}) = (v_1(\mathbf{x}), v_2(\mathbf{x})) \in L_2(Q_m) \mid \operatorname{div}\mathbf{v} \in L_2(Q_m); \quad (7.2.9) \right.$$

$$\left. \mathbf{v}(\mathbf{x})\mathbf{n} = 0 \text{ on } \partial Q_{\text{lat}} \right\}.$$

The condition $\mathbf{v}(\mathbf{x}) \in W_p$ makes it possible to determine the trace of the function $\mathbf{v}(\mathbf{x})\mathbf{n}$ on the surfaces $\{\partial D_i, i = 1, \ldots, N\}$ as elements of the functional space $H^{-1/2}(Q)$ and apply the formula of integration by parts to the functions under consideration (for details see Ekeland and Temam (1976); Lions and Magenes (1972)) and Section 1.2.3.

Relations (7.2.6) and (7.2.8) imply the two-sided estimate

$$J(\mathbf{v}) \le I(\phi^{\text{net}}) \le I(\phi), \quad (7.2.10)$$

for any $\phi \in V^{\text{net}}$ and $\mathbf{v} \in W_p$ such that $\operatorname{div}\mathbf{v}=0$.

In Sections 3.6.2 and 3.6.5 we constructed a trial function $\mathbf{v}(\mathbf{x}) \in W_p$, such that

(1) $\operatorname{div}\mathbf{v}(\mathbf{x}) = 0$ in Q_m (see formula (3.6.6));
(2) $\mathbf{v}(\mathbf{x}) = 0$ on ∂Q_{lat};
(3) the integral equations in (7.2.9) are satisfied, i.e. (see Section 3.6.2),

$$\int_{\partial D_i} \mathbf{v}(\mathbf{x})\mathbf{n}\,d\mathbf{x} = 0, \quad i \in I.$$

We denote this function by $\mathbf{v}_0(\mathbf{x})$.

In Sections 3.6.6–3.6.11, we constructed a trial function $\phi_0(\mathbf{x}) \in V_p$, such that $\phi(\mathbf{x}) = t_i^{\text{net}}$ on D_i. We denote this function by $\phi_0(\mathbf{x})$.

With these trial functions Theorem 3.14 was proved. Thus, for the mentioned functions $\phi_0(\mathbf{x})$ and $\mathbf{v}_0(\mathbf{x})$

$$|J(\mathbf{v}_0) - A_d| = O\left(\sqrt{\frac{\delta}{R}}\right) A_d, \quad (7.2.11)$$

$$|I(\phi_0) - A_d| = O\left(\sqrt{\frac{\delta}{R}}\right) A_d. \quad (7.2.12)$$

The conditions $\phi_0(\mathbf{x}) \in V$ and $\phi_0(\mathbf{x}) = t_i^{\text{net}}$ on D_i imply that $\phi_0(\mathbf{x}) \in V^{\text{net}}$. Then, $\phi_0(\mathbf{x})$ and $\mathbf{v}_0(\mathbf{x})$ can be used as trial functions in (7.2.10). Then, (7.2.5) follows from (7.2.10)–(7.2.12). $\qquad\square$

Relations (7.1.11) and (7.2.5) imply that

$$I(\phi^{\text{cont}}) - I(\phi^{\text{net}}) = O\left(\sqrt{\frac{\delta}{R}}\right) A_d. \quad (7.2.13)$$

Remark. By virtue of (7.1.11) and (7.2.5), it is possible to replace A_d by $A = I(\phi^{\text{cont}})$ or $I(\phi^{\text{net}})$ in the right-hand side of (7.2.13).

Now, we prove that (7.2.13) implies the closeness of the potentials of disks $\mathbf{t}^{\text{cont}}$ determined from the continuum problem and the potential of nodes $\mathbf{t}^{\text{net}}$ determined from the network model as $\delta \to 0$. Namely, we have the following theorem.

Theorem 7.6 *Let the disks $\{D_i, i = 1, \dots, N\}$ satisfy the uniform close-packing condition. Then $|\mathbf{t}^{\text{cont}} - \mathbf{t}^{\text{net}}| \to 0$ as $\delta \to 0$, where $\mathbf{t}^{\text{cont}} = \{t_i^{\text{cont}}, i \in I\}$ are the values of the potentials of the perfectly conducting bodies determined from the solution of problem (7.1.1)–(7.1.5) and $\mathbf{t}^{\text{net}} = \{t_i^{\text{net}}, i \in I\}$ are the potentials of the nodes determined from the solution of the network problem (7.1.9), (7.1.10).*

The next sections contain the proof of Theorem 7.6. The idea of the proof is the following. The quantities $I(\phi^{\text{cont}})$ and $I(\phi^{\text{net}})$ have the order of g_{ij}, and (7.2.13) means the asymptotic closeness of $I(\phi^{\text{cont}})$ and $I(\phi^{\text{net}})$. Generally, from the closeness of the energies, the closeness of solutions does not follow. In the considered case, in addition, $\phi^{\text{net}}(\mathbf{x}) \in V_p$ and the function $\phi^{\text{cont}}(\mathbf{x})$ gives the minimal value to the functional $I(\phi)$ on V_p. We will demonstrate that under this additional condition the closeness of the energies leads to the closeness of solutions $\mathbf{t}^{\text{cont}}$ and $\mathbf{t}^{\text{net}}$.

7.2.2 Step B. An auxiliary estimate for the energies

Lemma 7.7 *For any $\phi(\mathbf{x}) \in V_p$, the equality*

$$\int_{Q_m} |\nabla \phi^{\text{cont}}(\mathbf{x}) - \nabla \phi(\mathbf{x})|^2 dx = \int_{Q_m} |\nabla \phi(\mathbf{x})|^2 dx - \int_{Q_m} |\nabla \phi^{\text{cont}}(\mathbf{x})|^2 dx$$

(7.2.14)

holds.

Proof We take a function $\psi(\mathbf{x}) \in V_0$, where

$$V_0 = \left\{ \phi \in H^1(Q_m) \mid \phi(\mathbf{x}) = t_i^\phi \in \mathbb{R} \text{ in } D_i, \ i = 1, \dots, N; \right.$$

(7.2.15)

$$\left. \phi(\mathbf{x}) = 0 \text{ on } \partial Q^+, \phi(\mathbf{x}) = 0 \text{ on } \partial Q^- \right\}.$$

We can write the equality

$$\int_{Q_m} |\nabla \phi^{\text{cont}}(\mathbf{x}) + \nabla \psi(\mathbf{x})|^2 dx$$

(7.2.16)

$$= \int_{Q_m} |\nabla \phi^{\text{cont}}(\mathbf{x})|^2 dx + 2 \int_{Q_m} \nabla \phi^{\text{cont}}(\mathbf{x}) \nabla \psi(\mathbf{x}) dx + \int_{Q_m} |\nabla \psi(\mathbf{x})|^2 dx.$$

For the second integral on the right-hand side of (7.2.16), after integration by parts, we obtain

$$\int_{Q_m} \nabla \phi^{\text{cont}}(\mathbf{x}) \nabla \psi(\mathbf{x}) d\mathbf{x} \tag{7.2.17}$$

$$= - \int_{Q_m} \Delta \phi^{\text{cont}}(\mathbf{x}) \psi(\mathbf{x}) d\mathbf{x} + \sum_{i \in I} \int_{\partial D_i} \frac{\partial \phi^{\text{cont}}}{\partial \mathbf{n}} \psi(\mathbf{x}) d\mathbf{x}$$

$$+ \int_{\partial Q^+} \frac{\partial \phi^{\text{cont}}}{\partial \mathbf{n}} \psi(\mathbf{x}) d\mathbf{x} + \int_{\partial Q^-} \frac{\partial \phi^{\text{cont}}}{\partial \mathbf{n}} \psi(\mathbf{x}) d\mathbf{x}.$$

Due to $\psi(\mathbf{x}) \in V_0$, the last two integrals on the right-hand side of (7.2.17) are equal to zero. For the integrals in the sum on the right-hand side of (7.2.17) we have

$$\int_{\partial D_i} \frac{\partial \phi^{\text{cont}}}{\partial \mathbf{n}} \psi(\mathbf{x}) d\mathbf{x} = t_i^{\phi} \int_{\partial D_i} \frac{\partial \phi^{\text{cont}}}{\partial \mathbf{n}} d\mathbf{x} = 0.$$

Thus, we have

$$\int_{Q_m} |\nabla \phi^{\text{cont}}(\mathbf{x}) + \nabla \psi(\mathbf{x})|^2 d\mathbf{x} = \int_{Q_m} |\nabla \phi^{\text{cont}}(\mathbf{x})|^2 d\mathbf{x} + \int_{Q_m} |\nabla \psi(\mathbf{x})|^2 d\mathbf{x}. \tag{7.2.18}$$

For an arbitrary function $\phi \in V_p$, we consider $\psi(\mathbf{x}) = \phi^{\text{cont}}(\mathbf{x}) - \phi(\mathbf{x})$. Since ϕ^{cont}, $\phi \in V_p$, the function $\psi(\mathbf{x}) \in V_0$. For that function $\psi(\mathbf{x})$, (7.2.18) implies (7.2.14).

7.2.3 Step C. Estimation of the difference between the solutions to the original and network problems

The problem (7.1.1)–(7.1.5) is equivalent to the problem

$$\Delta \phi = 0 \text{ in } Q_m; \tag{7.2.19}$$

$$\phi(\mathbf{x}) = t_i^{\text{cont}} \text{ on } D_i, i = 1, \dots, N; \tag{7.2.20}$$

$$\frac{\partial \phi}{\partial \mathbf{n}}(\mathbf{x}) = 0 \text{ on } \partial Q_{\text{lat}}; \tag{7.2.21}$$

$$\phi(\mathbf{x}) = 1 \text{ on } \partial Q^+, \quad \phi(\mathbf{x}) = -1 \text{ on } \partial Q^-. \tag{7.2.22}$$

Consider the function $\psi(\mathbf{x}) = \phi^{\text{net}}(\mathbf{x}) - \phi^{\text{cont}}(\mathbf{x})$, where $\phi^{\text{net}}(\mathbf{x})$ and $\phi^{\text{cont}}(\mathbf{x})$ are the solutions to problems (7.2.19)–(7.2.22) and (7.2.1)–(7.2.4), respectively. The function $\psi(\mathbf{x})$ is the solution to the following problem (see (7.2.19)–(7.2.22) and (7.2.1)–(7.2.4)):

$$\Delta \psi = 0 \text{ in } Q_m; \tag{7.2.23}$$

$$\psi(\mathbf{x}) = t_i^{\text{cont}} - t_i^{\text{net}} = \tau_i \text{ on } D_i, i \in I; \tag{7.2.24}$$

$$\frac{\partial \psi}{\partial \mathbf{n}}(\mathbf{x}) = 0 \text{ on } \partial Q_{\text{lat}}; \tag{7.2.25}$$

$$\psi(\mathbf{x}) = 0 \text{ on } \partial Q^+, \quad \psi(\mathbf{x}) = 0 \text{ on } \partial Q^-. \tag{7.2.26}$$

Consider the Dirichlet integral

$$I(\psi) = \frac{1}{2} \int_{Q_m} |\nabla \psi(\mathbf{x})|^2 d\mathbf{x} = \frac{1}{2} \int_{Q_m} |\nabla \phi^{\text{net}}(\mathbf{x}) - \nabla \phi^{\text{cont}}(\mathbf{x})|^2 d\mathbf{x} \quad (7.2.27)$$

corresponding to the problem (7.2.23)–(7.2.26).

Relation (7.2.14) for $\phi(\mathbf{x}) = \phi^{\text{net}}(\mathbf{x})$ implies

$$|I(\phi^{\text{net}}) - I(\phi^{\text{cont}})| = \left| \int_{Q_m} (|\nabla \phi^{\text{net}}(\mathbf{x})|^2 - |\nabla \phi^{\text{cont}}(\mathbf{x})|^2) d\mathbf{x} \right| \geq I(\psi). \quad (7.2.28)$$

The rest of the proof is clear from the physical point of view. On one hand, in accordance with (3.14), $I(\phi^{\text{cont}})$ is relatively close to $I(\phi^{\text{net}})$. On the other hand, for non-vanishing τ_i the quantity $I(\psi)$ is sufficiently large. A possible way of resolving this contradiction is to assume that τ_i vanishes. The appropriate calculations are given below.

Due to (7.2.13) and the remark after (7.2.13), we have

$$I(\phi^{\text{cont}}) - I(\phi^{\text{net}}) = O\left(\sqrt{\frac{\delta}{R}}\right) A_d. \quad (7.2.29)$$

From (7.2.28) and (7.2.29) we have

$$I(\psi) \leq O\left(\sqrt{\frac{\delta}{R}}\right) A_d. \quad (7.2.30)$$

Now, we demonstrate that

$$I(\psi) \geq C\sqrt{\frac{R}{\delta}} \min |\tau_i - \tau_j|^2, \quad (7.2.31)$$

where the minimum is taken over the neighbor disks and the constant $C < \infty$ does not depend on τ_i, $i = 1, \ldots, K$ (K means the number of disks and pseudo-disks).

We use the lower estimate from (7.2.10), which, in the case under consideration, takes the form

$$I(\psi) \geq -\frac{1}{2} \int_{Q_m} \mathbf{v}^2(\mathbf{x}) d\mathbf{x} + \sum_{i \in I} \tau_i \int_{\partial D_i} \mathbf{v}(\mathbf{x}) \mathbf{n} d\mathbf{x} \quad (7.2.32)$$

$$+ \int_{\partial Q^+} \mathbf{v}(\mathbf{x}) \mathbf{n} d\mathbf{x} - \int_{\partial Q^-} \mathbf{v}(\mathbf{x}) \mathbf{n} d\mathbf{x},$$

for any $\mathbf{v} \in W_0 = \{\mathbf{v} \in L_2(Q_m) \mid \text{div} \mathbf{v} = 0, \, \mathbf{v}(\mathbf{x})\mathbf{n} = 0 \text{ on } \partial Q_{\text{lat}}\}$.

Consider two neighboring disks D_i and D_j such that $\tau_i \neq \tau_j$ and Π_{ij} is the channel between them (see Figure 7.1). Choose Π_{ij} narrow enough so that it does not intersect the channels between other disks for a dense disk packing. Such a choice is always possible. We consider the function

$$\mathbf{v}(\mathbf{x}) = \begin{cases} \left(0, \dfrac{\tau_i - \tau_j}{H(\mathbf{x})}\right) & \text{in } \Pi_{ij}, \\ (0, 0) & \text{outside } \Pi_{ij}, \end{cases} \quad (7.2.33)$$

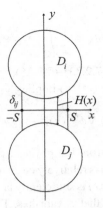

Figure 7.1 Channel between two neighbor disks.

where $H_{ij}(x) = \delta_{ij} + 2R - 2\sqrt{R^2 - x^2}$ is the distance between the disks (a rotation of the coordinate system to the position shown in Figure 7.1 is possible without changing the Dirichlet integrals) and x is the first coordinate of $\mathbf{x} = (x, y)$, see Figure 7.1.

The function (7.2.33) belongs to W_0 and it is divergence-free (see the comment on the formulas (3.6.5) and (3.6.6)). Therefore, it can be used in (7.2.32) as a trial function. For the function (7.2.33), we have (∂D^+ and ∂D^- are lateral boundaries of Π_{ij})

$$\int_{\partial Q^+} \mathbf{v}(\mathbf{x})\mathbf{n}\,dx + \int_{\partial Q^-} \mathbf{v}(\mathbf{x})\mathbf{n}\,dx = 0 \qquad (7.2.34)$$

and

$$-\frac{1}{2}\int_{Q_m} \mathbf{v}^2(\mathbf{x})\,dx = -\frac{1}{2}\int_{\Pi_{ij}} \mathbf{v}^2(\mathbf{x})\,dx \qquad (7.2.35)$$

$$= -\frac{1}{2}\int_{-S}^{S} \frac{(\tau_i - \tau_j)^2}{H(x)}\,dx = -\frac{(\tau_i - \tau_j)^2}{2}\int_{-S}^{S} \frac{dx}{H(x)}$$

($-S$ and S are shown in Figure 7.1).

For the function (7.2.33),

$$\sum_{i \in I} \tau_i \int_{\partial D_i} \mathbf{v}(\mathbf{x})\mathbf{n}\,dx = \tau_i \int_{\partial D_i} \mathbf{v}(\mathbf{x})\mathbf{n}\,dx + \tau_j \int_{\partial D_j} \mathbf{v}(\mathbf{x})\mathbf{n}\,dx.$$

Due to (7.2.34)

$$\int_{\partial D_i} \mathbf{v}(\mathbf{x})\mathbf{n}\,dx = -\int_{\partial D_j} \mathbf{v}(\mathbf{x})\mathbf{n}\,dx.$$

Then for the function (7.2.33), we have

$$\sum_{i \in I} \tau_i \int_{\partial D_i} \mathbf{v}(\mathbf{x})\mathbf{n}d\mathbf{x} = (\tau_i - \tau_j) \int_{\partial D_i} \mathbf{v}(\mathbf{x})\mathbf{n}d\mathbf{x}. \tag{7.2.36}$$

For the function (7.2.33) on the base of the neck Π_{ij} (on $\Pi_{ij} \cap D_i$), we have

$$\mathbf{v}(\mathbf{x})\mathbf{n} = \frac{\tau_i - \tau_j}{H(x)}\mathbf{e}_2\mathbf{n}, \tag{7.2.37}$$

where $\mathbf{e}_2 = (0, 1)$. Substituting (7.2.37) into (7.2.36), we obtain

$$\sum_{i \in I} \tau_i \int_{\partial D_i} \mathbf{v}(\mathbf{x})\mathbf{n}d\mathbf{x} = (\tau_i - \tau_j)^2 \int_{-S}^{S} \frac{\mathbf{e}_2\mathbf{n}s(x)dx}{H(x)}. \tag{7.2.38}$$

where $s(x)dx$ is the measure of length along a circular arc – the base of Π_{ij}.

Taking into account (7.2.34), (7.2.35), and (7.2.38), we see that the inequality (7.2.32) for $\mathbf{v}(\mathbf{x})$ given by (7.2.33) becomes

$$I(\psi) \geq (\tau_i - \tau_j)^2 \int_{-S}^{S} \left(-\frac{1}{2} + \mathbf{e}_2\mathbf{n}s(x) \right) \frac{dx}{H(x)}. \tag{7.2.39}$$

If the channel is chosen to be sufficiently narrow (i.e., S is taken sufficiently small), the inequality

$$-\frac{1}{2} + \mathbf{e}_2\mathbf{n}s(x) \geq c > 0 \tag{7.2.40}$$

is valid for any $x \in [-S, S]$. In what follows, we assume that Π_{ij} is chosen in such a way that (7.2.40) is satisfied. The integral $\displaystyle\int_{-S}^{S} \frac{dx}{H(x)}$ for a pair of disks is of the order of $\sqrt{\dfrac{R}{\delta}}$ (see Section 2.5.3). In view of this, (7.2.39) and (7.2.40) imply

$$I(\psi) \geq C(\tau_i - \tau_j)^2 \sqrt{\frac{R}{\delta_{ij}}}, \tag{7.2.41}$$

where $0 < C < \infty$ is independent of $\{\tau_i\}$.

Combining (7.2.41) and (7.2.30), we obtain

$$C(\tau_i - \tau_j)^2 \sqrt{\frac{R}{\delta_{ij}}} \leq O\left(\sqrt{\frac{\delta}{R}} \right) \sqrt{\frac{R}{\delta_{ij}}}$$

or

$$C(\tau_i - \tau_j)^2 \leq O\left(\sqrt{\frac{\delta}{R}} \right) \tag{7.2.42}$$

for the neighboring disks.

For the boundary disks (including quasi-disks), $\tau_i = 0$. Since the system of disks is connected, from (7.2.42) it follows that

$$\tau_i = O\left(\sqrt{\frac{\delta}{R}}\right) \tag{7.2.43}$$

for all $i = 1, \ldots, K$. $\qquad\qquad\square$

8

Application of the method of complex variables

The method of complex variables is a powerful and effective tool for the analysis of two-dimensional time-independent problems. It is widely applied in a great variety of planar problems (Lu, 1995; Mityushev and Rogosin, 2000), including problems related to composite materials, see, e.g., Berlyand and Mityushev (2001); Grigolyuk and Filshtinskij (1972); Mityushev (1997a, 2001, 2005); Mityushev and Adler (2002a); Mityushev (1997); Pesetskaya (2005). Applications of the method of complex variables for computing the overall properties of composite materials can be found in, e.g., Craster and Obnosov (2004); Makaruk *et al.* (2006); Mityushev (1997a, 2001); Mityushev and Adler (2002a); Mityushev *et al.* (2008); Movchan *et al.* (2002); Pesetskaya (2005).

8.1 $\mathbb{R}$-linear problem and functional equations

In the present section we study the transport properties of infinite unidirectional circular cylinders arbitrarily distributed in a uniform medium by the method of functional equations. Asymptotic formulas for the flux are deduced for densely-packed inclusions. These formulas involve two small parameters: the ratio of the distance between the disks to their radius $\frac{\delta}{R}$ and the degree of a perfect conductor or isolator ε. Using these formulas we amplify the method of functional equations by the structural approximation discussed in the previous chapters to compute the flux between densely-packed disks. Therefore, a combination of these methods yields analytical formulas for the flux in all scales of the distances between the disks for an arbitrary ratio of conductivities of components.

8.1.1 $\mathbb{R}$-linear problem

We start our considerations describing the geometry of composites presented in Section 3.1 (see, in particular, Figure 3.1) in terms of complex variables.

Following Mityushev (1993, 2005) consider N mutually disjoint disks $D_k = \{z \in \mathbb{C} : |z - \zeta_k| < R\}$ on the complex plane $\mathbb{C} \cong \mathbb{R}^2$ of the complex variable $z = x + iy$. Let Q_m be the complement of the closure of all disks to the extended complex plane. Hereafter, the letters z and t are used for a variable inside the domain, and on the boundary of the domain, respectively. We study the conduction of the composite, when the domains Q_m and D_k are occupied by materials of conductivities, respectively

$$a_m = 1, \quad a_i = \lambda > 0. \tag{8.1.1}$$

The potential $u(z)$ satisfies the Laplace equation

$$\Delta u = 0 \text{ in } \left(\bigcup_{k=1}^{N} D_k \right) \bigcup Q_m, \tag{8.1.2}$$

with the conjugation conditions:

$$u^+ = u^-, \quad \frac{\partial u^+}{\partial \mathbf{n}} = \lambda \frac{\partial u^-}{\partial \mathbf{n}} \text{ on } \partial D_k, \ k = 1, 2, \ldots, N, \tag{8.1.3}$$

where $\dfrac{\partial}{\partial \mathbf{n}}$ is the outward normal derivative and

$$u^+(t) = \lim_{\substack{z \to t, \\ z \in Q_m}} u(z), \ u^-(t) = \lim_{\substack{z \to t, \\ z \in D_k}} u(z),$$

$t \in \partial D_k, \ k = 1, 2, \ldots, N$.

We consider the external field applied in the Ox-direction, i.e.,

$$u \sim x = \Re z \text{ as } z \to \infty, \tag{8.1.4}$$

where $\Re$ stands for the real part.

We introduce the complex potentials $\varphi(z)$ and $\varphi_k(z)$ analytic in Q_m and D_k, respectively, and continuously differentiable in the closures of Q_m and D_k. The harmonic and analytic functions are related by the equalities

$$\varphi(z) + z = u(z) + iv(z), \ z \in Q_m,$$

$$\varphi_k(z) = \frac{1+\lambda}{2} (u_k(z) + iv_k(z)), \ z \in D_k,$$

where v and v_k are harmonic conjugate to u and u_k, respectively.

Let $\partial/\partial s$ be the derivative in the natural parameter s (i.e., $\partial/\partial \tau$, τ – unit tangent vector). Using the Cauchy–Riemann equations (Gakhov, 1966) $\dfrac{\partial u}{\partial \mathbf{n}} = -\dfrac{\partial v}{\partial s}$ and $\dfrac{\partial u_k}{\partial \mathbf{n}} = -\dfrac{\partial v_k}{\partial s}$ we can write the second relation (8.1.3) in the form

$$\frac{\partial v}{\partial s} = \lambda \frac{\partial v_k}{\partial s} \text{ on } \partial D_k. \tag{8.1.5}$$

Integration of (8.1.5) in s yields $v = \lambda v_k + c_k$ on ∂D_k, where c_k is an arbitrary real constant. We put $c_k = 0$, since a complex potential is determined up to a purely imaginary additive constant. Hence (8.1.3) becomes

$$u = u_k, \quad v = \lambda v_k \text{ on } \partial D_k. \tag{8.1.6}$$

Add the first relation (8.1.6) and the second relation (8.1.6) multiplied by i:

$$u + iv = u_k + i\lambda v_k \text{ on } \partial D_k. \tag{8.1.7}$$

Introduce the contrast parameter following Bergman *et al.* (1990)

$$\rho = \frac{a_i - a_m}{a_i + a_m} = \frac{\lambda - 1}{\lambda + 1}. \tag{8.1.8}$$

Substituting

$$u_k = \frac{1}{\lambda + 1}(\varphi_k + \overline{\varphi_k}), \quad v_k = \frac{1}{\lambda + 1} \cdot \frac{1}{i}(\varphi_k - \overline{\varphi_k}), \quad u + iv = \varphi(t) + t$$

in (8.1.7) we obtain the ℝ-linear conjugation problem

$$\varphi(t) = \varphi_k(t) - \rho\overline{\varphi_k(t)} - t, \quad |t - \zeta_k| = R, \quad k = 1, 2, \ldots, N. \tag{8.1.9}$$

Equation (8.1.9) ℝ-linearly relates the boundary values of the functions analytic on different sides of the curve.

In order to determine the current $\nabla u(x, y)$ we need the derivatives

$$\psi(z) := \frac{\partial \varphi}{\partial z} = \frac{\partial u}{\partial x} - i\frac{\partial u}{\partial y} - 1, \quad z \in Q_m,$$

$$\psi_k(z) := \frac{\partial \varphi_k}{\partial z} = \frac{1 + \lambda}{2}\left(\frac{\partial u}{\partial x} - i\frac{\partial u}{\partial y}\right), \quad z \in D_k. \tag{8.1.10}$$

Differentiating (8.1.9) we arrive at the following problem

$$\psi(t) = \psi_k(t) + \rho\left(\frac{R}{t - \zeta_k}\right)^2 \overline{\psi_k(t)} - 1, \ |t - \zeta_k| = R, \ k = 1, 2, \ldots, N. \tag{8.1.11}$$

Here, we use the following relation

$$\frac{\partial}{\partial t}\overline{\varphi_k(t)} = \frac{\partial}{\partial t}\overline{\varphi_k\left(\frac{R^2}{t - \zeta_k} + \zeta_k\right)} = -\left(\frac{R}{t - \zeta_k}\right)^2 \overline{\frac{\partial \varphi_k}{\partial t}(t)}, \ |t - \zeta_k| = R.$$

The ℝ-linear problems (8.1.9) and (8.1.11) are closely related to the scalar Riemann–Hilbert problem for multiply connected domains solved in Mityushev (1994)–Mityushev (1998) by the method presented below.

8.1.2 Functional equations

Following Mityushev (1994, 1998) consider the inversion with respect to the circle $|z - \zeta_k| = R$

$$z_{(m)}^* = \frac{R^2}{z - \zeta_k} + \zeta_k. \tag{8.1.12}$$

If a function $f(z)$ is analytic in the disk $|z - \zeta_k| < R$, the function $\overline{f\left(z^*_{(m)}\right)}$ is analytic outside of this disk.

Introduce the function

$$
\Phi\left(z\right) = \begin{cases}
\varphi_k(z) + \rho \sum\limits_{m \neq k} \overline{\varphi_m\left(z^*_{(m)}\right)} - z, & |z - \zeta_k| \leq R, \\[2ex]
\varphi\left(z\right) + \rho \sum\limits_{m=1}^{N} \overline{\varphi_m\left(z^*_{(m)}\right)}, & z \in Q_m
\end{cases}
$$

analytic in every disk D_k and in the domain Q_m. Calculate the jump across the circle $|z - \zeta_k| = R$

$$
\Delta_k := \Phi^+\left(t\right) - \Phi^-\left(t\right),
$$

where $\Phi^+\left(t\right) = \lim\limits_{z \to t \; z \in Q_m} \Phi\left(z\right)$, $\Phi^-\left(t\right) = \lim\limits_{z \to t \; z \in D_k} \Phi\left(z\right)$.

Using (8.1.9) we get $\Delta_k = 0$. It follows from the analytic continuation principle (Ahlfors, 1979) that $\Phi\left(z\right)$ is analytic in the extended complex plane. Then Liouville's theorem implies that $\Phi\left(z\right)$ is a constant, say c. The definition of $\Phi\left(z\right) \equiv c$ in $|z - \zeta_k| \leq R$ yields the following system of functional equations

$$
\varphi_k(z) = -\rho \sum_{m \neq k} \overline{\varphi_m\left(z^*_{(m)}\right)} + z + c. \tag{8.1.13}
$$

Differentiation of (8.1.13) by z and use of (8.1.10) yield the functional equations

$$
\psi_k(z) = \rho R^2 \sum_{m \neq k} \frac{1}{(z - \zeta_m)^2} \, \overline{\psi_m\left(z^*_{(m)}\right)} + 1, \tag{8.1.14}
$$

$$
|z - \zeta_k| \leq R, \; k = 1, 2, \ldots, N.
$$

The functional equations (8.1.13), (8.1.14) do not contain integral terms and contain compositions of functions. This fact essentially simplifies the solution to the equations discussed below.

Let $\psi_k(z)$ be a solution of (8.1.14). Then

$$
\psi\left(z\right) = \rho R^2 \sum_{m=1}^{N} \frac{1}{(z - \zeta_m)^2} \, \overline{\psi_m\left(z^*_{(m)}\right)}, \; z \in Q_m. \tag{8.1.15}
$$

It is convenient to write the flux $q = \left(\dfrac{\partial u}{\partial x}, \dfrac{\partial u}{\partial y}\right)$ as the complex value $q = \dfrac{\partial u}{\partial x} + i \dfrac{\partial u}{\partial y}$. Using (8.1.10) we have

$$
q(z) = \begin{cases}
\overline{\psi\left(z\right)} + 1, & z \in Q_m, \\[2ex]
\dfrac{2}{1 + \lambda} \overline{\psi_k(z)}, & z \in D_k, \; k = 1, 2, \ldots, N.
\end{cases} \tag{8.1.16}
$$

The above formulas have the following physical interpretation. Formula (8.1.14) means that the flux at the k-th inclusions is canceled by the fluxes at all other inclusions ($m = 1, 2, \ldots, N$, $m \neq k$) and by the external flux. Formulas (8.1.15) and (8.1.16) show that the flux at the matrix is canceled by the flux at all inclusions and by the external flux (compare to Rayleigh's approach of the same facts in discrete form presented by Rayleigh (1892); McPhedran *et al.* (1988); McPhedran and Milton (1987); Rylko (2000)).

Theorem 8.7 (Mityushev (1993)) *The system of functional equations (8.1.14) has a unique solution. This solution can be found by the method of successive approximations:*

$$\psi_k(z) = \sum_{p=0}^{\infty} (\rho R^2)^p \psi_k^{(p)}(z), \qquad (8.1.17)$$

where

$$\psi_k^{(0)}(z) = 1, \qquad (8.1.18)$$

$$\psi_k^{(p+1)}(z) = \sum_{m \neq k} \frac{1}{(z - \zeta_m)^2} \, \overline{\psi_m^{(p)}\left(z_{(m)}^*\right)}, \quad p = 0, 1, 2, \ldots.$$

The series (8.1.17) converges absolutely and uniformly and the iterations (8.1.18) uniformly in $|z - \zeta_k| \leq R$ for any fixed $|\rho| \leq 1$.

8.1.3 Flux between almost touching disks

Let $\delta_{km} = |\zeta_k - \zeta_m| - 2R$ be the distance between the disks D_k and D_m, and $\delta = \min_{k \neq m} \delta_{km}$. In this section, we investigate the problem when the dimensionless parameter $\dfrac{\delta}{R}$ is small, i.e., some disks are almost touching.

Consider two disks for which the minimum δ is attained. For definiteness we take $\delta = \delta_{12}$. Our goal is to estimate the flux (8.1.16) between the disks D_1 and D_2 at the point $z = \dfrac{\zeta_1 + \zeta_2}{2}$. Pair-interactions between these disks produce the main part of the local field. But the influence of all other disks to the field between the marked disks D_1 and D_2 is also taken into account.

In order to investigate the functions ψ_1 and ψ_2 satisfying the system (8.1.14) introduce the composition of two inversions with respect to the circles $|z - \zeta_1| = R$ and $|z - \zeta_2| = R$:

$$\alpha(z) = (z_{(1)}^*)_{(2)}^* = \frac{R^2(z - \zeta_1)}{R^2 + (\overline{\zeta_1 - \zeta_2})(z - \zeta_1)} + \zeta_2. \qquad (8.1.19)$$

The Möbius mapping (8.1.19) has two fixed points which can be found from the quadratic equation $\alpha(z) = z$:

$$z_1 = \frac{\zeta_1 + \zeta_2}{2} - \frac{a}{2}\sqrt{1 - \frac{4R^2}{|a|^2}}, \quad z_2 = \frac{\zeta_1 + \zeta_2}{2} + \frac{a}{2}\sqrt{1 - \frac{4R^2}{|a|^2}}, \quad (8.1.20)$$

where $a = \zeta_2 - \zeta_1$. Assuming that R is fixed and $\dfrac{\delta}{R}$ is a small parameter we get

$$z_1 = \frac{\zeta_1 + \zeta_2}{2} - e^{i\theta}\sqrt{R\delta} + O(\delta), \quad z_2 = \frac{\zeta_1 + \zeta_2}{2} + e^{i\theta}\sqrt{R\delta} + O(\delta), \quad (8.1.21)$$

where $\theta = \arg a$ is the argument of the complex number a.

Using (8.1.15) and (8.1.16) we estimate the flux between the disks at the point $z = \dfrac{\zeta_1 + \zeta_2}{2}$. We first consider the value from (8.1.15):

$$\psi\left(\frac{\zeta_1 + \zeta_2}{2}\right) = \rho R^2 \sum_{m=1}^{n} \frac{1}{\left(\frac{\zeta_1+\zeta_2}{2} - \zeta_m\right)^2} \overline{\psi_m\left(\frac{R^2}{\frac{\zeta_1 + \zeta_2}{2} - \zeta_m} + \zeta_m\right)}. \quad (8.1.22)$$

The term with $m = 1$ in (8.1.22) is equal to $\rho R^2 \dfrac{4}{a^2}\overline{\psi_1(w)}$, where $w = \dfrac{2R^2}{\overline{a}} + \zeta_1$. The distance between the points w and z_1 is of order $O(\sqrt{\delta})$, since

$$w - z_1 = \frac{2R^2}{\overline{a}} + \zeta_1 - \frac{\zeta_1 + \zeta_2}{2} + e^{i\theta}\sqrt{R\delta} + O(\delta) + O(\delta) = e^{i\theta}\sqrt{\delta R} + O(\delta). \quad (8.1.23)$$

It follows from (8.1.23) that

$$\overline{\psi_1\left(\frac{2R^2}{\overline{a}} + \zeta_1\right)} = \overline{\psi_1(z_1)} + O(\sqrt{\delta}). \quad (8.1.24)$$

The term with $m = 2$ in (8.1.22) is estimated by the same method:

$$\overline{\psi_2\left(-\frac{2R^2}{\overline{a}} + \zeta_1\right)} = \overline{\psi_2(z_2)} + O(\sqrt{\delta}). \quad (8.1.25)$$

The terms with $m \neq 1, 2$ from (8.1.22) are of order $O(1)$:

$$\rho R^2 \sum_{m \neq 1,2} \frac{R^2}{\left(\frac{\zeta_1 + \zeta_2}{2} - \zeta_m\right)^2} \overline{\psi_m\left(\frac{R^2}{\frac{\zeta_1 + \zeta_2}{2} - \zeta_m} + \zeta_m\right)} = C + O(\delta), \quad (8.1.26)$$

as $\delta \to 0$, where C is a constant independent of δ. Using (8.1.24)–(8.1.26) we obtain from (8.1.22)

$$\psi\left(\frac{\zeta_1 + \zeta_2}{2}\right) = \rho R^2 \frac{4}{a^2}[\overline{\psi_1(z_1)} + \overline{\psi_2(z_2)}] + C + O(\sqrt{\delta}). \quad (8.1.27)$$

We now proceed to estimate $\psi_1(z_1)$ and $\psi_2(z_2)$. Substitute $z = z_1$ in the first equation ($k = 1$) of (8.1.14):

$$\psi_1(z_1) = \rho R^2 \left(\frac{1}{(z_1 - \varsigma_2)^2} \, \overline{\psi_2 \left(\frac{R^2}{\overline{z_1 - \varsigma_2}} + \varsigma_2 \right)} \right. \tag{8.1.28}$$
$$\left. + \sum_{m \neq 1,2} \frac{1}{(z_1 - \varsigma_m)^2} \, \overline{\psi_m \left(z_{(m)}^* \right)} \right) + 1.$$

Using (8.1.21) one can check the relations

$$\frac{1}{(z_1 - \varsigma_2)^2} = \frac{4}{a^2} \left(1 - 2\sqrt{\frac{\delta}{R}} \right) + O(\delta), \tag{8.1.29}$$

$$\frac{R^2}{\overline{z_1 - \varsigma_2}} + \varsigma_2 = z_2 - \frac{2R^2}{\overline{a} \left(1 + \sqrt{\frac{\delta}{R}} \right)} + \frac{a}{2} \left(1 - \sqrt{\frac{\delta}{R}} \right) + O(\delta) = z_2 + O(\delta),$$
$$\tag{8.1.30}$$

since $|a| = 2R - \delta$. It follows from (8.1.30) that

$$\overline{\psi_2 \left(\frac{R^2}{\overline{z_1 - \varsigma_2}} + \varsigma_2 \right)} = \overline{\psi_2(z_2)} + O(\delta). \tag{8.1.31}$$

Substitution of (8.1.29) and (8.1.30) into (8.1.28) yields

$$\psi_1(z_1) = \rho e^{-2i\theta} \left(1 - 2\sqrt{\frac{\delta}{R}} \right) \overline{\psi_2(z_2)} + c_1 + O(\delta), \tag{8.1.32}$$

where $c_1 < \infty$ is a constant not depending on δ. Here, the following relation is used

$$\sum_{m \neq 1,2} \frac{1}{(z_1 - \varsigma_m)^2} \, \overline{\psi_m \left(z_{(m)}^* \right)} + 1 = c_1 + O(\delta).$$

Application of the same arguments to the second functional equation (8.1.14) yields

$$\psi_2(z_2) = \rho e^{-2i\theta} \left(1 - 2\sqrt{\frac{\delta}{R}} \right) \overline{\psi_1(z_1)} + c_2 + O(\delta). \tag{8.1.33}$$

In order to estimate (8.1.27) we add the relations (8.1.32) and (8.1.33):

$$X = \rho e^{-2i\theta} \left(1 - 2\sqrt{\frac{\delta}{R}} \right) \overline{X} + c + O(\delta), \tag{8.1.34}$$

where $X = \psi_1(z_1) + \psi_2(z_2)$. Solving equation (8.1.34) we have

$$X = \frac{c + \rho e^{-2i\theta} \left(1 - 2\sqrt{\frac{\delta}{R}} \right) \overline{c}}{1 - \rho^2 \left(1 - 4\sqrt{\frac{\delta}{R}} \right)} + O(\delta). \tag{8.1.35}$$

Using the relation $|a| = 2R + \delta$ and the definition of X we rewrite (8.1.27) in the form

$$\psi\left(\frac{\zeta_1 + \zeta_2}{2}\right) = \frac{4}{|a|^2}\rho R^2 e^{-2i\theta}\overline{X} + c + O(\sqrt{\delta}). \qquad (8.1.36)$$

Formulas (8.1.35)–(8.1.36) determine the behavior of $\psi\left(\dfrac{\zeta_1 + \zeta_2}{2}\right)$ for small δ. Below we discuss the limit cases of (8.1.35)–(8.1.36) when $\rho = 1$ (high conducting inclusions) and $\rho = -1$ (poorly conducting inclusions). The more complicated cases when $\rho = \pm(1 - \varepsilon)$, as ε tends to zero, are also discussed.

Let $\rho = 1$. Then (8.1.35)–(8.1.36) and (8.1.16) imply that the flux at the point $\dfrac{\zeta_1 + \zeta_2}{2}$ has the form

$$q_1 = q\left(\frac{\zeta_1 + \zeta_2}{2}\right) = \frac{2}{|a|^2}\sqrt{\frac{R}{\delta}}e^{i\theta}R^2\Re e^{i\theta}c + O(1). \qquad (8.1.37)$$

Let $\rho = -1$. Then

$$q_{-1} = q\left(\frac{\zeta_1 + \zeta_2}{2}\right) = -i\frac{2}{|a|^2}\sqrt{\frac{R}{\delta}}e^{i\theta}R^2\Im e^{i\theta}c + O(1), \qquad (8.1.38)$$

where $\Im$ stands for the imaginary part. Formulas (8.1.37) and (8.1.38) show that the fluxes q_{-1} and q_1 have in general a singularity of order $\sqrt{\dfrac{R}{\delta}}$ as δ tends to zero. The coefficients $\Re e^{i\theta}c$ and $\Im e^{i\theta}c$ are of order $O(1)$ and can vanish. The constant c can be determined by matching the fluxes computed in Chapter 4 and by (8.1.37) (for details see Section 8.1.4). Another consequence of formulas (8.1.37) and (8.1.38) is that the main terms of the vectors q_{-1} and q_1 are perpendicular.

A dimensionless small parameter ε is introduced to characterize the degree of perfect conductor/isolator

$$\varepsilon = \begin{cases} 1 - \rho = \dfrac{2a_m}{a_i + a_m} = \dfrac{2}{\lambda + 1}, & \text{as } a_i = \lambda \to \infty, \\[4mm] 1 + \rho = \dfrac{2a_i}{a_i + a_m} = \dfrac{2\lambda}{\lambda + 1}, & \text{as } a_i = \lambda \to 0, \end{cases} \qquad (8.1.39)$$

where the normalization (8.1.1) is used.

We now consider the asymptotic behavior of the flux when not only $\dfrac{\delta}{R}$ tend to zero, but also $\varepsilon \to 0$ where $\rho = 1 - \varepsilon$. Then (8.1.16), (8.1.35) and (8.1.36) imply the following formula

$$q_1 = e^{i\theta}\frac{2\Re ce^{i\theta} - \varepsilon(ce^{i\theta} + 2\bar{c}e^{-i\theta}) - 2\bar{c}e^{-i\theta}\sqrt{\frac{\delta}{R}}}{4\sqrt{\frac{\delta}{R}}(1 - 2\varepsilon) + 2\varepsilon} + \Delta, \qquad (8.1.40)$$

where $\Delta = O(1)$, as $\dfrac{\delta}{R}$ and ε tend to zero. The flux q_1 depends on two small parameters $\dfrac{\delta}{R}$ and ε and on one constant c which is determined by the field induced by the disks D_k, $k = 3, 4, \ldots, N$.

The following formula holds for poorly conducting inclusions

$$q_{-1} = -ie^{i\theta} \frac{2\Im ce^{i\theta} - \varepsilon(ce^{i\theta} + 2\bar{c}e^{-i\theta}) - 2\bar{c}e^{-i\theta}\sqrt{\frac{\delta}{R}}}{4\sqrt{\frac{\delta}{R}}(1 - 2\varepsilon) + 2\varepsilon} + \Delta, \qquad (8.1.41)$$

when $\rho = -1 + \varepsilon$, ε is a small positive parameter.

Formulas (8.1.40), (8.1.41) can be simplified for various scales of the small parameters $\sqrt{\dfrac{\delta}{R}}$ and ε. Consider an example when $\sqrt{\dfrac{\delta}{R}} \ll \varepsilon \ll 1$. Then (8.1.40) and (8.1.41) become, respectively,

$$q_1 = e^{i\theta}\frac{\Re ce^{i\theta}}{\varepsilon} + \Delta, \quad q_{-1} = -ie^{i\theta}\frac{\Im ce^{i\theta}}{\varepsilon} + \Delta. \qquad (8.1.42)$$

The asymptotic formulas (8.1.35)–(8.1.42) are consistent with the theory presented in Chapter 3, e.g., one of the results on the densely-packed disks of high conductivity ($\rho = 1$). The continuous composite is approximated by a network connecting the centers of the disks chosen in accordance with the Voronoi tessellation, see Section 3.2.5. The specific flux between the disks with numbers i and j is given by the asymptotic formula (2.5.8)

$$q_{ij} = \pi\sqrt{\frac{R}{\delta_{ij}}}. \qquad (8.1.43)$$

Below, we compare this formula with the results obtained in Section 8.1.3.

It is worth noting that (8.1.35) and (8.1.36) imply that the flux between closed disks does not form a singularity at δ if $|\rho|$ is away from one. Therefore, the structural approximation (see Chapter 2) cannot be applied in this case.

The asymptotic formula (8.1.37) confirms the formula (8.1.43) obtained for $\rho = 1$ (for details see Section 8.1.4). It follows from (8.1.38) that the structural approximation can be applied for $\rho = -1$. This fact can also be deduced from the following general consideration. The flux $q(z)$ has the form (8.1.16), where the complex potentials $\psi(z)$ and $\psi_k(z)$ satisfy the ℝ-linear problem (8.1.11). Let us multiply (8.1.11) by $-i$:

$$-i\psi(t) = -i\psi_k(t) + \rho\left(\frac{R}{t - \zeta_k}\right)^2 \overline{(-i)\psi_k(t)} + i, \ t \in \partial D_k, \qquad (8.1.44)$$

$$k = 1, 2, \ldots, N.$$

One can see that $-i\psi(z)$ and $-i\psi_k(z)$ satisfy the ℝ-linear problem (8.1.44) which describes the flux in another composite with the same geometry, but the external field is rotated by 90° and the conductivities of inclusions and the matrix are interchanged (ρ is replaced by $-\rho$). As a result of these changes, the flux in the second composite is equal to the flux in the original one, but it is rotated by 90°. Similar arguments were used by Keller (1964) to deduce his famous duality relation for the effective conductivity. Therefore, the structural approximation can be applied to poorly conducting inclusions ($\rho = -1$) for 2D composites.

If $\sqrt{\dfrac{\delta}{R}} \ll \varepsilon \ll 1$, the flux is estimated only via ε. As it follows from (8.1.42), the flux between the disks is proportional to $\dfrac{1}{\varepsilon}$. Thus, the flux is sensitive to deviations of ρ near the limit points $\rho = \pm1$, when ε exceeds $\sqrt{\dfrac{\delta}{R}}$. Hence, the structural approximation developed in Chapters 3 and 4 can also be applied, but with the asymptotics (8.1.42) depending on ε.

8.1.4 Combination of functional equations and structural approximation

The theoretical results presented in the previous sections can be used not only to separate out the cases when the method of functional equations or structural approximation can be applied to dilute or densely-packed composites, respectively; they are also useful to combine both approaches to composites with all scales of the distances between the disks.

The flux in the matrix is given by formulas (8.1.15) and (8.1.16), where ψ_k are calculated in Theorem 8.7. These calculations can be expensive for the flux in the gaps between close disks (see Rylko (2008b)). While the formula (8.1.43) describes the total specific flux between two disks, it does not give any estimates on the flux locally in the gaps. To this end, we will use the method of structural approximation to find the flux in the gaps between close disks when $\rho = 1$, and the method of functional equations to compute the flux far away from the gaps. As we will see below, the results obtained by these methods match.

We estimate the flux at the mid point between the disks. Introduce a local coordinates near the two chosen disks in such a way that the centers of the disks $-\dfrac{a}{2}, \dfrac{a}{2}$ lie on the real axis. Introduce the positive constant

$$S = \frac{\dfrac{R}{a} + \dfrac{1}{2}\left(1 + \sqrt{1 - 4\left(\dfrac{R}{a}\right)^2}\right)}{\dfrac{R}{a} + \dfrac{1}{2}\left(1 - \sqrt{1 - 4\left(\dfrac{R}{a}\right)^2}\right)}. \tag{8.1.45}$$

The fixed point z_2 defined by (8.1.20) takes the form

$$z_2 = \frac{a}{2}\sqrt{1 - 4\left(\dfrac{R}{a}\right)^2}. \tag{8.1.46}$$

This allows us to give an explicit formula for the potential

$$u(z) = -\frac{u_0}{2 \ln S} \ln \left| \frac{z - z_2}{z + z_2} \right|.$$ (8.1.47)

In order to prove (8.1.47) consider the conformal mapping

$$w = \frac{z - z_2}{z + z_2}$$ (8.1.48)

of $D = \{z \in \mathbb{C} : |z \pm a/2| > R\}$ onto the annulus $D' = \{w \in \mathbb{C} : 1/S < |w| < S\}$. The value of the function $u(z)$ is equal to $\pm u_0/2$ on the circles $|z \pm a/2| = R$, since $u(z) = -\frac{u_0}{2 \ln S} \ln |w|$.

According to (8.1.16) the flux can be expressed as the complex value

$$q(z) = \overline{\psi(z)} + 1 = -\frac{u_0}{2 \ln S} \frac{4z_2}{z^2 - z_2^2} + 1.$$ (8.1.49)

The flux at the point $z = 0$ is equal to

$$q(0) = \frac{u_0}{z_2 \ln S} + 1.$$ (8.1.50)

Assuming that $\delta \to 0$ and using (8.1.45), (8.1.46) we obtain from (8.1.50)

$$q(0) = \frac{u_0}{2\delta} + O\left(\frac{1}{\sqrt{\delta}}\right).$$ (8.1.51)

One can see that $q(0)$ depends on the difference of the potentials on the disks u_0 and the specific flux δ^{-1} at the mid point which differs from the total specific flux (8.1.43). In order to estimate u_0 in the external field we consider the following problem

$$\hat{u}(z) = -\frac{u_0}{2}, \quad \left| z + \frac{a}{2} \right| = R, \quad \hat{u}(z) = \frac{u_0}{2}, \quad \left| z - \frac{a}{2} \right| = R,$$ (8.1.52)

where the function $\hat{u}(z)$ is harmonic in D except at infinity where $\hat{u}(z) \sim -x$.

Rewrite the problem (8.1.52) in terms of the variable w given by (8.1.48):

$$v(w) = -\frac{u_0}{2} + z_2\Re\frac{1+w}{1-w}, \quad |w| = \frac{1}{S}, \quad v(w) = \frac{u_0}{2} + z_2\Re\frac{1+w}{1-w}, \quad |w| = S,$$ (8.1.53)

where $v(w) = \hat{u}(z) + \Re z$ is harmonic in D', $B_1 = \{|w| < 1/S\}$ and $B_2 = \{|w| > S\}$. The boundary-value problem (8.1.53) can be solved by the method of functional equations. Introduce the complex potentials $\varphi(w) := v(w) + i\tilde{v}(w)$, $\varphi_1(w) := v_1(w) + i\tilde{v}_1(w)$ and $\varphi_2(w) := v_2(w) + i\tilde{v}_2(w)$ analytic in D', B_1 and B_2, respectively. In terms of the potentials, the equations (8.1.53) become

$$\Re\varphi(w) = \Re\varphi_1(w) = -\frac{u_0}{2} + z_2\Re\frac{1+w}{1-w}, \quad \text{on } \partial B_1,$$ (8.1.54)

$$\Re\varphi(w) = \Re\varphi_2(w) = \frac{u_0}{2} + z_2\Re\frac{1+w}{1-w}, \quad \text{on } \partial B_2,$$

which are satisfied by the solutions of the following $\mathbb{R}$-linear problem

$$\varphi(w) = \varphi_1(w) - \overline{\varphi_1(w)} - \frac{u_0}{2} + z_2 \frac{1+w}{1-w}, \quad \text{on } \partial B_1, \qquad (8.1.55)$$

$$\varphi(w) = \varphi_2(w) - \overline{\varphi_2(w)} + \frac{u_0}{2} + z_2 \frac{1+w}{1-w}, \quad \text{on } \partial B_2.$$

Mityushev (1994) has established that the problems (8.1.53), (8.1.55) are equivalent. Proceeding as in Section 8.1.2, introduce

$$\Phi(w) = \begin{cases} \varphi_1(w) + \overline{\varphi_2\left(\dfrac{S^2}{\overline{w}}\right)} - \dfrac{u_0}{2} + z_2 \dfrac{1+w}{1-w}, & w \in B_1, \\[3ex] \varphi_2(w) + \overline{\varphi_1\left(\dfrac{1}{S^2\overline{w}}\right)} + \dfrac{u_0}{2} + z_2 \dfrac{1+w}{1-w}, & w \in B_2, \\[3ex] \varphi(w) + \overline{\varphi_1\left(\dfrac{1}{S^2\overline{w}}\right)} + \overline{\varphi_2\left(\dfrac{S^2}{\overline{w}}\right)}, & w \in D'. \end{cases}$$

The conditions (8.1.54) imply that $\Phi(w)$ is continuous across the interfaces. By the analytic continuation principle, we conclude that $\Phi(w)$ is analytic in the extended complex plane. Moreover, since it is bounded, by Liouville's theorem, $\Phi(w)$ is constant c.

Thus, we arrive at the following functional equations

$$\varphi_1(w) = -\overline{\varphi_2\left(\frac{S^2}{\overline{w}}\right)} + \frac{u_0}{2} - z_2 \frac{1+w}{1-w} + c, \quad |w| \leq \frac{1}{S}, \qquad (8.1.56)$$

$$\varphi_2(w) = -\overline{\varphi_1\left(\frac{1}{S^2\overline{w}}\right)} - \frac{u_0}{2} - z_2 \frac{1+w}{1-w} + c, \quad |w| \geq S. \qquad (8.1.57)$$

It is possible to solve equations (8.1.56)–(8.1.57). But for our purposes it is sufficient to determine u_0. It is easy to get

$$u_0 = 2z_2 \qquad (8.1.58)$$

by substitution of $w = 0$ in (8.1.56) and $w = \infty$ in (8.1.57).

The flux at the mid point $q(0)$ from (8.1.50) becomes

$$q(0) = \frac{2}{\ln S} + 1. \qquad (8.1.59)$$

Assuming that $\dfrac{\delta}{R} = \dfrac{a}{R} - 2$ tends to zero we obtain from (8.1.59) and (8.1.45)

$$q(0) = 2\sqrt{\frac{R}{\delta}} + O(1). \qquad (8.1.60)$$

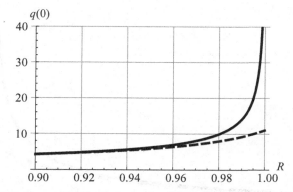

Figure 8.1 Dependence of the flux between the disks on R. Data are for: five iterations of the functional equations (dashed), 200 iterations of the functional equations (long dashes), and formula (8.1.60) (solid). The long dashes and the solid curve coincide.

This is the main formula to estimate the flux between close disks. One can see from (8.1.58) and (8.1.21) that the difference of the potentials between the disks is

$$u_0 = 2\sqrt{R\delta} + O(\delta). \tag{8.1.61}$$

The total flux between the disks is calculated by substitution of $z = iy$ in (8.1.49) and by calculation of the integral

$$\frac{u_0}{\ln S} \int_{-\infty}^{\infty} \frac{z_2}{y^2 + z_2^2} dy = \frac{\pi}{\ln S} u_0 = \pi \sqrt{\frac{R}{\delta}} u_0 + O(1). \tag{8.1.62}$$

Hence, the total specific flux is equal to $\dfrac{\pi}{\ln S} = \pi\sqrt{\dfrac{R}{\delta}} + O(1)$. This coincides with (8.1.43) (see (2.5.8)) and confirms that asymptotic formulas deduced by functional equations coincide with the formulas of the structural approximation.

The difference between the results obtained by the two methods and the possibility to match these results are demonstrated in the following example. Consider two perfectly conducting disks with centers $\zeta_1 = -1$ and $\zeta_2 = 1$. The flux q is calculated at the mid point $z = 0$. It follows from Figure 8.1 that five iterations in the method of functional equations (see Theorem 8.7) are not enough for $R \sim 0.99$. The results obtained by 200 iterations coincide with the results obtained by formula (8.1.60). However, one can see from Figure 8.2 that 200 iterations are not enough for $R \sim 0.9999$. Hence, the flux can be precisely computed by functional equations exterior to a sufficiently small domain $|z| < \eta$. The value of η is chosen in such a way that the flux is matched with the flux given by (8.1.60) in $|z| < \eta$ along the circle $|z| = \eta$. It follows from Figures 8.1 and 8.2 that η can be taken as 0.05 for five iterations and as 0.00005 for 200 iterations.

Based on the constructive method presented in Chapter 3 and on the asymptotic formulas (8.1.37)–(8.1.42) it is possible to apply the following method to

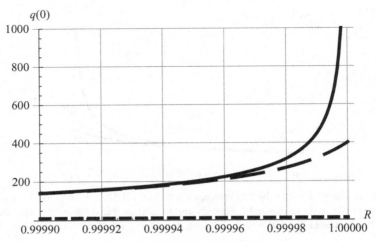

Figure 8.2 Notation as in Figure 8.1.

determine the local flux for arbitrary fixed locations of the disks. First, using the method of functional equations with a moderate number of iterations yields an analytic approximate formula for the flux out of the gaps between closed disks. Second, the method presented in Chapter 3 yields exact expressions for the differences of the potentials Δu_{ij} between the closed disks with the numbers i and j for any fixed location of the disks. Then, one can determine the constants $C_{ij} = \dfrac{R^2}{2} \Re e^{i\theta} c$ from (8.1.37) by the relation $C_{ij} = \dfrac{\Delta u_{ij}}{\sqrt{R\delta_{ij}}}$ and obtain the flux (8.1.37) in the gaps between the disks. Note that (8.1.61) implies that Δu_{ij} is of order $\sqrt{R\delta_{ij}}$.

The same constants c_{ij} are used in (8.1.38) for $\rho = -1$ and in (8.1.42) for
$$\sqrt{\frac{\delta_{ij}}{R}} \ll \varepsilon \ll 1.$$

8.2 Doubly-periodic problems

The method of complex potentials and functional equations is applied to conductivity problems with a finite number of inclusions in the whole plane in Section 8.1. This approach can be useful to get analytical formulas for the effective conductivity tensor A for dilute composites. It is worth noting that the concentration of a finite number of inclusions in the plane is equal to zero. In order to obtain formulas for A in the case of higher concentrations we have to investigate periodic structures (Bensoussan *et al.*, 1975; Oleinik *et al.*, 1962) or structures reduced to a cell which can be periodically continued onto the whole plane (see Chapter 3).

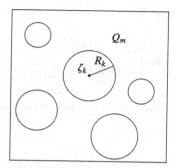

Figure 8.3 Periodicity cell of a doubly-periodic composite.

In this section, we extend and apply the method of Section 8.1 to boundary-value problems for periodic composites by use of elliptic function theory outlined in Section 1.5.

8.2.1 Formulation of the boundary-value problem

Following Berlyand and Mityushev (2001) consider a periodic two-dimensional lattice $\mathcal{L}$ which is defined by the two fundamental translation vectors 1 and i in the complex plane $\mathbb{C}$. The $(0, 0)$-th cell $Q_{(0,0)}$ is the unit square

$$Q_{(0,0)} = \{z = x + iy \in \mathbb{C} : -1/2 < x, y < 1/2\}.$$

The periodic lattice $\mathcal{L}$ consists of the cells $Q_{(m_1,m_2)} = Q_{(0,0)} + m_1 + im_2$ obtained by periodic repetition by the cell $Q_{(0,0)}$. Assume that the periodicity cell $Q_{(0,0)}$ contains mutually non-overlapping disks D_k (particles) of radii R_k centered at ζ_k, that is

$$D_k = \{z \in \mathbb{C} : |z - \zeta_k| < R_k\}, k = 1, 2, \ldots, N.$$

We also introduce the notation for the circles of the disks $T_k = \{z \in \mathbb{C} : |z - \zeta_k| = R_k\}$, and for the part of the unit square occupied by the matrix

$$Q_m = P \backslash \left(\bigcup_{k=1}^{N} (D_k \bigcup T_k) \right).$$

We study the conductivity of the doubly-periodic composite material, when the domains Q_m (matrix) and D_k (particles) are occupied by materials of conductivities a_m and a_i, respectively, see Figure 8.3. The potential $u(x, y)$ satisfies the Laplace equation

$$\Delta u = 0 \text{ in } \left(\bigcup_{k=1}^{N} D_k \right) \bigcup Q_m, \qquad (8.2.1)$$

with the conjugation conditions:

$$[u] = 0 \text{ on } T_k, \tag{8.2.2}$$

$$\left[a(\mathbf{x}) \frac{\partial u}{\partial \mathbf{n}} \right] = 0 \text{ on } T_k, \ k = 1, 2, \dots, N, \tag{8.2.3}$$

where [ ] stands for the "jump", the difference of the values of a function across the interface curve T_k.

The function

$$a(\mathbf{x}) = \begin{cases} a_m = 1 & \text{in matrix,} \\ a_i = \lambda & \text{in disks.} \end{cases}$$

describes the spatial distribution of the conductivity coefficient in the composite. For the sake of definiteness we assume in this section that $a_i > a_m$, i.e., $\lambda > 1$.

We also impose on $u(x, y)$ the quasi-periodicity conditions which correspond to an external field applied in the Ox-direction

$$u(x + 1, y) = u(x, y) + 1, \tag{8.2.4}$$

$$u(x, y + 1) = u(x, y).$$

Our main objective is to find the analytical dependence of the effective conductivity for macroscopically isotropic composites

$$A = \int_P a(x, y) |\nabla u(x, y)|^2 dx dy \tag{8.2.5}$$

on geometrical and physical parameters. We compute A for an arbitrary deterministic array of particles and on this basis in the next sections we consider analogous problems for random locations of ζ_k and random distributions of the disk's radii R_k.

We rewrite problem (8.2.1)–(8.2.4) in terms of complex potentials $\psi(z)$ and $\psi_k(z)$ which are analytic in Q_m and D_k, respectively, and continuous in the closures of Q_m and D_k (compare to (8.1.10))

$$\psi(z) = \frac{\partial u}{\partial x} - i \frac{\partial u}{\partial y} - 1, \ z \in Q_m,$$

$$\psi_k(z) = \frac{a_i + a_m}{2a_m} \left(\frac{\partial u}{\partial x} - i \frac{\partial u}{\partial y} \right), \ z \in D_k. \tag{8.2.6}$$

Straightforward calculations imply from (8.2.4) and (8.2.6) that $\psi(z)$ is doubly-periodic:

$$\psi(z + 1) = \psi(z) = \psi(z + i). \tag{8.2.7}$$

Repeating the arguments presented in Section 8.1.1 we rewrite the two real conditions (8.2.3) as one complex condition (compare to (8.1.11))

$$\psi(t) = \psi_k(t) + \rho \left(\frac{R_k}{t - \zeta_k} \right)^2 \overline{\psi_k(t)} - 1,$$
$$|t - \zeta_k| = R_k, \quad k = 1, 2, \ldots, N, \tag{8.2.8}$$

where ρ is introduced in (8.1.8).

Following Section 8.1.2 one can reduce the boundary-value problem (8.2.8) to a system of functional equations (compare to (8.1.14))

$$\psi_q(z) = \rho \sum_{k=1}^{N} \sum_{m_1, m_2}^{*} (W_{m_1 m_2 k} \psi_k)(z) + 1,$$
$$|z - \zeta_q| \leq R_q, \quad q = 1, 2, \ldots, N, \tag{8.2.9}$$

where $\displaystyle\sum_{m_1, m_2}^{*}$ is taken over all integers m_1 and m_2 except $m_1 = m_2 = 0$ for $k = q$.
The operator $W_{m_1 m_2 k}$ is defined as follows

$$(W_{m_1 m_2 k} \psi_k)(z) = \left(\frac{R_k}{z - \zeta_k - m_1 - i m_2} \right)^2 \overline{\psi_k \left(\frac{R_k^2}{z - \zeta_k - m_1 - i m_2} + \zeta_k \right)}. \tag{8.2.10}$$

The properties of this operator and the convergence of the infinite sum in the right-hand part of (8.2.9) are outlined in Mityushev (1997b). The main difficulty of the reduction of (8.2.8) to (8.2.9) is based on the conditional convergence of the infinite sum which must be well defined by fixing the order of summation. To overcome this difficulty we use the Eisenstein summation method presented in Section 1.5 and below.

A formal solution of the functional equations (8.2.9) can be sought in the form of power series in the maximal radius $R_l = \max_{1 \leq k \leq N} R_k$

$$\psi_m(z) = \sum_{q=0}^{\infty} \psi_m^{(q)}(z) R_l^{2q}, \quad m = 1, 2, \ldots, N. \tag{8.2.11}$$

Then the coefficients $\psi_m^{(q)}(z)$ are analytic in z. Expansion in any other fixed R_k^2 yields the same result.

We next write each function $\psi_m^{(q)}(z)$ in the form of Taylor series

$$\psi_m^{(q)}(z) = \sum_{n=0}^{\infty} \psi_{nm}^{(q)} (z - \zeta_m)^n, \quad m = 1, 2, \ldots, N. \tag{8.2.12}$$

Then we formally have

$$\sum_{m_1,m_2} (W_{m_1 m_2 m} \psi_m^{(q)})(z) = \sum_{m_1,m_2} \sum_{n=0}^{\infty} \overline{\psi_{nm}^{(q)}} \frac{R_m^{2(n+1)}}{(z - \zeta_m - m_1 - im_2)^{n+2}} \quad (8.2.13)$$

$$= \sum_{n=0}^{\infty} \overline{\psi_{nm}^{(q)}} R_m^{2(n+1)} \sum_{m_1,m_2} \frac{1}{(z - \zeta_m - m_1 - im_2)^{n+2}}.$$

The latter expression contains the sums

$$E_{n+2}(Z) = \sum_{m_1,m_2} \frac{1}{(Z - m_1 - im_2)^{n+2}}, \quad (8.2.14)$$

where $Z = z - \zeta_m$. For any $n > 1$, (8.2.14) converges absolutely for any $z \neq m_1 + im_2$ for integers m_1 and m_2. This is the Eisenstein series of order $n + 2$ (compare to (1.5.19)). The conditionally convergent sum for $n = 0$ can be expressed as the iterated limit

$$E_2(Z) = \sum_{m_1,m_2} \frac{1}{(Z - m_1 - im_2)^2} \quad (8.2.15)$$

$$= \lim_{M_2 \to \infty} \sum_{m_2=-M_2}^{M_2} \lim_{M_1 \to \infty} \sum_{m_1=-M_1}^{M_1} \frac{1}{(Z - m_1 - im_2)^2}.$$

The limit determines the order of summation: first sum along the Ox-axis and then stretch the infinite strip in the Oy-direction. It is hard to compute the Eisenstein functions by definition, since the series (8.2.14)–(8.2.15) converge slowly. Formulas for more efficient computations are described at the end of Section 1.5.

Introduce the ratios

$$\nu_k = R_k^{-2} R_l^2, \quad k = 1, 2, \dots, N. \quad (8.2.16)$$

Substituting (8.2.11), (8.2.12) in (8.2.9) and using (8.2.14), (8.2.16) we obtain the relation

$$\sum_{q=0}^{\infty} \psi_m^{(q)}(z) R_l^{2q} = \rho \sum_{k=1}^{N} \sum_{n=0}^{\infty} \sum_{q=0}^{\infty} \overline{\psi_{nk}^{(q)}} R_l^{2(q+n+1)} \nu_k^{n+1} E_{n+2}(z - \zeta_k) + 1,$$

$$|z - \zeta_k| \leq R_m, \quad m = 1, 2, \dots, N. \quad (8.2.17)$$

Equating the coefficients in front of R_l^2 in (8.2.17) we arrive at the following proposition in which we determine $\psi_k^{(q)}(z)$ recursively and thus obtain the solution $\psi_k(z)$ of the functional equations (8.2.9).

Proposition 8.8 (Berlyand and Mityushev, 2001; Mityushev, 1997b, 1999)
Let $\psi_k(z)$ be a solution of the functional equations (8.2.9). Then it admits the representation (8.2.11), where

$$\psi_m^{(0)}(z) = 1, \tag{8.2.18}$$

$$\psi_m^{(q+1)}(z) = \rho \sum_{k=1}^{N} [\overline{\psi_{0k}^{(q)}} v_k E_2(z - \zeta_k) + \overline{\psi_{1k}^{(q-1)}} v_k^2 E_3(z - \zeta_k) + \ldots$$

$$+ \overline{\psi_{q-1,k}^{(1)}} v_k^q E_{q+1}(z - \zeta_k)], \quad q = 0, 1, 2, \ldots.$$

Here, $\psi_{lk}^{(q)}$ is the l-th coefficient of the Taylor expansion of $\psi_k^{(q)}(z)$ (see 8.2.12)). The function $\psi_k(z)$ is represented in the form of the absolutely and uniformly convergent series (8.2.11) in the closure of D_k. The iterations (8.2.18) converges uniformly in the closures of D_k.

Remark. In the formula (8.2.17) for a disk D_m ($m = 1, 2, \ldots, N$) and later in the chapter we introduce the function

$$E_l^{(m)}(z - \zeta_k) = \begin{cases} E_l(z - \zeta_k) & \text{if } k \neq m, \\ E_l(z - \zeta_m) - (z - \zeta_m)^{-l} & \text{if } k = m, \end{cases} \tag{8.2.19}$$

where $l = 2, 3, \ldots$. For simplicity we write E_l instead of $E_l^{(m)}$. The use of the designation (8.2.19) simplifies the final formula for the effective conductivity. For instance, (8.2.19) implies

$$E_n(0) = S_n, \quad n = 2, 3, \ldots, \tag{8.2.20}$$

where S_n are the Eisenstein–Rayleigh lattice sums (1.5.14).

From (8.2.18) we obtain the explicit formulas

$$\psi_m^{(1)}(z) = \rho \sum_{k=1}^{N} v_k E_2(z - \zeta_k), \tag{8.2.21}$$

$$\psi_m^{(2)}(z) = \rho^2 \sum_{k,k_1=1}^{N} v_k v_{k_1} \overline{E_2(\zeta_k - \zeta_{k_1})} E_2(z - \zeta_{k_1}),$$

$$\psi_m^{(3)}(z) = \rho^3 \sum_{k,k_1,k_2=1}^{N} v_k v_{k_1} v_{k_2} \overline{E_2(\zeta_k - \zeta_{k_1})} E_2(\zeta_{k_1} - \zeta_{k_2}) E_2(z - \zeta_{k_2})$$

$$- 2\rho^2 \sum_{k,k_1=1}^{N} v_k^2 v_{k_1} \overline{E_3(\zeta_k - \zeta_{k_1})} E_3(z - \zeta_k)$$

and so on.

Thus, we have determined the functions $\psi_k(z)$ in (8.2.11) recursively; it is shown in the next section that these functions determine the effective conductivity, so that it is not necessary to find $\psi(z)$.

8.2.2 The complex variable formula for effective conductivity

The local relation between the flux $\mathbf{J}$ and the gradient $\nabla\phi$ in the unit cell $Q_{(0,0)}$ is given by the Fourier law in the thermal conductivity or Ohm's law in electric problems

$$
\mathbf{J} = \begin{cases} a_m\nabla u = \nabla u & \text{in } Q_m, \\ a_i\nabla u = \lambda\nabla u & \text{in } \bigcup_{\ell=1}^{N} D_\ell. \end{cases}
\tag{8.2.22}
$$

Generally, the effective medium is anisotropic. The effective conductivity tensor of an anisotropic medium in two dimensions has the form

$$
A = a_m \begin{pmatrix} A^x & A^{xy} \\ A^{xy} & A^y \end{pmatrix}.
\tag{8.2.23}
$$

It relates the average values over the unit cell of the flux and the gradient

$$
\mathbf{J} = A\nabla u.
\tag{8.2.24}
$$

The multiplier a_m is kept in this representation to stress that the components A^x, A^{xy}, A^y are dimensionless, though $a_m = 1$ in the normalized form (8.1.1).

In order to determine A it is sufficient to solve two cell problems (see Jikov *et al.* (1994); Bensoussan *et al.* (1978); Kalamkarov and Kolpakov (1997); Bakhvalov and Panasenko (1989)) when the external field is applied either in the Ox- (see (8.2.1)–(8.2.3)) or Oy-direction.

Fix for definiteness the Ox-direction. Then

$$
a_m A^x = a_m \int_{Q_m} \frac{\partial u}{\partial x} dxdy + a_i \sum_{\ell=1}^{N} \int_{D_\ell} \frac{\partial u}{\partial x} dxdy,
\tag{8.2.25}
$$

$$
a_m A^{xy} = a_m \int_{Q_m} \frac{\partial u}{\partial y} dxdy + a_i \sum_{\ell=1}^{N} \int_{D_\ell} \frac{\partial u}{\partial y} dxdy,
\tag{8.2.26}
$$

where $u(x, y)$ is the solution of the problem (8.2.1)–(8.2.4).

The physical meaning of $a_m A^x$ is the total flux in the Ox-direction through the composite under the action of unit difference of potential applied to the composite in the Ox-direction, i.e., a component of the effective conductivity tensor. In an anisotropic composite, application of unit difference of potential in the Ox-direction may cause a flux in the Oy-direction. This flux is equal to $a_m A^{xy}$, i.e., also a component of the effective conductivity tensor.

Using Green's formula

$$
\int_{G} \left(\frac{\partial g}{\partial x} - \frac{\partial f}{\partial y} \right) dx\,dy = \int_{\partial G} f(x, y)dx + g(x, y)dy
\tag{8.2.27}
$$

we obtain

$$\int_{Q_m} \frac{\partial u}{\partial x} dx dy = \int_{\partial Q_m} u(x, y) dy = \int_{\partial Q_{(0,0)}} u(x, y) dy - \sum_{\ell=1}^{N} \int_{\partial D_\ell} u(x, y) dy.$$

Here we assume that the boundary of the cell $\partial Q_{(0,0)}$ and the curves ∂D_ℓ are oriented in the counterclockwise direction. Since the jump of u on the unit cell along the x-direction is equal to 1, we get

$$\int_{Q_m} \frac{\partial u}{\partial x} dx dy = 1 - \sum_{\ell=1}^{N} \int_{D_\ell} \frac{\partial u}{\partial x} dx dy.$$

Along similar lines we have

$$\int_{Q_m} \frac{\partial u}{\partial y} dx dy = - \sum_{\ell=1}^{N} \int_{D_\ell} \frac{\partial u}{\partial y} dx dy.$$

Then (8.2.25)–(8.2.26) become

$$a_m (A^x - i A^{xy}) = a_m + (a_i - a_m) \sum_{\ell=1}^{N} \int_{D_\ell} \left(\frac{\partial u}{\partial x} - i \frac{\partial u}{\partial y} \right) dx dy.$$

Using (8.2.6) (see also (8.1.10)) we obtain

$$A^x - i A^{xy} = 1 + 2\rho \sum_{\ell=1}^{N} \int_{D_\ell} \psi_\ell(z) dx dy.$$

Due to the mean value theorem for harmonic functions we have

$$A^x - i A^{xy} = 1 + 2\rho \sum_{\ell=1}^{N} \pi R_\ell^2 \psi_\ell(\zeta_\ell). \tag{8.2.28}$$

Since $\psi_\ell(z)$ has been calculated in Proposition 8.8, formula (8.2.28) provides an exact formula for $A^x - i A^{xy}$.

For simplicity we assume that our composite material is macroscopically isotropic. Then the effective tensor A is of the form $A\mathcal{I}$, where A is the effective conductivity and $\mathcal{I}$ is the unit tensor. In this case $A^{xy} = 0$ and (8.2.28) becomes

$$A = 1 + 2\rho V \left[B_0 + B_1 V + B_2 V^2 + \cdots \right], \tag{8.2.29}$$

$$V = \pi R_l^2 N \tilde{v} \tag{8.2.30}$$

which is the total area fraction, $\tilde{v} = \sum_{\ell=1}^{N} v_\ell$,

$$B_q = \frac{1}{\pi^q \tilde{v}^{q+1}} \sum_{\ell=1}^{N} v_\ell \psi_\ell^{(q)}(\zeta_m), \quad q = 0, 1, \dots \tag{8.2.31}$$

In order to evaluate $\psi_\ell^{(q)}(\zeta_\ell)$ we introduce

$$Z_{p_1,p_2,...,p_M}^{k_0,k_1,...,k_M} = E_{p_1}(\zeta_{k_0} - \zeta_{k_1})\overline{E_{p_2}(\zeta_{k_1} - \zeta_{k_2})}\ldots \mathbf{C}^{M-1}E_{p_M}(\zeta_{k_{M-1}} - \zeta_{k_M}),$$
(8.2.32)

where $\mathbf{C}$ is the complex conjugation operator, $M = 1, 2, 3, \ldots$; k_j takes values $1, 2, \ldots N$; $p_j = 1, 2, 3, \ldots$. It is convenient to put $E_n(0) := S_n$ (see (8.2.20)). Let

$$P = p_1 + p_2 + \cdots + p_M - M + 1$$

and introduce

$$X_{p_1,p_2,...,p_M} = \frac{1}{\widetilde{\nu}^P} \sum_{k_0,k_1,...,k_M} \nu_{k_0} \nu_{k_1}^{p_1-1} \ldots \nu_{k_M}^{p_M-1} Z_{p_1,p_2,...,p_M}^{k_0,k_1,...,k_M}.$$
(8.2.33)

In accordance with (8.2.18), ν_{k_0} corresponds to ν_k and each function E_n from (8.2.18) has the coefficient ν_k^{q-1}. Combining (8.2.29)–(8.2.33) we obtain the coefficients in (8.2.29) (Szczepkowski *et al.*, 2003):

$$B_0 = 1, \quad B_1 = \frac{\rho}{\pi N^2}X_2,$$
(8.2.34)

$$B_2 = \frac{\rho^2}{\pi^2 N^3}X_{22}, \quad B_3 = \frac{1}{\pi^3 N^4}\left[-2\rho^2 X_{33} + \rho^3 X_{222}\right],$$

$$B_4 = \frac{1}{\pi^4 N^5}[3\rho^2 X_{44} - 2\rho^3(X_{332} + X_{233}) + \rho^4 X_{2222}],$$

$$B_5 = \frac{1}{\pi^5 N^6}[-4\rho^2 X_{55} + \rho^3(3X_{442} + 6X_{343} + 3X_{244})$$
$$- 2\rho^4(X_{3322} + X_{2332} + X_{2233}) + \rho^5 X_{22222}],$$

$$B_6 = \frac{1}{\pi^6 N^7}[5\rho^2 X_{66} - \rho^3(4X_{255} + 12X_{354} + 12X_{453} + 4X_{552})$$
$$+ \rho^4(3X_{2244} + +6X_{2343} + 4X_{3333} + 3X_{2442} + 6X_{3432} + 3X_{4422})$$
$$- 2\rho^5(X_{22233} + X_{22332} + +X_{23322} + X_{33222}) + \rho^6 X_{222222})].$$

It is possible to proceed with (8.2.34) and calculate the next coefficients $B_q, q > 6$. It follows from Proposition 8.8 and (8.2.31) that B_q has the following structure

$$B_q = \sum_{\substack{p_1,p_2,...,p_M \in S^q; \\ k_0,k_1,...,k_M}} \beta_{q,p_1,p_2,...,p_M} \frac{\nu_{k_0}\nu_{k_1}^{p_1-1}\ldots\nu_{k_M}^{p_M-1}}{\widetilde{\nu}^P}Z_{p_1,p_2,...,p_M}^{k_0,k_1,...,k_M},$$
(8.2.35)

where S^q is the set of all indices $2 \le p_1, \ldots, p_M \le q$ such that $p_1 + p_2 + \cdots + p_M = 2q$, $1 \le k_1, \ldots, k_M \le N$ and the constants $\beta_{q,p_1,p_2,...,p_M}$ depend only on ρ and N. For instance, B_1 involves only one term of the type (8.2.33) with the coefficient (see (8.2.34))

$$\beta_{12} = \frac{\rho}{\pi N^2}.$$

Note. For each given $q, p_1, p_2, \ldots, p_M$ the coefficient $\beta_{q,p_1,p_2,\ldots,p_M}$ can be obtained in explicit form. However, for the forthcoming analysis of random models a constructive algorithm for $\beta_{q,p_1,p_2,\ldots,p_M}$, which follows from (8.2.31), and recursive formulas (8.2.18), are sufficient.

It follows from (8.2.35) and (8.2.32) that B_q is a sum of $X_{p_1,p_2,\ldots,p_M}$ with coefficients depending only on ρ and N. Each term $X_{p_1,p_2,\ldots,p_M}$ is a sum of $Z^{k_0,k_1,\ldots,k_M}_{p_1,p_2,\ldots,p_M}$, which depend only on the locations of the centers $\zeta_{k_0}, \zeta_{k_1}, \ldots, \zeta_{k_M}$. The coefficients $v_{k_0} v_{k_1}^{p_1-1} \ldots v_{k_M}^{p_M-1} \widetilde{v}^{-P}$ depend only on the radii of the particles. This decomposition of B_q holds for each parameter ρ, ζ_k, R_k for which the particles D_k do not overlap.

8.3 Optimal design problem for monodispersed composites

The analytical formulas obtained in Section 8.2 contain the centers of inclusions ζ_k in symbolic form. In this section, we treat ζ_k as random variables. Thus, the deterministic results of Section 8.2 are developed here for random composites. Our constructive approach gives answers to the fundamental questions related to the optimal properties of composites. As examples we discuss here finite perturbations of a regular square array and the polydispersity problem in the framework of the "shaking model".

8.3.1 "Shaking model" for a random composite

In this section we apply the formulas from the previous one to evaluate the effective conductivity A of an isotropic random composite with identical inclusions, i.e., $R = R_k$ for all k.

In Chapter 4, the random distribution of disks was arbitrary and the only way to analyze the problem of such complexity was through numerical simulations. In this section, the randomness of the distribution of disks is modeled by introducing the so-called "shaking parameter" d, which characterizes the random locations of the centers of the disks over the periodicity cell. In other words, random configurations of the disks are obtained by "randomization" of a given periodic array such that this single scalar parameter (a real number) controls the randomness – this allows for an analytical solution (see the precise definition below) (Berlyand and Mityushev, 2005).

Consider the basic periodicity cell $Q_{(0,0)}$ and suppose that the cell contains $N = n^2$ particles, where n is an integer number. Partition $Q_{(0,0)}$ into N equal subcells so that their centers b_k form a square array. For example, if $N = 4$ we have

$$b_1 = \frac{1}{4}(1+i), \quad b_2 = \frac{1}{4}(-1+i), \quad b_3 = \frac{1}{4}(-1-i), \quad b_4 = \frac{1}{4}(1-i).$$

The probability distribution of ζ_k, $k = 1, 2, \ldots, N$, is defined as follows. Let $(\Omega, \mathcal{F}, P)$ be a probability space and $d > 0$ be a fixed number, called the "shaking parameter". In each subcell introduce the "shaking disk"

$$G_k := \{z \in \mathbb{C} : |z - b_k| < d\}$$

of radius d centered at b_k. The random variable $\zeta_k = \zeta_k(\omega)$ (center of the particle D_k; ω stands for an event in probability space) is randomly chosen within the disk G_k with the uniform density

$$f_k(z) = \begin{cases} \dfrac{1}{\pi d^2} & \text{if} \quad |z - b_k| < d, \\ 0 & \text{otherwise}. \end{cases}$$

Since we assume that each disk (particle) D_k lies in a subcell of size $\dfrac{1}{\sqrt{N}}$, d must satisfy the natural condition

$$2(d + R) < \frac{1}{\sqrt{N}}. \tag{8.3.1}$$

Recall that the effective conductivity A is given by the formula (8.2.29). Notice that now $A = A(\omega)$ is a random variable. Letting $\langle a \rangle$ be the expectation of a random variable a, we introduce the *effective conductivity in the random monodispersed case* as

$$\mathcal{A} := \langle A(\omega) \rangle.$$

Thus, from (8.2.29) we get

$$\langle A(\omega) \rangle = 1 + 2\rho V \left[\langle B_0 \rangle + \langle B_1 \rangle V + \langle B_2 \rangle V^2 + \cdots \right]. \tag{8.3.2}$$

where, in the monodispersed case,

$$\langle B_q \rangle = \sum_{\substack{p_1, p_2, \ldots, p_M \in S^q; \\ k_0, k_1, \ldots, k_M}} \beta_{q, p_1, p_2, \ldots, p_M} \langle Z^{k_0, k_1, \ldots, k_M}_{p_1, p_2, \ldots, p_M} \rangle. \tag{8.3.3}$$

The following result has been established in Berlyand and Mityushev (2001).

Proposition 8.9 *Suppose that the centers ζ_k are equi-probably located in the shaking disks G_k of equal radii d. Then $\langle Z^{k_0, k_1, \ldots, k_M}_{p_1, p_2, \ldots, p_M} \rangle$ is represented in the form of the power series in d^2*

$$\langle Z^{k_0, k_1, \ldots, k_M}_{p_1, p_2, \ldots, p_M} \rangle = \sum_{s=0}^{\infty} \gamma^{k_0, k_1, \ldots, k_M}_{p_1, p_2, \ldots, p_M}(s) d^{2s}. \tag{8.3.4}$$

Moreover, the following inequality holds

$$\beta_{q, p_1, p_2, \ldots, p_M} \gamma^{k_0, k_1, \ldots, k_M}_{p_1, p_2, \ldots, p_M}(s) \geq 0 \tag{8.3.5}$$

for all $q, p_1, p_2, \ldots, p_M$; $k_0, k_1, \ldots, k_M$; and $s = 0, 1, 2, \ldots$.

A formal proof of this statement is rather technical. One can see full details of the proof by induction in the papers of Berlyand and Mityushev (2001, 2005). The main idea of the proof becomes clear after a close look at the formulas (8.2.34). One can see that the terms X_2, X_{22}, X_{44}, X_{442} with their coefficients in (8.2.34) have a positive sign and the terms X_{33}, X_{332}, X_{233} have a negative sign. This rule can be precisely written and proved by induction for all terms $X_{p_1,p_2,...,p_M}$. Another observation concerns $Z_{p_1,p_2,...,p_M}^{k_1,k_2,...,k_M}$ introduced in (8.2.32). In order to calculate $\langle Z_{p_1,p_2,...,p_M}^{k_0,k_1,...,k_M} \rangle$ over a disk of radius d it is sufficient to expand the terms $E_p(\zeta_k - \zeta_m)$ into a series (see (1.5.24) and (8.2.19)–(8.2.20) for $k = m$) and to calculate this integral term by term. As a result we obtain the series

$$\langle Z_{p_1,p_2,...,p_M}^{k_0,k_1,...,k_M} \rangle = \sum_{s_0,s_1,...,s_M} \Gamma_{p_1,p_2,...,p_M}^{k_0,k_1,...,k_M} (s_0, s_1, \ldots, s_M) d^{2(s_0+s_1+...+s_M)-1}, \quad (8.3.6)$$

which can be written in the form (8.3.4) as a function analytic in d^2. Investigation of its coefficients yields an analogous rule for the signs of $\langle Z_{p_1,p_2,...,p_M}^{k_0,k_1,...,k_M} \rangle$. It is hard to precisely calculate the coefficients $\gamma_{p_1,p_2,...,p_M}^{k_0,k_1,...,k_M} (s)$, but it is possible to show by induction that $\gamma_{p_1,p_2,...,p_M}^{k_0,k_1,...,k_M} (s)$ are real and have the same sign as the corresponding $\beta_{q,p_1,p_2,...,p_M}$. This yields inequality (8.3.5).

Inequality (8.3.5) immediately implies that

$$\mathcal{A}(d = 0) \leq \mathcal{A}(d > 0). \quad (8.3.7)$$

This inequality means that the conductivity of the regular composite ($d = 0$) is always less than the conductivity of any "shaking composite". An analogous result for a 3D cubic lattice had been proved by Kozlov (1989) for infinitely small d for dilute composites, i.e., for infinitely small R. Kozlov's investigation into the infinitesimal statement (Kozlov, 1989) is based on an asymptotic formula which corresponds to our formula (8.3.4) with two terms, i.e., taken to $O(d^2)$. It is worth mentioning that our proof is valid for finite R and finite d, not infinitely small.

8.3.2 Optimal design problem for a deterministic monodisperced composite

In the present section, we demonstrate by an example from Drygaś and Mityushev (2009) that regular arrays have optimal properties also in a deterministic statement.

Consider a simple optimal design problem for a composite with two inclusion per unit cell ($N = 2$) with centers ζ_1, ζ_2 of radius R. In accordance with (8.2.29)–(8.2.34)

$$A^x - iA^{xy} = 1 + 2\rho V \left(1 + \frac{\rho}{4\pi}X_2 V + \frac{\rho^2}{8\pi^2}X_{22}V^2\right) + O(V^4), \quad (8.3.8)$$

where a_m is normalized to 1,

$$X_2 = \sum_{k,k_1=1,2} E_2(\zeta_{k_1} - \zeta_k),$$

$$X_{22} = \sum_{k,k_1,k_2=1,2} E_2(\zeta_{k_1} - \zeta_k)\overline{E_2(\zeta_{k_2} - \zeta_{k_1})}.$$

The value $A^y + iA^{xy}$ can be formally obtained from (8.3.8) by replacing of X_2 and X_{22} by

$$X_2^* = \sum_{k,k_1=1,2} E_2(i(\zeta_{k_1} - \zeta_k)),$$

$$X_{22}^* = \sum_{k,k_1,k_2=1,2} E_2(i(\zeta_{k_1} - \zeta_k))\overline{E_2(i(\zeta_{k_2} - \zeta_{k_1}))}.$$

Then,

$$A^y + iA^{xy} = 1 + 2\rho V\left(1 + \frac{\rho}{4\pi}X_2^*V + \frac{\rho^2}{8\pi^2}X_{22}^*V^2\right) + O(V^4). \tag{8.3.9}$$

This equation is obtained from (8.3.8) by rotation of the coordinates by 90° when x is replaced by $-y$ and y by x. Using (8.3.8)–(8.3.9), we calculate the invariant of the tensor A

$$I_1 = \frac{1}{2}(A^x + A^y) \tag{8.3.10}$$

$$= 1 + \rho V\left(1 + \frac{\rho}{4\pi}(X_2 + X_2^*)V + \frac{\rho^2}{8\pi^2}(X_{22} + X_{22}^*)V^2\right) + O(V^4).$$

It follow from the properties of the Eisenstein functions that $X_2 + X_2^* = 4\pi$ and $X_{22} = X_{22}^*$ (see (1.5.23)–(1.5.25)).

Consider the following optimal design problem. Given parameters ρ, V, that is, ρ, V are fixed in (8.3.10), find a distribution of the disks in the unit cell such that I_1 attains a maximum (minimum). X_{22} can be considered as a function of $a = \zeta_2 - \zeta_1$:

$$X_{22}(a) = 2|\wp(a)|^2 + 8\pi^2, \tag{8.3.11}$$

where (8.2.20), (1.5.23) and (1.5.2) are used. In order to find the extremism of I_1 up to $O(V^4)$, it is sufficient to investigate the function $X_{22}(a)$, which attains a minimum when $\wp(a) = 0$. Equation (8.3.11) implies that $\wp'(a) = 0$. Therefore, the point a has to be a double zero of $\wp(z)$. The order of $\wp(z)$ is equal to 2; hence, the equations $\wp(a) = \wp'(a) = 0$ can have only three candidates for a. It follows from (1.5.26) that $\wp(\frac{1}{2}) = -\wp(\frac{i}{2})$ Then, (1.5.28) yields

$$\wp\left(\frac{1+i}{2}\right) = \wp'\left(\frac{1+i}{2}\right) = 0. \tag{8.3.12}$$

Then the unique required value of a is $\dfrac{1+i}{2}$. This case corresponds to a regular array of inclusions when the disks are located at the vertices of the square array with the fundamental translation vectors $\dfrac{1}{2}(1 \pm i)$.

The maximal value of $X_{22}(a)$ in the unit cell is attained at the boundary of the domain $|a| > 2R$ since the function $\wp(a)$ is analytic in the unit cell except zero. Therefore, the maximum of $|\wp(a)|^2$ has to be found on the circle $a = 2R\exp(i\theta)$. Using (1.5.25), one can see that $|\wp(a)|^2$ does not exceed $|\wp(2R)|^2$. The latter value is attained at the following angles θ: $0, \pm\dfrac{\pi}{2}, \pi$. Therefore, the maximum of I_1 up to $O(V^4)$ is attained when two disks touch each other and their centers lie on the lines parallel to Ox or to Oy.

8.4 Random polydispersed composite

In this section, we present an analysis of the polydispersity effect with the complex variable method following the work of Berlyand and Mityushev (2001). The method is based on complex variable techniques and series representations of the solution obtained in Section 8.2. It allows us to investigate the effect of polydispersity qualitatively and obtain accurate quantitative results for the cases of small (close to zero) and large (close to one) values of the polydispersity parameter. This analysis complements the numerical results presented in Chapter 4.

Following the previous section we introduce two "shaking models":

(i) the "bumping model",
(ii) the "well-separated model".

Model (i) describes random geometrical arrays when particles are closely packed and not well-separated. The key ingredient of this model is the introduction of two shaking parameters d_l and d_s (depending on the sizes of the particles) for large and small disks, respectively. The only restriction on d_l and d_s is that the disks cannot overlap (but can be very close to each other). In the "well-separated model" the shaking parameter is the same for large and small disks.

Model (ii) clearly can be applied in the dilute case and it is not surprising that this model leads to the inequality

$$\mathcal{A} > \mathcal{A}_{\text{mono}}. \tag{8.4.1}$$

However, the central result of this section is that model (i) predicts regimes where either (8.4.1) or the opposite inequality

$$\mathcal{A} < \mathcal{A}_{\text{mono}} \tag{8.4.2}$$

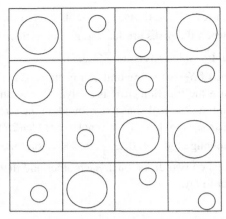

Figure 8.4 Shaking of inclusions within small squares.

hold, depending on the value of the polydispersity parameter p (the relative volume fractions of large and small disks). This result is consistent with the numerical example in Section 4.6.2.

We consider a polydispersed composite when circular particles D_k of conductivity λ are randomly embedded into a matrix of conductivity 1. The particles D_k are of two different sizes: large and small. We denote the radii of the large disks by R_l and the radii of the small disks by R_s, $R_l \geq R_s$. The local geometry of such a composite is presented in Figure 8.4.

The *polydispersity parameter* p, $0 \leq p \leq 1$, characterizes the proportion of small disks ($p = 0$ means all disks are large, $p = 1$ means all disks are small) for a fixed total volume fraction V, that is the total volume fraction (concentration) of the particles is

$$V = N\pi (R_s^2 p + R_l^2 (1 - p)). \tag{8.4.3}$$

Remark. In (8.4.3) the parameters R_l, R_s, p and V are given, and the number of particles N depends on them. Notice that for no choice of R_l, R_s, p and V does such an integer value of N exist. However, in what follows we may assume N to be sufficiently big. In other words, R_s and R_l can be arbitrarily small compared to the periodicity cell, since only the ratio

$$\chi = \left(\frac{R_s}{R_l}\right)^2 \tag{8.4.4}$$

is important. Therefore, the error due to the possible non-integer number of inclusions is negligible.

Our main objective is to obtain the analytical dependence of the effective conductivity $\mathcal{A}$ on the polydispersity parameter p. This allows for qualitative comparison of the effective conductivities of the polydispersed composites $\mathcal{A}(p)$ and the effective conductivities of the monodispersed composites $\mathcal{A}(p)$ and

$$\mathcal{A}_{\text{mono}} = \mathcal{A}(0).$$

Note that if V is fixed, $\mathcal{A}(0) = \mathcal{A}(1)$ since the cases $p = 0$ and $p = 1$ are essentially the same up to re-scaling.

We introduce the random polydispersed model in the following way. Similarly to Section 8.3.1, we partition the basic periodicity cell $Q_{(0,0)}$ into $N = n^2$ equal square subcells. The centers of these subcells b_k form a square array in $Q_{(0,0)}$. In the k-th subcell, introduce the "shaking disk" of radius d_k centered at b_k:

$$G_k = \{z \in \mathbb{C} : |z - b_k| < d_k\}.$$

The random variable $\zeta_k = \zeta_k(\omega)$, the center of the k-th inclusion, is randomly chosen within the disk G_k with the density

$$(\pi d_k^2)^{-1} = |G_k|^{-1}.$$

Since we assume that each disk (particle) D_k lies in a subcell of size $\dfrac{1}{\sqrt{N}}$ the shaking parameter d_k must satisfy the natural condition

$$2(d_k + R_k) < \frac{1}{\sqrt{N}}. \tag{8.4.5}$$

Along with the random variables $\{\zeta_k, k = 1, \ldots, N\}$, we introduce the random variable $\nu_k = \nu_k(\omega)$ (see (8.2.16)) which characterizes the size of the k-th particle:

$$\nu_k(\omega) = \begin{cases} \chi & \text{with the probability } p, \\ 1 & \text{with the probability } 1 - p. \end{cases} \tag{8.4.6}$$

In the two models described below inequality (8.4.5) represents the non-overlapping of the disks D_k which models impenetrable particles. We will study two models of polydispersity: the well-separated model in which all shaking parameters d_k are identical, and the bumping model in which the shaking parameters d_k are determined by the sizes of the particles.

Our main assumptions for the random model are the following:

(i) N is a sufficiently large fixed number;[1]
(ii) the random variables ν_k ($k = 1, 2, \ldots, N$) are independent identically distributed (iid);
(iii) the random variables ζ_k ($k = 1, 2, \ldots, N$) are independent and uniformly distributed within G_k;
(iv) the random variables ν_k and ζ_k are mutually independent.

Condition (iv) is assumed only for the same "shaking parameter" d.

---

[1] Doubling the cell in the Ox- and Oy-directions with normalization of the area increases N four times without changing the problem.

Below we consider functions $f(\zeta_1, \zeta_2, \ldots, \zeta_N; \nu_1, \nu_2, \ldots, \nu_N)$, where $\nu_1, \nu_2, \ldots, \nu_N$ are independent random variables defined above. Denote by

$$F(\nu_1, \nu_2, \ldots, \nu_N) = \langle f(\zeta_1, \zeta_2, \ldots, \zeta_N; \nu_1, \nu_2, \ldots, \nu_N) \rangle$$

the expectation of f as a function of the random variables $\zeta_1, \zeta_2, \ldots, \zeta_N$ (ν_k, $k = 1, 2, \ldots, N$ are parameters). More precisely, $\langle . \rangle$ means the expectation in the probability space defined as a product of the probability spaces of each ζ_k. Further, one can calculate the expectation

$$\mathbb{E}[F(\nu_1, \nu_2, \ldots, \nu_N)] = \mathbb{E}[\langle f(\zeta_1, \zeta_2, \ldots, \zeta_N; \nu_1, \nu_2, \ldots, \nu_N) \rangle] \qquad (8.4.7)$$

with respect to random variables $\nu_1, \nu_2, \ldots, \nu_N$. Note that it is possible to define a probability space which corresponds to a joint distribution of all ζ_k and ν_k and introduce a corresponding "double" expectation in all of these random variables. This double expectation leads to the same result as the consecutive expectations (8.4.7). This model is a generalization of the polydispersed case of the analogous shaking model introduced in Section 8.3.1 for monodispersed random composites where only the expectation $\langle . \rangle$ was used.

Notice that for given $N = n^2$, the total concentration is not fixed, but is a random variable instead. Indeed,

$$V = V_N(\omega) = \pi R_l^2 (\nu_1 + \cdots + \nu_N). \qquad (8.4.8)$$

Note that

$$\mathbb{E}[V_N] = N\pi R_l^2 \mathbb{E}[\nu_k] = N\pi (R_s^2 p + R_l^2(1 - p)),$$

which is consistent with (8.4.3). Moreover, for large N we can apply the law of large numbers to V_N to conclude that

$$\frac{V_N}{N} \to^P \pi R_l^2 \mathbb{E}[\nu_1],$$

where $\to^P$ stands for convergence in probability.

According to (8.4.7) we define the effective conductivity of a random polydispersed composite $\mathcal{A}$ by averaging of A (see (8.2.5)) over random locations and random radii

$$\mathcal{A} = \mathbb{E}[\langle A \rangle]. \qquad (8.4.9)$$

Take the expectation $\langle . \rangle$ and $\mathbb{E}[.]$ in (8.2.29). We have (see the first formula (8.2.34) for B_0)

$$\mathcal{A} = 1 + 2\rho V \left(1 + \mathbb{E}[\langle B_1 \rangle V] + \mathbb{E}[\langle B_2 \rangle V^2] + \cdots \right), \qquad (8.4.10)$$

where $0 \leq V \leq V_{\max}$ and

$$V_{\max} = \frac{\pi}{4}(p\chi + 1 - p)$$

corresponds to touching particles.

8.4.1 Well-separated model (identical shaking parameters)

In the present section we assume that all shaking parameters are identical, that is

$$d_1 = d_2 = \ldots = d_N = d, \tag{8.4.11}$$

where

$$d + R_l < \frac{1}{2\sqrt{N}}. \tag{8.4.12}$$

We calculate A using (8.4.10)–(8.4.11). In order to calculate $\mathbb{E}[\langle B_q \rangle V^q]$ we use (8.2.35)

$$\mathbb{E}[\langle B_q \rangle] \tag{8.4.13}$$

$$= \sum_{\substack{p_1, p_2, \ldots, p_M \in S^q; \\ k_0, k_1, \ldots, k_M}} \beta_{q, p_1, p_2, \ldots, p_M} \mathbb{E}\left[\frac{v_{k_0} v_{k_1}^{p_1 - 1} \ldots v_{k_M}^{p_M - 1} V^q}{K_1(\chi)^P} \right] \langle Z_{p_1, p_2, \ldots, p_M}^{k_0, k_1, \ldots, k_M} \rangle,$$

taking into account the following properties:

(i) $\mathbb{E}[\langle \beta_{q, p_1, p_2, \ldots, p_M} \rangle] = \beta_{q, p_1, p_2, \ldots, p_M}$, since $\beta_{q, p_1, p_2, \ldots, p_M}$ does not depend on v_k and ζ_k;

(ii) the values $Z_{p_1, p_2, \ldots, p_M}^{k_0, k_1, \ldots, k_M}$ do not depend on the random values v_k, $k = 1, 2, \ldots, N$, hence

$$\mathbb{E}[Z_{p_1, p_2, \ldots, p_M}^{k_0, k_1, \ldots, k_M}] = Z_{p_1, p_2, \ldots, p_M}^{k_0, k_1, \ldots, k_M}$$

and therefore

$$\mathbb{E}[\langle Z_{p_1, p_2, \ldots, p_M}^{k_0, k_1, \ldots, k_M} \rangle] = \langle Z_{p_1, p_2, \ldots, p_M}^{k_0, k_1, \ldots, k_M} \rangle;$$

(iii) $\langle v_{k_0} v_{k_1}^{p_1 - 1} \ldots v_{k_M}^{p_M - 1} \rangle = v_{k_0} v_{k_1}^{p_1 - 1} \ldots v_{k_M}^{p_M - 1}$, since the random variables v_k do not depend on ζ_k.

Recall that for the monodispersed case we have from (8.2.31)

$$A_{\text{mono}} = \langle A \rangle = 1 + 2\rho V_{\text{mono}} \left(1 + C_1 V_{\text{mono}} + C_2 V_{\text{mono}}^2 + \cdots \right), \tag{8.4.14}$$

where

$$C_q = \sum_{\substack{p_1, p_2, \ldots, p_M; \\ k_0, k_1, \ldots, k_M}} \beta_{q, p_1, p_2, \ldots, p_M} \langle Z_{p_1, p_2, \ldots, p_M}^{k_0, k_1, \ldots, k_M} \rangle \tag{8.4.15}$$

and

$$0 < V_{\text{mono}} < V_{\text{max}}.$$

Here $V_{\text{max}} = \frac{\pi}{4}$ corresponds to touching particles.

In order to compare A_{mono} and A defined by (8.4.14) and (8.4.10), respectively, we fix the shaking parameter d and the value of V_{mono} equal to the expectation of the total volume fraction of the polydispersed composite

$$V_{\mathrm{mono}} = \mathbb{E}[V] = \pi N(pR_s^2 + (1-p)R_l^2).$$

Introduce

$$w_{q,p_1,p_2,\ldots,p_M} = \frac{\mathbb{E}\left[v_{k_0} v_{k_1}^{p_1-1} \ldots v_{k_M}^{p_M-1} V^q\right]}{(\chi p + 1 - p)^P}. \tag{8.4.16}$$

For simplicity we first compute (8.4.16) assuming that all subscripts $k_0, k_1, \ldots, k_M$ in (8.4.16) are different. Then the corresponding random variables $v_{k_0}, v_{k_1}^{p_1-1}, \ldots, v_{k_M}^{p_M-1}$ are also different and therefore independent. From (8.4.8) we have

$$\mathbb{E}\left[v_{k_0} v_{k_1}^{p_1-1} \ldots v_{k_M}^{p_M-1} V^q\right] = \pi^q R_l^{2q} \mathbb{E}\left[v_{k_0} v_{k_1}^{p_1-1} \ldots v_{k_M}^{p_M-1} \left(\sum_{k=1}^{N} v_k\right)^q\right].$$

Expanding $\left(\sum_{k=1}^{N} v_k\right)^q$, we get the sum of the terms of the form $\mathbb{E}[v_{k_0}^{r_0} v_{k_1}^{p_1-1+r_1} \ldots v_{k_M}^{p_M-1+r_M}]$, $r_0 + \cdots + r_M = q$. Using the independence of v_k,

$$\mathbb{E}\left[v_{k_0}^{r_0} v_{k_1}^{p_1-1+r_1} \ldots v_{k_M}^{p_M-1+r_M}\right]$$
$$= K_{r_0}(\chi) K_{p_1-1+r_1}(\chi) K_{p_2-1+r_2}(\chi) \ldots K_{p_M-1+r_M}(\chi),$$

where the following moments are introduced

$$\mathbb{E}[v_k^q] = K_q(\chi) = \chi^q p + 1 - p, \quad q = 1, 2, \ldots. \tag{8.4.17}$$

Hence, (8.4.16) implies

$$w_{p_1,p_2,\ldots,p_M} = \frac{K_{p_1-1}(\chi) K_{p_2-1}(\chi) \ldots K_{p_M-1}(\chi)}{K_1^{p_1-1}(\chi) K_1^{p_2-1}(\chi) \ldots K_1^{p_M-1}(\chi)}. \tag{8.4.18}$$

If in (8.4.16) $k_i = k_j$ for some $i \neq j$, we also arrive at an analogous formula (8.4.16) which requires more cumbersome notation.

We make use of the elementary inequality (Jensen's inequality in the context of probability theory)

$$\mathbb{E}\left[v_k^q\right] = \chi^q p + 1 - p \geq (\chi p + 1 - p)^q \tag{8.4.19}$$

which holds for any $q \geq 1$, $\chi \in [0, 1]$ and $p \in [0, 1]$. From (8.4.19), we deduce

$$K_{r_0}(\chi) K_{p_1-1+r_1}(\chi) K_{p_2-1+r_2}(\chi) \ldots K_{p_M-1+r_M}(\chi) \tag{8.4.20}$$
$$\geq K_1^{r_0}(\chi) K_1^{p_1-1+r_1}(\chi) K_1^{p_2-1+r_2}(\chi) \ldots K_1^{p_M-1+r_M}(\chi).$$

Observing that $K_1(\chi)^P = K_1(\chi)K_1^{p_1-1}(\chi)K_1^{p_2-1}(\chi)\dots K_1^{p_M-1}$ and using (8.4.20), after simple algebraic manipulations we get

$$w_{q,p_1,p_2,\dots,p_M} \geq (\pi N(pR_s^2 + (1-p)R_l^2))^q = V_{\text{mono}}^q. \qquad (8.4.21)$$

Moreover, equality in (8.4.21) is attained in two cases (i) $\chi = 1$ that is $p = 0$ and no polydispersity present; (ii) $p_1 = p_2 = \dots = p_M = 2$ (see (8.4.18)).

It follows from inequality (8.4.21) that

$$\mathbb{E}[\langle B_q V^q \rangle] \geq C_q V_{\text{mono}}^q,$$

when C_q has the form (8.4.15). This yields the following

Proposition 8.10 *Let $\mathcal{A}_{\text{mono}}$ defined by (8.4.14) be the effective conductivity of a macroscopically isotropic monodispersed composite and $\mathcal{A}$ defined by (8.4.9) (see also (8.4.10)) be the effective conductivity of a macroscopically isotropic polydispersed composite. Then for any fixed d, satisfying the non-overlapping condition (8.4.5), the following inequality holds*

$$\mathcal{A}_{\text{mono}} \leq \mathcal{A}, \qquad (8.4.22)$$

for any relative concentration p of the small particles and if the fixed total volume fraction V_{mono} in the monodispersed case equals the expectation of V in the polydispersed case.

Note. Proposition 8.10 was proved for a fixed value of the shaking parameter d. However, it can be easily generalized to the case when the shaking parameters d_k, $k = 1, 2, \dots, N$, are iid's in the interval $[d_{\min}, d_{\max}]$. This is the most general case of identical shaking, since d_k do not depend on the radii of the disks, unlike in the bumping model (see the next section). This model describes the physical situation when particles are well-separated (not too close to touching) and, in particular, it applies to the dilute case.

8.4.2 Bumping model (two different shaking parameters)

In this section following Berlyand and Mityushev (2001) we introduce a model that describes highly packed composites with particles of two different sizes. In such composites particles are often not well-separated and percolation patterns dominate the effective behavior. That is why no approach based on the Maxwell–Garnett formula (and its refinements) will work here.

We now present a heuristic motivation for our model. In the high-concentration regime where particles are close to each other and there is not much room for them to move around, a large particle would bump into another one after a relatively small change in the location. However, a small particle is likely to have more room to move around before it bumps into a neighboring particle. Thus, while in a low-concentration (more generally well-separated) composite it is reasonable to introduce identical shaking parameters, as we did in the previous section, for

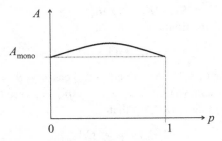

Figure 8.5 Dependence of the effective conductivity A on the relative volume fraction of the small particles p for monodispersed composites (broken line) and for polydispersed composites (solid line) in the dilute case.

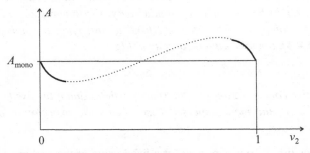

Figure 8.6 Dependence of the effective conductivity A_{mono} on the relative volume fraction of the small particles p for monodispersed composites, broken line; and for polydispersed composites A, solid and dot-dashed line, in the case of high concentrations of particles. The solid line is rigorously justified, the dot-dashed is supported by numerical results, and
$$\chi = (R_s/R_l)^2 \text{ is fixed.}$$

highly packed random composites it is natural to introduce two different shaking parameters d_s, d_l for small and large particles, respectively, such that

$$d_l < d < d_s, \tag{8.4.23}$$

where d is the shaking parameter for the monodispersed model introduced at the beginning of the section (at the same total volume fraction). In other words the inequality (8.4.23) models the geometric observation that small particles are more mobile than large ones. As we will see below our model leads to an S-shaped dependence of the effective conductivity on the polydispersity parameter (for a fixed total volume fraction), that is (8.4.1) holds for p close to one, and (8.4.2) holds for small p (see Figure 8.6). Also (8.4.2) agrees with the numerical computations from Section 4.6, in which a decrease in the effective conductivity was observed for a mixture of large and small particles when p is small.

In order to compare the effective conductivity A for the well-separated and for the bumping models we compare the expectation $\mathbb{E}[\langle . \rangle]$ of the corresponding coefficients (8.2.35) in the mono- and polydispersed cases. Recall that $\mathbb{E}[\langle . \rangle]$ corresponds to subsequent averaging over random locations of the centers ζ_k (shaking parameters) followed by averaging over random sizes v_k of the disks D_k. In both cases we begin with (8.4.13).

We first calculate $\langle Z_{p_1,p_2,...,p_M}^{k_0,k_1,...,k_M} \rangle$. The well-separated model coincides with the "shaking model" discussed in Section 8.3.1. In this case $\langle Z_{p_1,p_2,...,p_M}^{k_0,k_1,...,k_M} \rangle$ is calculated by (8.3.6). For the bumping model the difference in calculations is based on the replacement of $d^{2(s_0+...+s_M)}$ in (8.3.6) by

$$\frac{v_{k_0} v_{k_1}^{p_1-1} \ldots v_{k_M}^{p_M-1}}{K_1(\chi)^P} d_{k_0} d_{k_1}^{p_1-1} \ldots d_{k_M}^{p_M-1}, \tag{8.4.24}$$
$$k_j = 1, 2, \ldots, N; \quad M = 1, 2, 3, \ldots$$

Next, we take the expectation $\mathbb{E}[.]$ in (8.4.24)

$$\mathbb{E}\left[\frac{v_{k_0} v_{k_1}^{p_1-1} \ldots v_{k_M}^{p_M-1}}{K_1(\chi)^P} d_{k_0} d_{k_1}^{p_1-1} \ldots d_{k_M}^{p_M-1}\right]. \tag{8.4.25}$$

Since each index $k_0, k_1, \ldots, k_M$ takes one of the values $1, 2, \ldots, N$ and M is any natural number, there may be repeated indices among $k_0, k_1, \ldots, k_M$. We select the repeated indices and using the independence of the random variables v_k $(d_k(v_k))$ decompose (8.4.25) into a product of terms in the following form

$$\mathbb{E}\left[\frac{v_k^j}{K_1^j(\chi)} d_k^j\right] = \frac{p\chi^j d_s^j + (1-p)d_l^j}{(p\chi + (1-p))^j}. \tag{8.4.26}$$

Therefore, comparison of $\mathcal{A}$ and $\mathcal{A}_{\text{mono}}$ amounts to comparison of the terms

$$\omega(d_s, d_l) = \frac{p\chi^j d_s^j + (1-p)d_l^j}{(p\chi + (1-p))^j} \text{ and } \omega(d, d) = d^j. \tag{8.4.27}$$

We do this for two limiting cases: for sufficiently small p ($p \approx 0$, i. e., almost all disks are large) and for p close to 1 ($p \approx 1$, i.e., almost all disks are small) for fixed χ. Due to (8.4.23) we obtain

$$\omega(d_s, d_l) = d_l^j < d^j \text{ as } p \approx 0, \tag{8.4.28}$$
$$\omega(d_s, d_l) = d_s^j > d^j \text{ as } p \approx 1.$$

In order to compare $\mathcal{A}$ and $\mathcal{A}_{\text{mono}}$ for the same total concentration V, we first fix all parameters corresponding to the monodispersed composite, i.e., either the radius R of the disks, or the number of the subcells (disks) per unit cell, N. These values depend on each other through the relation $V = N\pi R^2$.

For the polydispersed composite we use the same V (more precisely, $\mathbb{E}[V]$) and N and choose the relative volume fraction p ($0 \leq p \leq 1$) and the parameter χ. The radii R_s and R_l are then determined by solving the system of equations (8.4.4) and (8.4.3).

Secondly, we chose the following shaking parameters

$$d = \frac{1}{2\sqrt{N}} - R, \tag{8.4.29}$$

$$d_s = \frac{1}{2\sqrt{N}} - R_s, \tag{8.4.30}$$

$$d_l = \frac{1}{2\sqrt{N}} - R_l. \tag{8.4.31}$$

These relations mean that we take the maximum possible shaking parameters within the framework of the non-overlapping model. Our parameters are consistent in the following sense. If p goes to 0 or 1, or if $\chi \to 1$, then a polydispersed composite becomes a monodispersed one.

In Proposition 8.11 below, we compare $\mathcal{A}$ with the constant $\mathcal{A}_{mono}$ for any p and χ. This proposition follows from the comparison of the values (8.4.27) which for $p \approx 0$ and for $p \approx 1$ satisfy inequality (8.4.28) established above. Figure 8.6 illustrates this comparison for a fixed χ and shaking parameters (8.4.29) and (8.4.30).

Proposition 8.11 *Let $\mathcal{A}_{mono}$ defined by (8.4.14) be the effective conductivity of a macroscopically isotropic monodispersed composite and $\mathcal{A}$ defined by (8.4.9) be the effective conductivity of a macroscopically isotropic polydispersed composite (see also (8.4.10)). Then for a fixed V (same total volume fraction for mono- and polydispersed composites) and any shaking parameters d_s, d_l, d satisfying the conditions (8.4.5) and (8.4.23), the following inequalities hold*

$$\mathcal{A} < \mathcal{A}_{mono} \text{ for } p \approx 0, \tag{8.4.32}$$

$$\mathcal{A} > \mathcal{A}_{mono} \text{ for } p \approx 1.$$

The analysis presented demonstrates that $\mathcal{A}$ has different behavior depending on the conditions on the mobility of particles (8.4.23) and (8.4.11). Condition (8.4.23) arises from natural restrictions that at high concentrations disks bump into each other, since they are hard and cannot penetrate. This condition implies the behavior displayed in Figure 8.6, which agrees with the numerical analysis presented above. The case $N = 2$ was investigated analytically by Rylko (2008a) by the method described in Section 8.2. It was demonstrated by Rylko (2008a) that both inequalities (8.4.32) take place also in the deterministic statement for different locations of inclusions.

For the low-concentration regime it is natural that the probability of bumping is close to zero. Therefore, the shaking parameters should be equal as in (8.4.11). This leads to the behavior described in Figure 8.5 and agrees with rigorous asymptotical analysis for small concentration (Torquato, 2002; Thovert and Acrivos, 1989; Thovert et al., 1990).

The same curve (see Figure 8.5) was obtained by Robinson and Friedman (2001) using a clever two-step application of the Maxwell–Garnett formula. However, their analysis (Robinson and Friedman, 2001) tacitly uses assumption of small concentration, because the Maxwell–Garnett formula is derived for small concentration only. In particular, no use of this formula is capable of capturing percolation effects. Note that for concentrations above the percolation threshold V_{cr} connectivity patterns (percolation patterns) dominate the behavior of the effective conductivity while for $V < V_{cr}$ the percolation effects are non-essential and the behavior presented in Figure 8.5, though rigorously justified only for small concentrations $V \ll 1$, is expected to be valid for $V < V_{cr}$.

We remark that the shaking model has a restriction that each disk is confined in its own subcell which is why the number of disks in the basic cell $Q_{(0,0)}$ is always the square of an integer. This restriction is introduced in order to obtain an analytical solution (a series in which all coefficients are evaluated recursively) which is only possible for very few specific models. A prominent example of a model, which can be treated analytically, is the Keller–Dykhne random checkerboard (Keller, 1964; Dykhne, 1971), where the effective conductivity is given by a simple formula. It is known that the qualitative behavior of the effective properties predicted by such restricted models is usually valid for a much wider class of physical problems.

References

Abbot, J.R., N. Tetlow, A.L. Graham, S.A. Altobell, E. Fukushima, L.A. Mondy and T.A. Stephens. (1991). Experimental observations of particle migration in concentrated suspensions: Couette flow. *J. Rheol.*, **35**:773–795.

Aboudi, J. (1991). *Mechanics of Composite Material: A Unified Micromechanics Approach.* Elsevier Science, Amsterdam.

Acrivos, A. and E. Chang. (1986). A model for estimating transport quantities in two-phase materials. *Phys. Fluids*, **29**(3):3–4.

Adams, R.A. (1975). *Sobolev Spaces.* Academic Press, New York.

Ahlfors, L. (1979). *Complex Analysis*, 3rd ed. McGraw–Hill, New York.

Akhiezer, N.I. (1990). *Elements of the Theory of Elliptic Functions.* American Mathematical Society, Providence, RI.

Almgren, R.F. (1985). An isotropic three-dimensional structure with Poisson's ratio $= -1$. *J. Elasticity*, **15**:427–430.

Ambegaokar, V., B.I. Halperin and J.S. Langer. (1971). Hopping conductivity in disordered systems. *Phys. Rev.*, B, **4**(8):2612–2620.

Andrianov, I.V., V.V. Danishevs'kyy and A.L. Kalamkarov. (2002). Asymptotic analysis of effective conductivity of composite materials with large rhombic fibres. *Composite Struct.*, **56**(33):229–234.

Andrianov, I.V., V.V. Danishevs'kyy and S. Tokarzewski. (1996). Two-point quasifractional approximants for effective conductivity of a simple cubic lattice of spheres. *Int. J. Heat Mass Transfer*, **39**(11):2349–2352.

Andrianov, I.V., G.A. Starushenko, V.V. Danishevskiy and S. Tokarzewski. (1999). Homogenization procedure and Padé approximants for effective heat conductivity of a composite material with cylindrical inclusions having square cross-section. *Proc. R. Soc. London*, A, **455**:3401–3413.

Annin, B.D., A.L. Kalamkarov, A.G. Kolpakov and V.Z. Parton. (1993). *Computation and Design of Composite Material and Structural Elements (in Russian).* Nauka, Novosibirsk.

Aurenhammer, F. and R. Klein. (1999). Voronoi diagrams. In: *Handbook of Computational Geometry* (J.R. Sack and J. Urrutia, eds.), North Holland, Amsterdam, pp. 201–291.

Avellaneda, M. (1987). Optimal bounds and microgeometries for an elastic two-phase composite. *SIAM J. Appl. Math.*, **47**(6):1216–1228.

Babůshka, I., B. Anderson, P. Smith and K. Levin. (1999). Damage analysis of fiber composites. Part I. Statistical analysis on fiber scale. *Comput. Methods Appl. Mech. Engng.*, **172**:27–77.

Bakhvalov, N.S. and G.P. Panasenko. (1989). *Homogenization: Averaging Processes in Periodic Media*. Kluwer Academic Publishers, Dordrecht.

Balberg, I. (1987). Recent developments in continuum percolation. *Phil. Mag.*, B **30**:991–1003.

Batchelor, G.K. and R.W. O'Brien. (1977). Thermal or electrical conduction through a granular material. *Proc. R. Soc. London*, A, **335**:313–333.

Batchelor, G.K. and C.S. Wen. (1972). Sedimentation in a dilute dispersion of spheres. *J. Fluid Mech.*, **52**:245–268.

Bendsøe, M.P. and N. Kikuchi. (1988). Generating optimal topologies in structural design using a homogenization method. *Comp. Meth. Appl. Mech. Engng*, **71**:95–112.

Bendsøe, M.P. and O. Sigmund. (2004). *Topology Optimization*. Springer-Verlag, Berlin.

Bensoussan, A., J.-L. Lions and G. Papanicolaou. (1975). Sur quelques phénomènes asymptotiques d'évolution. *Compt. Rend. Acad. Sci. Paris*, Ser. A-B, **281**(10):A317–A322.

Bensoussan, A., J.-L. Lions and G. Papanicolaou. (1978). *Asymptotic Analysis for Periodic Structures*. North Holland, Amsterdam.

Beran, M.J. (1968). *Statistical Continuum Theories*. John Wiley, New York.

Beran, M.J. and J. Molyneux. (1966). Use of classical variational principles to determine bounds for the effective bulk modulus in heterogeneous media. *Quart. Appl. Math.*, **24**:107–118.

Berdichevsky, V.L. (2009). *Variational Principles of Continuum Mechanics*. Springer-Verlag, Berlin.

Bergman, D.J. (1983). The dielectric constant of a composite material – a problem in classical physics. *Phys. Reports*, C**43**:378–407.

Bergman, D.J., E. Duering and M. Murat. (1990). Discrete network models for the low-field Hall effect near a percolation threshold: Theory and simulation. *J. Stat. Phys.*, **1**(58):1–43.

Bergman, D.J. and K.J. Dunn. (1992). Bulk effective dielectric constant of a composite with periodic micro-geometry. *Phys. Rev. B*, **45**:13262–13271.

Berlyand, L., L. Borcea and A. Panchenko. (2005). Network approximation for effective viscosity of concentrated suspensions with complex geometries. *SIAM J. Math. Anal.*, **36**(5):1580–1628.

Berlyand, L., Y. Gorb and A. Novikov. (2005). Discrete network approximation for highly-packed composites with irregular geometry in three dimensions. In: *Multiscale Methods in Science and Engineering* (B. Engquist, P. Lotstedt and O. Runborg, eds.), Springer-Verlag, Berlin, pp. 21–58.

Berlyand, L., Y. Gorb and A. Novikov. (2009). Fictitious fluid approach and anomalous blow-up of the dissipation rate in a two-dimensional model of concentrated suspensions. *Arch. Rational Mech. Anal.*, **193**(3):585–622.

Berlyand, L. and A. Kolpakov. (2001). Network approximation in the limit of small interparticle distance of the effective properties of a high-contrast random dispersed composite. *Arch. Rational Mech. Anal.*, **159**(3):179–227.

Berlyand, L. and S. Kozlov. (1992). Asymptotics of the homogenized moduli for the elastic chess-board composite. *Arch. Rational Mech. Anal.*, **118**(2): 95–112.

Berlyand, L. and V. Mityushev. (2001). Generalized Clausius–Mossotti formula for random composite with circular fibers. *J. Stat. Phys.*, **102**(1/2):115–145.

Berlyand, L. and V. Mityushev. (2005). Increase and decrease of the effective conductivity of a two phase composite due to polydispersity. *J. Stat. Phys.*, **118**(3/4):479–507.

Berlyand, L. and A. Novikov. (2002). Error of the network approximation for densely packed composites with irregular geometry. *SIAM J. Math. Anal.*, **34**(2):385–408.

Berlyand, L. and K. Promislow. (1995). Effective elastic moduli of a soft medium with hard polygonal inclusions and extremal behavior of effective Poisson's ratio. *J. Elasticity*, **40**(1):45–73.

Berlyand, L.V. and A. Panchenko. (2007). Strong and weak blow up of the viscous dissipation rates for concentrated suspensions. *J. Fluid Mech.*, **578**:1–34.

Bhattacharya, K., R.V. Kohn and S. Kozlov. (1999). Some examples of nonlinear homogenization involving nearly degenerate energies. *Proc. R. Soc. London*, A, **455**: 567–583.

Bollobás, B. (1998). *Modern Graph Theory*. Springer-Verlag, New York.

Bonnecaze, R.T. and J.F. Brady. (1991). The effective conductivity of random suspensions of spherical particles. *Proc. R. Soc. London*, Ser. A, **432**:445–465.

Borcea, L. (1998). Asymptotic analysis of quasi-static transport in high contrast conductive media. *SIAM J. Appl. Math.*, **2**(59):597–635.

Borcea, L., J.G. Berryman and G. Papanicolaou. (1999). Matching pursuit for imaging high-contrast conductivity. *Inverse Problems*, **15**:811–849.

Borcea, L. and G. Papanicolaou. (1998). Network approximation for transport properties of high contrast conductivity. *Inverse Problems*, **4**(15):501–539.

Born, M. and K. Huang. (1954). *Dynamical Theory of Crystal Lattices*. Oxford University Press, Oxford.

Bourgeat, A., A. Mikelic and S. Wright. (1994). Stochastic two-scale convergence in the mean and applications. *J. Reine Angew. Math.*, **456**:19–51.

Bourgeat, A. and A. Piatnitski. (2004). Approximations of effective coefficients in stochastic homogenization. *Ann. Inst. H. Poincaré*, **40**:153–165.

Brady, J.F. (1993). The rheological behavior of concentrated colloidal suspensions. *J. Chem. Phys.*, **99**:567–581.

Brady, J.F. and G. Bossis. (1985). The rheology of concentrated suspensions of spheres in simple shear flow by numerical simulation. *J. Fluid Mech.*, **155**:105–129.

Brodbent, S.R. and J.M. Hammerslay. (1957). Percolation processes I. Crystals and mazes. *Math. Proc. Cambridge Phil. Soc.*, **53**:629–641.

Broutman, L.J. and R.H. Krock, eds. (1974). *Composite Materials*. Vol. 1–8. Academic Press, New York.

Brown, W.F. (1956). *Dielectrics*. Springer-Verlag, Berlin.

Bruno, O. (1991). The effective conductivity of strongly heterogeneous composites. *Proc. R. Soc. London*, A, **433**:353–381.

Bürger, R. and W.L. Wendland. (2001). Sedimentation and suspension flows: historical perspective and some recent developments. *J. Engng. Math.*, **41**(2/3):101–116.

Burkill, J.C. (2004). *The Lebesgue Integral*. Cambridge University Press, Cambridge.

Caillerie, D. (1978). Sur la comportement limite d'une inclusion mince de grande rigidité dans un corps élastique. *Compt. Rend. Acad. Sci. Paris*, Ser. A., **287**:675–678.

Carreau, P.J. and F. Cotton. (2002). Rheological properties of concentrated suspensions. In: *Transport Processes in Bubbles, Drops and Particles* (D. De Kee and R.P. Chhabra, eds.), Taylor & Francis, London.

Chang, Ch. and R.L. Powell. (1994). Effect of particle size distribution on the rheology of a concentrated bimodal suspension. *J. Rheol.*, **38**:85–98.

Chen, H.-S. and A. Acrivos. (1978). The effective elastic moduli materials containing spherical inclusions at non-dilute concentration. *Int. J. Solids Struct.*, **14**:349–364.

Cheng, H. and L. Greengard. (1997). On the numerical evaluation of electrostatic fields in a dense random dispersions of cylinders. *J. Comput. Phys.*, **136**:626–639.

Cheng, H. and L. Greengard. (1998). A method of images for the evaluation of electrostatic fields in a system of closely spaced conducting cylinders. *SIAM J. Appl. Math.*, **50**: 122–141.

Cherkaev, A.V. (2000). *Variational Methods for Structural Optimization*. Springer-Verlag, Berlin.

Chinh, Ph.D. (1997). Overall properties of planar quasisymmetric randomly inhomogeneous media: Estimates and cell models. *Phys. Rev. E*, **56**:652–660.

Chou, T.-W. and F.K. Ko, eds. (1989). *Textile Structural Composites*. Elsevier Science, Amsterdam.

Christensen, R.M. (1979). *Mechanics of Composite Materials*. John Wiley, New York.

Chung, J.W., J.Th.M. De Hosson and E. van der Giessen. (1996). Fracture of a disordered 3-D spring network: A computer simulation methodology. *Phys. Rev. B*, **54**:15094–15100.

Clerc, J.P., G. Giraud, J.M. Laugier and J.M. Luck. (1990). The electrical conductivity of binary disordered systems, percolation clusters, fractals and related models. *Adv. Phys.*, **39**(3):191–309.

Courant, R.S. and D. Hilbert. (1953). *Methods of Mathematical Physics*. John Wiley, New York.

Coussot, P. (2002). Flows of concentrated granular mixtures. In: *Transport Processes in Bubbles, Drops and Particles* (R.P. Chhabra and D. De Kee, eds.), Taylor & Francis, London, pp. 291–315.

Craster, R.V. and Yu.V. Obnosov. (2004). A three-phase tessellation: Solution and effective properties. *Proc. R. Soc. London*, A, **460**:1017–1037.

Curtin, W.A. and H. Scher. (1990a). Brittle fracture in disordered materials: A spring network model. *J. Mater. Res.*, **5**:535–553.

Curtin, W.A. and H. Scher. (1990b). Mechanical modeling using a spring network. *J. Mater. Res.*, **5**:554–562.

Del Maso, G. (1993). *An Introduction to Γ-Convergence*. Birkhäuser, Boston.

Diaz, A.R. and N. Kikuchi. (1992). Solutions to shape and topology eigenvalue optimization problems using a homogenization method. *Int. J. Num. Meth. Engng*, **35**:1487–1502.

Dieudonne, J.A. (1969). *Treatise on Analysis*. Academic Press, New York.

Ding, J., H.E. Warriner and J.A. Zasadzinski. (2002). Viscosity of two-dimensional suspensions. *Phys. Rev. Lett.*, **88**(16):168102.1–168102.4.

Dobrodumov, A.M. and A.M. El'yashevich. (1973). Simulation of brittle fracture of polymers by a network model in the Monte Carlo method. *Sov. Solid State Phys.*, **15**: 1259–1260.

Doyle, W.T. (1978). The Clasius–Mossotti problem for cubic arrays of spheres. *J. Appl. Phys.*, **49**:795–797.

Drummon, J.E. and M.I. Tahir. (1984). Laminar viscous flow through regular arrays of parallel solid cylinders. *Int. J. Multiphase Flow*, **10**:515–540.

Drygaś, P. and V. Mityushev. (2009). Effective conductivity of unidirectional cylinders with interfacial resistance. *Quarterly J. Mech. Appl. Math.*, **62**(3):235–262.

Dykhne, A.M. (1971). Conductivity of a two-dimensional two-phase system. *Sov. Phys.*, **32**(63):63–65.

Einstein, A. (1906). Eine neue Bestimmung der Molekuldimensionen. *Ann. Phys.*, **19**:289–306.

Ekeland, I. and R. Temam. (1976). *Convex Analysis and Variational Problems*. North Holland, Amsterdam.

Evans, L.C. and W. Gangbo. (1999). Differential equation methods for the Monge–Kantorovich mass transfer problem. *Mem. Amer. Math. Soc.*, **137**(653):viii+66.

Evans, L.C. and R.F. Gariepy. (1992). *Measure Theory and Fine Properties of Functions*. CRC Press, Boca Raton, FL.

Feng, N.A. (1985). Percolation properties of granular elastic networks in two dimensions. *Phys. Rev. B*, **32**(1):510–513.

Feng, N.A. and A. Acrivos. (1985). On the viscosity of concentrated suspensions of solid spheres. *Chem. Engng Sci.*, **22**:847–853.

Flaherty, J.E. and J.B. Keller. (1973). Elastic behavior of composite media. *Comm. Pure Appl. Math.*, **26**:565–580.

Flory, P.J. (1941). Molecular size distribution in three dimensional polymers. I. Gelation. *J. Amer. Chem. Soc.*, **63**:3083–3090.

Fox, L. (1964). *An Introduction to Numerical Linear Algebra*. Clarendon Press, Oxford.

Francfort, G.A. and F. Murat. (1986). Homogenization and optimal bounds in linear electricity. *Arch. Rational Mech. Anal.*, **94**(4):307–334.

Frenkel, N.A. and A. Acrivos. (1967). On the viscosity of concentrated suspension of solid spheres. *Chem. Engng Sci.*, **22**:847–853.

Friis, E.A., R.S. Lakes and J.B. Park. (1988). Negative Poisson's ratio polymeric and metallic foams. *J. Mater. Sci.*, **23**:4406–4414.

Gakhov, F.D. (1966). *Boundary Value Problems*. Pergamon Press, Oxford.

Garboczi, E.J. and J.F. Douglas. (1996). Intrinsic conductivity of objects having arbitrary shape and conductivity. *Phys. Rev. E*, **53**(6):6169–6180.

Gaudiello, A. and A.G. Kolpakov. (2011). Influence of non degenerated joint on the global and local behavior of joined rods. *Int. J. Engng. Sci.*, **49**(3):295–309.

Good, I.J. (1949). The number of individuals in a cascade process. *Math. Proc. Cambridge Phil. Soc.*, **45**:360–363.

Goto, H. and H. Kuno. (1984). Flow of suspensions containing particles of two different sizes through a capillary tube. II. Effect of the particle size ratio. *J. Rheol.*, **28**:197–205.

Graham, A.L. (1981). On the viscosity of a suspension of solid particles. *Appl. Sci. Res.*, **37**:275–286.

Greengard, L. and J.-Y. Lee. (2006). Electrostatics and heat conduction in high contrast composite materials. *J. Comput. Phys.*, **211**(1):64–76.

Greengard, L. and M. Moura. (1994). On the numerical evaluation of electrostatic fields in composite materials. *Acta Numerica*, **3**:379–410.

Grigolyuk, E.I. and L.A. Filshtinskij. (1972). *Periodical Piecewise Homogeneous Elastic Structures (in Russian)*. Nauka, Moscow.

Grimet, G. (1992). *Percolation*. Springer-Verlag, Berlin.

Gupta, P.K. and A.R. Cooper. (1990). Topologically disordered networks of rigid polytopes. *J. Non-Crystal. Solids*, **123**(14):14–21.

Halperin, B.I., S. Feng and P.N. Sen. (1985). Difference between lattice and continuum percolation transport exponents. *Phys. Rev. Lett.*, **54**:2391–2394.

Happel, J. (1959). Viscous flow relative to arrays of cylinders. *AIChE J.*, **5**:174–177.

Hasimoto, H. (1959). On the periodic fundamental solutions of the Stokes equations and their application to viscous flow past a cubic array of spheres. *J. Fluid Mech.*, **5**: 317–328.

Haug, E.J., K.K. Choi and V. Komkov. (1986). *Design Sensitivity Analysis of Structural Systems*. Academic Press, Orlando, FL.

Herrmann, H.J., A. Hansen and S. Roux. (1989). Fracture of disordered, elastic lattices in two dimensions. *Phys. Rev. B*, **39**:637–648.

Hill, R. (1963). Elastic properties of reinforced solids: Some theoretical principles. *J. Mech. Phys. Solids*, **11**:357–372.

Hill, R. (1996). Characterization of thermally conductive epoxy composite fillers. *Proc. Technical Program "Emerging Packing Technology"*, Surface Mount Tech. Symp., pp. 125–131.

Hill, R.F. and P.H. Supancic. (2002). Thermal conductivity of platelet-filled polymer composite. *J. Am. Cer. Soc.*, **85**:851–857.

Hinsen, K. and B.U. Felderhof. (1992). Dielectric constant of a suspension of uniform spheres. *Phys. Rev. B*, **46**(20):12955–12963.

Hrennikoff, A. (1941). Solution of a problem of elasticity by the framework method. *J. Appl. Mech.*, **8**:169–175.

Jabin, P.-E. and F. Otto. (2004). Identification of the dilute regime in particle sedimentation. *Commun. Math. Phys.*, **250**:415–432.

Jeffrey, D.J. and A. Acrivos. (1976). The rheological properties of suspensions of rigid particles. *AIChE J.*, **22**:417–432.

Jikov, V.V., S.M. Kozlov and O.A. Oleinik. (1994). *Homogenization of Differential Operators and Integral Functionals*. Springer-Verlag, Berlin.

Kalamkarov, A.L. and A.G. Kolpakov. (1996). On the analysis and design of fiber reinforced composite shells. *Trans. ASME. J. Appl. Mech.*, **63**(4):939–945.

Kalamkarov, A.L. and A.G. Kolpakov. (1997). *Analysis, Design and Optimization of Composite Structures*. John Wiley, Chichester.

Karal Jr., F.C. and J.B. Keller. (1966). Effective dielectric constant, permeability, and conductivity of a random medium and the velocity and attenuation coefficient of coherent waves. *J. Math. Phys.*, **7**:661–670.

Kato, T. (1976). *Perturbation Theory for Linear Operators*. Springer-Verlag, New York.

Keller, J.B. (1963). Conductivity of a medium containing a dense array of perfectly conducting spheres or cylinders or nonconducting cylinders. *J. Appl. Phys.*, 4(34):991–993.

Keller, J.B. (1964). A theorem on the conductivity of a composite medium. *J. Math. Phys.*, 5:548–549.

Keller, J.B. (1987). Effective conductivity of a periodic composite composed of two very unequal conductors. *J. Math. Phys.*, 10(28):2516–2520.

Keller, J.B. and D. Sachs. (1964). Calculations of conductivity of a medium containing cylindrical inclusions. *J. Appl. Phys.*, 35:537–538.

Kellomaki, M., J. Astrom and J. Timonen. (1996). Rigidity and dynamics of random spring networks. *Phys. Rev. Lett.*, 77:2730–2733.

Kelly, A. and Yu.N. Rabotnov, eds. (1988). *Handbook of Composites*. North Holland, Amsterdam.

Kesten, H. (1992). *Percolation Theory for Mathematicians*. Birkhäuser, Boston.

Kolmogorov, A.N. and S.V. Fomin. (1970). *Introductory Real Analysis*. Prentice Hall, Englewood Cliffs, NJ.

Kolpakov, A.A. (2007). Numerical verification of existence of the energy-concentration effect in a high-contrast high-filled composite material. *J. Engng Phys. Thermophys.*, 80(4):812–819.

Kolpakov, A.A. and A.G. Kolpakov. (2007). Asymptotics of the capacity of a system of closely placed bodies. Tamm's shielding effect and network models. *Doklady Phys.*, 415(2):188–192.

Kolpakov, A.A. and A.G. Kolpakov. (2010). *Capacity and Transport in Contrast Composite Structures: Asymptotic Analysis and Applications*. CRC Press, Boca Raton, FL.

Kolpakov, A.G. (1987). Averaged characteristics of thermoelastic frames. Izvestiay of the Academy of Science of the USSR. *Mechanics of Solids*, 22(6):53–61.

Kolpakov, A.G. (1985). Determination of the average characteristics of elastic frameworks. *J. Appl. Math. Mech.*, 49:739–745.

Kolpakov, A.G. (1988). Asymptotics of the first boundary value problem for an elliptic equation in a region with a thin covering. *Siberian Math. J.*, 6:74–84.

Kolpakov, A.G. (1992). Glued bodies. *Differential Equations*, 28(8):1131–1139.

Kolpakov, A.G. (2004). *Stressed Composite Structures: Homogenized Models for Thin-Walled Nonhomogeneous Structures with Initial Stresses*. Springer-Verlag, Berlin.

Kolpakov, A.G. (2005). Asymptotic behavior of the conducting properties of high-contrast media. *J. Appl. Mech. Tech. Phys.*, 46(3):412–422.

Kolpakov, A.G. (2006a). The asymptotic screening and network models. *J. Engng Phys. Thermophys.*, 2:39–47.

Kolpakov, A.G. (2006b). Convergence of solutions for a network approximation of the two-dimensional Laplace equation in a domain with a system of absolutely conducting disks. *Comp. Math. Math. Phys.*, 46(9):1682–1691.

Kolpakov, A.G. (2011). Influence of non degenerated joint on the global and local behavior of joined plates. *Int. J. Engng. Sci.*, 49(11):1216–1231.

Koplik, J. (1982). Creeping flow in two-dimensional networks. *J. Fluid Mech.*, 119: 219–247.

Kozlov, S.M. (1978). Averaging of random structures (in Russian). *Doklady Acad. Nauk SSSR*, **241**(5):1016–11019.

Kozlov, S.M. (1980). Averaging of random operators. *Math. USSR Sbornik*, **37**:167–180.

Kozlov, S.M. (1989). Geometric aspects of averaging. *Russian Math. Surv.*, **2**(44):91–144.

Kozlov, S.M. (1992). On the domain of variations of apparent added masses, polarization and effective characteristics of composites. *J. Appl. Math. Mech.*, **56**(1):102–107.

Kuchling, H. (1980). *Physics*. VEB Fachbuchverlag, Leipzig.

Kun, F. and H. Herrmann. (1996). A study of fragmentation processes using a discrete element method. *Comput. Meth. Appl. Mech. Engng*, **138**:3–18.

Ladd, A.J.C. (1997). Sedimentation of homogeneous suspensions of non-Brownian spheres. *Phys. Fluids*, **9**(3):491–499.

Ladyzhenskaya, O.A. and N.N. Ural'tseva. (1968). *Linear and Quasilinear Elliptic Equations*. Academic Press, New York.

Lakes, R. (1991). Deformation mechanisms of negative Poisson's ratio materials: Structural aspects. *J. Mater. Sci.*, **26**:2287–2292.

Lamb, H. (1991). *Hydrodynamics*. Dover, New York.

Landauer, R. (1978). Electrical conductivity in inhomogeneous media. In: *Electrical Transport and Optical Properties of Inhomogeneous Media* (J.C. Garland, D.B. Tanner, eds.), American Institute of Physics. Woodbury, New York, pp. 2–43.

Leal, G. (1992). *Laminar Flow and Convective Transport Processes: Scaling Principles and Asymptotic Analysis*. Butterworth–Heinemann, Amsterdam.

Leighton, D. and A. Acrivos. (1987). Measurement of shear-induced self-diffusion in concentrated suspensions of spheres. *J. Fluid Mech.*, **177**:109–131.

Lenczner, M. (1997). Homogénéisation d'un circuit électrique. *C.R. Acad. Sci. Paris, Série II B*, **324**(9):537–542.

Lieberman, G.M. (1988). Boundary regularity for solutions of degenerate elliptic equations. *Nonlinear Anal.*, **12**(11):1203–1219.

Limat, L. (1988). Percolation and Cosserat elasticity: Exact results on a deterministic fractal. *Phys. Rev., B*, **37**:672–675.

Lions, J.-L. (1978). Notes on some computational aspects of the homogenization method in composite materials. In: *Computational Methods in Mathematics, Geophysics and Optimal Control*, Nauka, Novosibirsk, pp. 5–19.

Lions, J.-L. and E. Magenes. (1972). *Non-Homogeneous Boundary Value Problems and Applications*, Vol. 1, 2. Springer-Verlag, Berlin.

Lipton, R. (1994). Optimal bounds on the effective elastic tensor for orthotropic composites. *Proc. R. Soc. London*, A, **444**:399–410.

Love, A.E.H. (1929). *A Treatise on the Mathematical Theory of Elasticity*. Oxford University Press, Oxford.

Lu, J.-K. (1995). *Complex Variable Methods in Plane Elasticity*. World Scientific, Singapore.

Lévy, T. (1986). Application of homogenization to the study of a suspension of force-free particles. In: *Trends in Applications of Pure Mathematics to Mechanics*. Lecture Notes in Physics 249, Springer-Verlag, Berlin, pp. 349–353.

Makaruk, S.F., V.V. Mityushev and S.V. Rogosin. (2006). An optimal design problem for two-dimensional composite materials. A constructive approach. In: *Analytic Methods of Analysis and Differential Equations.* AMADE 2003 (A.A. Kilbas and S.V. Rogosin, eds.). Cambridge Scientific, Cottenham, Cambridge, pp. 153–167.

Markov, K.Z. (2000). Elementary micromechanics of heterogeneous media. In: *Heterogeneous Media: Micromechanics Modeling Methods and Simulation* (K. Markov and L. Preziosi, eds.), Birkhäuser, Basel, pp. 1–162.

Maury, B. (1999). Direct simulations of 2D fluid–particle flows in biperiodic domains. *J. Comput. Phys.*, **156**(2):325–351.

Maxwell, J.C. (1873). *Treatise on Electricity and Magnetism.* Clarendon Press, Oxford.

McAllister, L.E. and W.L. Lachman. (1983). Multidirectional carbon–carbon composites. In: *Handbook of Composites, Vol. 4. Fabrication of Composites* (A. Kelly and S.T. Mileiko, eds.), North Holland, Amsterdam, pp. 109–176.

McKenzie, D.R., R.C. McPhedran and G.H. Derrik. (1978). The conductivity of a lattice of spheres II. The body centered and face centered lattices. *Proc. R. Soc. London, A,* **362**:211–232.

McPhedran, R. (1986). Transport property of cylinder pairs and of the square array of cylinders. *Proc. R. Soc. London, A,* **408**:31–43.

McPhedran, R., L. Poladian and G.W. Milton. (1988). Asymptotic studies of closely spaced, highly conducting cylinders. *Proc. R. Soc. London, A,* **415**:195–196.

McPhedran, R.C. and D.R. McKenzie. (1978). The conductivity of a lattice of spheres I. The simple cubic lattice. *Proc. R. Soc. London, A,* **359**:45–63.

McPhedran, R.C. and G.W. Milton. (1987). Transport properties of touching cylinder pairs and of a square array of touching cylinders. *Proc. R. Soc. London,* **A411**:313–326.

Meester, R. and R. Roy. (1992). *Continuum Percolation.* Cambridge University Press, Cambridge.

Melrose, D.B. and R.C. McPhedran. (1991). *Electromagnetic Processes in Dispersive Media.* Cambridge University Press, Cambridge.

Meredith, R.E. and C.W. Tobias. (1960). Resistance to potential flow through a cubical array of spheres. *J. Appl. Physics,* **31**:1270–1273.

Mertensson, E. and U. Gafvert. (2003). Three-dimensional impedance networks for modeling frequency dependent electrical properties of composite materials. *J. Phys. D: Appl. Phys.,* **36**:1864–1872.

Mertensson, E. and U. Gafvert. (2004). A three-dimensional network model describing a non-linear composite material. *J. Phys. D: Appl. Phys.,* **37**:112–119.

Michel, J.C., H. Moulinec and P. Suquet. (2000). A computational method based on augmented Lagrangians and fast Fourier transforms for composites with high contrast. *Comp. Model. Engng Sci.,* **1**(2):79–88.

Michel, J.C., H. Moulinec and P. Suquet. (2002). A computational scheme for linear and non-linear composites with arbitrary phase contrast. *Int. J. Numer. Meth. Engng.,* **52**:139–160.

Milton, G.M. (1992). Composite materials with Poisson's ratios close to -1. *J. Mech. Phys. Solids,* **40**:1105–1137.

Milton, G.W. (2002). *The Theory of Composites.* Cambridge University Press, Cambridge.

Mityushev, V. (1993). Plane problem for the steady heat conduction of a material with circular inclusions. *Arch. Mech.*, **45**(2):211–215.

Mityushev, V. (1994). Solution of the Hilbert boundary value problem for a multiply connected domain. *Slupskie Prace Mat.-Przyr.*, **9a**:33–67.

Mityushev, V. (1997a). A functional equation in a class of analytic functions and composite materials. *Demostratio Math.*, **30**:63–70.

Mityushev, V. (1997b). Functional equations and their applications in the mechanics of composites. *Demonstratio Math.*, **30**(1):64–70.

Mityushev, V. (1998). Hilbert boundary value problem for multiply connected domains. *Complex Variables*, **35**:283–295.

Mityushev, V. (1999). Transport properties of two-dimensional composite materials with circular inclusions. *Proc. R. Soc. London*, **A455**:2513–2528.

Mityushev, V. (2001). Transport properties of doubly periodic arrays of circular cylinders and optimal design problems. *Appl. Math. Optim.*, **44**:17–31.

Mityushev, V. (2005). *R*-linear problem on the torus and its application to composites. *Complex Variables*, **50**(7–10):621–630.

Mityushev, V. (2009). Conductivity of a two-dimensional composite containing elliptical inclusions. *Proc. R. Soc. A*, **465**:2991–3010.

Mityushev, V. and P.M. Adler. (2002a). Longitudinal permeability of a doubly periodic rectangular array of cylinders. I. *Z. Angew. Math. Mech.*, **82**:335–345.

Mityushev, V. and P.M. Adler. (2002b). Longitudinal permeability of a doubly periodic rectangular array of cylinders. II. An arbitrary distribution of cylinders inside the unit cell. *Z. Angew. Math. Phys.*, **53**:486–517.

Mityushev, V., E. Pesetskaya and S. Rogosin. (2008). Analytical methods for heat conduction in composites and porous media in cellular and porous materials. In: *Cellular and Porous Materials: Thermal Properties Simulation and Prediction* (A. Ochsner, G. Murch and M. de Lemos, eds.), Wiley-VCH, Weinheim.

Mityushev, V. and S.V. Rogozin. (2000). *Constructive Methods for Linear and Nonlinear Boundary Value Problems of Analytic Function Theory*. Chapman & Hall/CRC, Boca Raton, FL.

Mityushev, V.V. (1997). Transport properties of doubly-periodic arrays of circular cylinders. *Z. Angew. Math. Mech.*, **77**:115–120.

Mizohata, S. (1973). *The Theory of Partial Differential Equations*. Cambridge University Press, Cambridge.

Molchanov, S. (1994). Lectures on random media. In: *Lectures on Probability Theory* (D. Bakry, R.D. Gill and S.A. Molchanov, eds.), Springer-Verlag, Berlin, pp. 242–411.

Molyneux, J. (1970). Effective permittivity of a polycrystalline dielectric. *J. Math. Phys.*, **11**(4):1172–1184.

Movchan, A.B., N.V. Movchan and C.G. Poulton. (2002). *Asymptotic Models of Fields in Dilute and Densely Packed Composites*. Imperial College Press, London.

Nemat-Nasser, S. and M. Hori. (1993). *Micromechanics*. Elsevier Science, Amsterdam.

Nettelblad, B., E. Mårtensson, C. Önneby, U. Gäfvert and A. Gustafsson. (2003). Two percolation thresholds due to geometrical effects: Experimental and simulated results. *J. Phys. D: Appl. Phys.*, **36**(4):399–405.

Newman, M.E.J. (2003). The structure and functions of complex networks. *SIAM Rev.*, **45**(2):167–256.

Nicorovici, N.A. and R.C. McPhedran. (1996). Transport properties of arrays of elliptical cylinders. *Phys. Rev.* E, **54**:1945–1957.

Noor, A.K. (1988). Continuum modeling for repetitive structures. *Appl. Mech. Rev.*, **41**(7): 285–296.

Nott, P.R. and J.F. Brady. (1994). Pressure-driven flow of suspensions: Simulation and theory. *J. Fluid Mech.*, **275**:157–199.

Novikov, A. (2009). A discrete network approximation for effective conductivity of non-ohmic high-contrast composites. *Commun. Math. Sci.*, **7**(3):719–740.

Novikov, V.V. and Chr. Friedrich. (2005). Viscoelastic properties of composite materials with random structure. *Phys. Rev. E*, **72**:021506–1–021506–9.

Novozilov, V.V. (1970). On the relationship between average values of the stress tensor and strain tensor in statistically isotropic elastic bodies. *Appl. Math. Mech.*, **34**(1):67–74.

Nunan, K.C. and J.B. Keller. (1984a). Effective elasticity tensor for a periodic composite. *J. Mech. Phys. Solids*, **32**:259–280.

Nunan, K.C. and J.B. Keller. (1984b). Effective velocity of a periodic suspension. *J. Fluid Mech.*, **142**:269–287.

Oleinik, O.A., A.S. Shamaev and G.A. Yosifian. (1962). *Mathematical Problems in Elasticity and Homogenization*. North Holland, Amsterdam.

Ostoja-Starzewski, M. (2006). Material spatial randomness – from statistical to representative volume element. *Probab. Eng. Mech.*, **21**(2):112–132.

Panasenko, G.P. (2005). *Multi–Scale Modeling for Structures and Composites*. Springer-Verlag, Berlin.

Panasenko, G.P. and G. Virnovsky. (2003). Homogenization of two-phase flow: high contrast of phase permeability. *C.R. Mecanique*, **331**:9–15.

Papanicolaou, G.C. (1995). Diffusion in random media. In: *Surveys in Applied Mathematics* (J.B. Keller, D. McLaughlin and G. Papanicolaou, eds.), Plenum Press, New York, pp. 205–255.

Papanicolaou, G.C. and S.R.S. Varadhan. (1981). Boundary value problems with rapidly oscillating random coefficients. *Seria Coll. Janos Bolyai*, **27**:835–873.

Perrins, W.T., R.C. McPhedran and D.R. McKenzie. (1979). Transport properties of regular arrays of cylinders. *Proc. R. Soc. London, A*, **369**:207–225.

Pesetskaya, E.V. (2005). Effective conductivity of composite materials with random positions of cylindrical inclusions: Finite number inclusions in the cell. *Applic. Anal.*, **84**(8):843–865.

Peterseim, D. (2010). Triangulating a system of disks. In: *Proc. 26th Eur. Workshop Comp. Geometry*, pp. 241–244.

Peterseim, D. (2012). Robustness of finite elements simulation in densely packed random particle composites. *Networks Meter. Media*, **7**(1):113–126.

Pham Huy, H. and E. Sanchez-Palencia. (1974). Phénomènes de transmission à travers des couches minces de conductivité élevée. *J. Math. Anal. Appl.*, **47**:284–309.

Phillips, R.J., R.C. Armstrong, R.A. Brown, A.L. Graham and J.R. Abbot. (1992). A constitutive equation for concentrated suspensions that accounts for shear–induced particle migration. *Phys. Fluids, A*, **4**:30–40.

Poincaré, H. (1886). Sur les integrals irregulieres des equations lineaires. *Acta Math.*, **8**: 295–344.

Poslinski, A.J., M.E. Ryan, R.K. Gupta, S.G. Seshadri and F.J. Frechette. (1988). Rheological behavior of filled polymeric systems II. The effect of bimodal size distribution of particulates. *J. Rheol.*, **32**:751–771.

Prager, S. (1963). Diffusion and viscous flow in concentrated suspension. *Physica*, **29**: 129–139.

Pshenichnov, G.I. (1993). *A Theory of Latticed Plates and Shells*. World Scientific, Singapore.

Rayleigh, Lord (J.W. Strutt). (1892). On the influence of obstacles arranged in rectangular order upon the properties of the medium. *Phil. Mag.*, **34**(241):481–491.

Reuss, A. (1929). Berechnung der Fließ grenze von Mischkristallen auf Grund der Plastizitätsbedingung für Einkristalle. *Z. Angew. Math. Mech.*, **9**:49–58.

Robinson, D.A. and S.F Friedman. (2001). Effect of particle size distribution on the effective dielectric permittivity of saturated granular media. *Water Resour. Res.*, **37**(1):33–40.

Rockafellar, R.T. (1969). *Convex Functions and Duality in Optimization Problems and Dynamics*. Springer–Verlag, Berlin.

Rockafellar, R.T. (1970). *Convex Analysis*. Princeton University Press, Princeton, NJ.

Roux, S. and E. Guyon. (1985). Mechanical percolation: A small beam lattice study. *J. Physique Lett.*, **46**:999–1004.

Rudin, W. (1964). *Principles of Mathematical Analysis*. McGraw–Hill, New York.

Rudin, W. (1992). *Functional Analysis*. McGraw–Hill, New York.

Runge, I. (1925). Zur elektrischen Leitfähigkeit metallischer Aggregate. *Z. Tech. Physic*, **61**(6):61–68.

Rylko, N. (2000). Transport properties of a rectangular array of highly conducting cylinders. *J. Engng. Math.*, **38**:1–12.

Rylko, N. (2008a). Effect of polydispersity in conductivity of unidirectional cylinders. *Arch. Mater. Sci. Engng*, **29**:45–52.

Rylko, N. (2008b). Structure of the scalar field around unidirectional circular cylinders. *Proc. R. Soc. London, A*, **464**:391–407.

Sab, K. (1992). On the homogenization and the simulation of random materials. *Eur. J. Mech., A/Solids*, **11**(5):585–607.

Sahimi, M. (2003). *Heterogeneous Materials*, Vol. 1, 2. Springer-Verlag, New York.

Sanchez-Palencia, E. (1974). Problèmes de perturbations liés aux phénomènes de conduction à travers des couches minces de grande résistivité. *J. Math. Pure Appl.*, **53**: 251–270.

Sanchez-Palencia, E. (1980). *Non-Homogeneous Media and Vibration Theory*. Springer-Verlag, Berlin.

Sangani, A.S. and A. Acrivos. (1983). The effective conductivity of a periodic array of spheres. *Proc. R. Soc. London, A*, **386**:263–275.

Schwartz, L. (1966). *Theorie des Distributions*. Hermann, Paris.

Schwartz, L.M., D.L. Johnson and S. Feng. (1984). Vibration modes in granular materials. *Phys. Rev. Lett.*, **52**(831):831–834.

Shermergor, T.D. (1977). *Elasticity Theory of Micro–Inhomogeneous Media (in Russian)*. Nauka, Moscow.

Shook, C.A and M.C. Rocko. (1991). *Slurry Flow, Principles and Practice.* Butterworth–Heinemann, Boston, MA.

Sierou, A. and J.F. Brady. (2002). Rheology and microstructure in concentrated noncolloidal suspensions. *J. Rheol.*, **46**(5):1031–1056.

Simonenko, I.B. (1974). Electrostatics problems for an inhomogeneous medium: A case of thin dielectric with high dielectric constant: I. *Differential Equations*, **10**:301–309.

Simonenko, I.B. (1975a). Electrostatics problems for an inhomogeneous medium: A case of thin dielectric with high dielectric constant: II. *Differential Equations*, **11**:1870–1878.

Simonenko, I.B. (1975b). Limit problem of conductivity in an inhomogeneous medium. *Siberian Math. J.*, **16**:1291–1300.

Smythe, W.R. (1950). *Static and Dynamical Electricity*, 2nd ed. McGraw–Hill, New York.

Sobolev, S.L. (1937). On the boundary value problem for polyharmonic functions (in Russian). *Matem. Zbornik*, **2**(3):465–499.

Sobolev, S.L. (1950). *Some Applications of Functional Analysis to Mathematical Physics* (in Russian). Leningrad State University, Leningrad.

Stauffer, D. and A. Aharony. (1992). *Introduction to Percolation Theory.* Taylor & Francis, London.

Stockmayer, W.H. (1943). Theory of molecular size distribution and gel formation in branched-chain polymers. *J. Chem. Phys.*, **11**:45–55.

Subia, S., M.S. Ingber, L.A. Mondy, S.A. Altobelli and A.L. Graham. (1998). Modeling of concentrated suspensions using a continuum constitutive equation. *J. Fluid Mech.*, **373**:193–219.

Szczepkowski, J., A.E. Malevich and V. Mityushev. (2003). Macroscopic properties of similar arrays of cylinders. *Quart. J. Appl. Math. Mech.*, **56**(4):617–628.

Tamm, I.E. (1979). *Fundamentals of the Theory of Electricity.* Mir Publishers, Moscow.

Temam, R. (1979). *Navier–Stokes Equations.* North Holland, Amsterdam.

Thovert, J.F. and A. Acrivos. (1989). The effective thermal conductivity of a random polydispersed suspension of spheres to order c^2. *Chem. Eng. Comm.*, **82**:177–191.

Thovert, J.F., I.C. Kim, S. Torquato and A. Acrivos. (1990). Bounds on the effective properties of polydispersed suspensions of spheres: An evaluation of two relevant morphological parameters. *J. Appl. Phys.*, **67**:6088–6098.

Timoshemko, S. and J.N. Goodier. (1951). *Theory of Elasticity.* McGraw–Hill, New York.

Torquato, S. (2002). *Random Heterogeneous Materials.* Springer-Verlag, Berlin.

van Lint, J.H. and R.M. Wilson. (2001). *A Course in Combinatorics*, 2nd ed. Cambridge University Press, Cambridge.

Vinogradov, V. and G.W. Milton. (2005). An accelerated fast Fourier transform algorithm for nonlinear composites. *Advances Comp. Experim. Enging. Sci. Proc. ICCES'05.* Available at www.math.utah.edu/vladim/papers/publications.html.

Voigt, W. (1910). *Lehrbuch der Kristallphysik.* Teubner, Stuttgart.

Voronoi, G. (1908). Nouvelles applications des paramètres continus à la théorie des formes quadratiques. Deuxieme Memoire. Recherches sur les parallelloedres primitifs. *J. Reine Angew. Math.*, **134**(198):198–287.

Walpole, L.J. (1966). On bounds for the overall elastic moduli of inhomogeneous systems. *J. Mech. Phys. Solids*, **14**:151–162.

Weil, A. (1976). *Elliptic Functions According to Eisenstein and Kronecker*. Springer-Verlag, Berlin.

Wermer, J. (1974). *Potential Theory*. Springer-Verlag, Berlin.

West, B.W. (2000). *Introduction to Graph Theory*. Prentice Hall, NJ.

Willis, J.R. (2002). Lectures on mechanics of random media. In: *Mechanics of Random and Multiscale Microstructures, CISM Lecture Notes* (D. Jeulin and M. Ostoja-Starzewski, eds.), Springer–Verlag, Vienne, pp. 221–267.

Yan, Y., J. Li and L.M. Sander. (1989). Fracture growth in 2-D elastic networks with the Born model. *Europhys. Lett.*, **10**:7–13.

Yang, C.S. and P.M. Hui. (1991). Effective nonlinear response in random nonlinear resistor networks: Numerical studies. *Phys. Rev. B*, **44**:12559–12561.

Yardley, J.G., A.J. Reuben and R.C. McPhedran. (2001). The transport properties of layers of elliptical cylinders. *Proc. R. Soc. London, A*, **457**:395–423.

Yeh, R.H.T. (1970a). Variational principles of elastic moduli of composite materials. *J. Appl. Phys.*, **41**(8):3353–3356.

Yeh, R.H.T. (1970b). Variational principles of transport properties of composite materials. *J. Appl. Phys.*, **41**(1):224–226.

Yosida, K. (1971). *Functional Analysis*. Springer-Verlag, Berlin.

Yurinski, V.V. (1980). Average of an elliptic boundary problem with random coefficients. *Siberian Math. J.*, **21**:470–482.

Yurinski, V.V. (1986). Averaging of symmetric diffusion in a random medium. *Siberian Math. J.*, **27**(4):603–613.

Zeidler, E. (1995). *Applied Functional Analysis: Applications to Mathematical Physics*. Springer-Verlag, Berlin.

Zuzovsky, M. and H. Brenner. (1977). Effective conductivities of composite materials composed of a cubic arrangement of spherical particles embedded in an isotropic matrix. *Z. Angew. Math. Phys.*, **28**(6):979–992.

Index

Printed in the United States
by Baker & Taylor Publisher Services